Lecture Notes in Physics

The Editorial Policy for Proceedings

The series Lecture Notes in Physics reports new developments in physical research and teaching – quickly, informally, and at a high level. The proceedings to be considered for publication in this series should be limited to only a few areas of research, and these should be closely related to each other. The contributions should be of a high standard and should avoid lengthy redraftings of papers already published or about to be published elsewhere. As a whole, the proceedings should aim for a balanced presentation of the theme of the conference including a description of the techniques used and enough motivation for a broad readership. It should not be assumed that the published proceedings must reflect the conference in its entirety. (A listing or abstracts of papers presented at the meeting but not included in the proceedings could be added as an appendix.)
When applying for publication in the series Lecture Notes in Physics the volume's editor(s) should submit sufficient material to enable the series editors and their referees to make a fairly accurate evaluation (e.g. a complete list of speakers and titles of papers to be presented and abstracts). If, based on this information, the proceedings are (tentatively) accepted, the volume's editor(s), whose name(s) will appear on the title pages, should select the papers suitable for publication and have them refereed (as for a journal) when appropriate. As a rule discussions will not be accepted. The series editors and Springer-Verlag will normally not interfere with the detailed editing except in fairly obvious cases or on technical matters.
Final acceptance is expressed by the series editor in charge, in consultation with Springer-Verlag only after receiving the complete manuscript. It might help to send a copy of the authors' manuscripts in advance to the editor in charge to discuss possible revisions with him. As a general rule, the series editor will confirm his tentative acceptance if the final manuscript corresponds to the original concept discussed, if the quality of the contribution meets the requirements of the series, and if the final size of the manuscript does not greatly exceed the number of pages originally agreed upon.
The manuscript should be forwarded to Springer-Verlag shortly after the meeting. In cases of extreme delay (more than six months after the conference) the series editors will check once more the timeliness of the papers. Therefore, the volume's editor(s) should establish strict deadlines, or collect the articles during the conference and have them revised on the spot. If a delay is unavoidable, one should encourage the authors to update their contributions if appropriate. The editors of proceedings are strongly advised to inform contributors about these points at an early stage.
The final manuscript should contain a table of contents and an informative introduction accessible also to readers not particularly familiar with the topic of the conference. The contributions should be in English. The volume's editor(s) should check the contributions for the correct use of language. At Springer-Verlag only the prefaces will be checked by a copy-editor for language and style. Grave linguistic or technical shortcomings may lead to the rejection of contributions by the series editors.
A conference report should not exceed a total of 500 pages. Keeping the size within this bound should be achieved by a stricter selection of articles and not by imposing an upper limit to the length of the individual papers.
Editors receive jointly 30 complimentary copies of their book. They are entitled to purchase further copies of their book at a reduced rate. As a rule no reprints of individual contributions can be supplied. No royalty is paid on Lecture Notes in Physics volumes. Commitment to publish is made by letter of interest rather than by signing a formal contract. Springer-Verlag secures the copyright for each volume.

The Production Process

The books are hardbound, and quality paper appropriate to the needs of the authors is used. Publication time is about ten weeks. More than twenty years of experience guarantee authors the best possible service. To reach the goal of rapid publication at a low price the technique of photographic reproduction from a camera-ready manuscript was chosen. This process shifts the main responsibility for the technical quality considerably from the publisher to the authors. We therefore urge all authors and editors of proceedings to observe very carefully the essentials for the preparation of camera-ready manuscripts, which we will supply on request. This applies especially to the quality of figures and halftones submitted for publication. In addition, it might be useful to look at some of the volumes already published.
As a special service, we offer free of charge LATEX and TEX macro packages to format the text according to Springer-Verlag's quality requirements. We strongly recommend that you make use of this offer, since the result will be a book of considerably improved technical quality.
To avoid mistakes and time-consuming correspondence during the production period the conference editors should request special instructions from the publisher well before the beginning of the conference. Manuscripts not meeting the technical standard of the series will have to be returned for improvement.

For further information please contact Springer-Verlag, Physics Editorial Department V, Tiergartenstrasse 17, W-6900 Heidelberg, FRG

Erik Balslev (Ed.)

Schrödinger Operators
The Quantum Mechanical Many-Body Problem

Proceedings of a Workshop
Held at Aarhus, Denmark
15 May - 1 August 1991

Springer-Verlag Berlin Heidelberg GmbH

Editor

Erik Balslev
Matematisk Institut, Universitetsparken
Ny Munkegade, 8000 Aarhus C, Denmark

ISBN 978-3-662-13888-5 ISBN 978-3-540-47107-3 (eBook)
DOI 10.1007/978-3-540-47107-3

Originally published by Springer-Verlag Berlin Heidelberg New York in 1992
Softcover reprint of the hardcover 1st edition 1992

Typesetting: Camera ready by author/editor

58/3140-543210 - Printed on acid-free paper

Preface

The theory of Schrödinger operators has undergone an impressive development in recent years. V. Enss extended his time-dependent method to prove asymptotic completeness for the 3–body problem, including also long-range pair potentials with a decay rate of at least $|x|^{-(\sqrt{3}-1)-\varepsilon}$. I. M. Sigal and S. Soffer proved completeness for n-body systems with short-range pair potentials decaying faster than $|x|^{-1}$. Further progress was made by J.M. Graf, who gave a new, simpler proof of completeness of n-body systems in the short-range case.

This development as well as the progress made in this area in many other directions gave the impetus for the organization of a workshop on Schrödinger operators with many-body systems as a central theme. Such a workshop took place in Aarhus, 15 May - 1 August 1991

Many specialists participated in the workshop and contributed in many ways to making it a rewarding experience, involving individual research, informal discussions and stimulating lectures. The secret hope for new results on the long-range case was fulfilled, when J. Derezinski developed the major new ideas of a proof of completeness for n-body operators with pair potentials decaying faster than $|x|^{-(\sqrt{3}-1)}$. This decay rate seems in fact to be a natural boundary for the problem.

Many other interesting lines of research were pursued and are reflected in the wide range of subjects covered in this volume. Some articles are based on lectures given during the workshop, while others present work partly carried out there.

I want to thank the Danish Science Foundation and the Aarhus University Research Foundation for financial support, which made this workshop possible.

Aarhus, February 1992

Erik Balslev

Contents

PERTURBATIONS OF GENERALIZED SCHRÖDINGER OPERATORS IN STOCHASTIC SPECTRAL ANALYSIS

Michael Demuth, Karl-Weierstrass-Institut für Mathematik
Mohrenstrasse 39
1086 Berlin-Mitte, Germany

J.A. van Casteren, Department of Math. and Comp. Science
University of Antwerp (UIA)
Universiteitsplein 1
2610 Wilrijk/Antwerp, Belgium

The objective of stochastic spectral analysis is explained. It is used to study regular perturbations for a general class of generators of Feller semigroups, also called generalized Schrödinger operators. Upon introducing the Kato-Feller norm, the asymptotic behaviour of several spectral data can be studied. In the present article mainly the convergence of scattering matrices is considered.

1. BASIC ASSUMPTIONS IN STOCHASTIC SPECTRAL ANALYSIS.

There are two ways of introducing Schrödinger semigroups. In the purely analytic setting one starts with the self-adjoint realization H_0 of the Laplace operator $-\frac{1}{2}\Delta$ in $L^2(\mathbf{R}^n)$. It generates a semigroup which consists of integral operators with integral kernels given by the Wiener density function

$$p_W(t,x,y) = (2\pi t)^{-n/2} \exp\left(-\frac{|x-y|^2}{2t}\right), \; t>0, \;; x,\; y \in \mathbf{R}^n \tag{1}$$

Moreover one assumes the presence of a potential operator V, also generating a semigroup and chosen in such a way that $H_0 \dot{+} V$ is a well-defined (self-adjoint) operator. Then one can apply the Trotter product formula to establish a Feynman-Kac representation of the semigroup $\exp\left(-t\left(H_0 \dot{+} V\right)\right)$ in $L^2(\mathbf{R}^n)$.

From the stochastic point of view this introduction of Schrödinger semigroups is somewhat artificial, because one needs several a priori assumptions on the potential V. In this context the fact can be used that H_0 generates the Wiener process. Therefore the Wiener integral is available, meaning that we can start the perturbation problem of H_0 by requiring that the right-hand-side of the Feynman-Kac formula is finite, or more precisely that it defines a (strongly continuous) semigroup in $L^p(\mathbf{R}^n)$, $1 \le p \le \infty$. In fact one is interested in the properties of operators of the form

$$[P(t)f](x) = \int_{\Omega_x} \exp\left(-\int_0^t V(\omega(s))\right) f(\omega(t)) \mathbf{P}_W(d\omega), \tag{2}$$

where $t \ge 0$, where f belongs to $L^p(\mathbf{R}^n)$, $1 \le p \le \infty$, where $\mathbf{P}_W$ is the Wiener measure on the path space Ω_x consisting of continuous paths on $[0,\infty)$ with values in

$\mathbf{R}^n$ and starting at x, i.e. $\omega(0) = x$. It turns out that Kato's potentials are optimal for $P(t)$ to be a bounded operator from $L^\infty(\mathbf{R}^n)$ to $L^\infty(\mathbf{R}^n)$. Moreover for such potentials it follows that $\lim_{\lambda\to\infty} \left\| (H_0 + \lambda I)^{-1} |V| \right\|_{\infty,\infty} = 0$, which implies that the form bound of V with respect to H_0 is zero. Hence Kato's class is the most natural class for studying semigroups with self-adjoint generators $H_0 \dot{+} V$ in $L^2(\mathbf{R}^n)$.

But both ways of introducing Schrödinger semigroups have the same disadvantage: one starts with a single operator H_0, which is often the Laplacian. This is somewhat inconsistent if one employs the second way to introduce potentials stochastically via the Wiener integral (2).

Thus we suggest a new, a third way to introduce semigroups, called Feller semigroups. Its perturbed generators can be called *generalized Schrödinger operators.* We shall use a purely stochastic way of introducing these semigroups. Notice the fact that the central link between operator theory and stochastic analysis is the transition function or the transition density function. This provides, on one side, a semi-group with a corresponding generator. On the other side, one may use the underlying Markov process for proving several results.

For a concise notation we confine ourselves to formulate all the conditions in terms of the density function, although at the start we could also do this for the transition function. In order to include a wide range of Feller generators, it is our aim to impose conditions on the density function, that are as general as possible. Since these conditions also have to be manageable, there will be some restrictions, like continuity and certain boundedness properties of the density. As a consequence we are able to study the spectral properties of the generators using the stochastic representation of their resolvents. Let us call this branch of spectral theory: Stochastic Spectral Analysis. Therefore we suppose that the following

Basic assumptions of stochastic spectral analysis (BASSA)

are verified.

(1) **Existence** Let $p : \mathbf{R}_+ \times \mathbf{R}^n \times \mathbf{R}^n \to \mathbf{R}_+$ be a continuous function. This function is supposed to be Markov transition density in the sense that it satisfies

$$\int_{\mathbf{R}^n} p(s,x,u)p(t,u,y)du = p(s+t,x,y)$$

and

$$\int_{\mathbf{R}^n} p(s,x,y)dy \le 1.$$

The corresponding Markov process is denoted by $(\mathbf{R}_+; \Omega, \mathcal{F}, \mathbf{P}_x; \omega(t))$ with right continuous paths $\omega(\cdot)$.

(2) **Feller property** For every $f \in C_\infty(\mathbf{R}^n)$ (continuous functions vanishing at infinity) the function $x \mapsto \int f(y)p(t,x,y)dy$ also belongs to $C_\infty(\mathbf{R}^n)$.

(3) **Continuity** For every $f \in C_\infty(\mathbf{R}^n)$ and for any $x \in \mathbf{R}^n$ the following equality is valid:

$$\lim_{t\downarrow 0} \int_{\mathbf{R}^n} f(y)p(t,x,y)dy = f(x).$$

(4) **Symmetry** The identity

$$p(t,x,y) = p(t,y,x)$$

holds for all $t \in \mathbf{R}_+$ and for all $x,\ y \in \mathbf{R}^n$.

A density function $p(t,x,y)$, verifying (1), (2), (3) and (4) corresponds to a semi-group. Because of the symmetry such a semi-group is self-adjoint with a generator K_0, which is self-adjoint and which verifies

$$\left[e^{-tK_0}f\right](x) := \int_{\mathbf{R}^n} f(y)p(t,x,y)dy \tag{3}$$

$$= \int_{\Omega} f(\omega(t))\mathbf{P}_x(d\omega). \tag{4}$$

2. ADMISSABLE GENERALIZED SCHRÖDINGER OPERATORS.

The perturbations of such free generators K_0 are defined in the same way as for Kato's class potentials.

Definition 2.1. Let $W : \mathbf{R}^n \to [0,\infty]$ be a Borel measurable function. The function W is said to belong to KF if

$$\limsup_{\tau \downarrow 0} \left\| \int_0^\tau e^{-sK_0} W ds \right\|_{\infty,\infty} = 0. \tag{5}$$

The function W is said to belong to $\mathrm{KF}_{\mathrm{loc}}$ if $1_B W$ belongs to KF for every compact subset B of $\mathbf{R}^n$. A potential $V = V_+ - V_-$ is called a *Kato-Feller* potential if V_+ belongs $\mathrm{KF}_{\mathrm{loc}}$ and if V_- belongs to KF.

It was shown by the second author [Cas-1], [Cas-2] that $K_0 \dot{+} V$ is a generator of a strongly continuous semigroup in $L^p(\mathbf{R}^n)$, $1 \le p < \infty$. In $L^2(\mathbf{R}^n)$ it is a self-adjoint, positivity preserving semi-group with a self-adjoint generator. In [De Cas-2] the authors give abstract conditions in order that this Feynman-Kac generator is the Friedrichs' extension of $K_0 + V$. Using the short-hand notation of a conditional measure we get

$$\left(e^{-t(K_0 \dot{+} V)}\right)(x,y) = \lim_{\tau \uparrow t} \mathbf{E}_x \left\{ e^{-\int_0^\tau V(\omega(s))ds} p(t-\tau,\omega(\tau),y) \right\} \tag{6}$$

$$=: \int_{\Omega_x^{y,t}} e^{-\int_0^t V(\omega(s))ds} \mathbf{P}_x^{y,t}(d\omega), \tag{7}$$

where $\Omega_x^{y,t} = \{\omega \in \Omega : \omega(0) = 0, \omega(t) = y\}$. The function $e^{-t(K_0 \dot{+} V)}(x,y)$ is continuous, it satisfies the Chapman-Kolmogorov equation and it is symmetric in x and y. Moreover, this kernel of the integral operator $e^{-t(K_0 \dot{+} V)}$ can be estimated by the free kernel (see e.g. [Cas-2]):

$$\left(e^{-t(K_0 \dot{+} V)}\right)(x,y) \le c e^{At} \left[\sup_{x,y} p(t,x,y)\right]^{1/q} [p(t,x,y)]^{1/r} \tag{8}$$

with $\frac{1}{q}+\frac{1}{r}=1$, $1<r,\ q<\infty$, and c, A positive constants. Notice that bounded potentials are always Kato-Feller potentials. In particular this is of interest for scattering spectral results, where the behaviour of the potentials in finite regions is of no great importance, only the potential part at infinity has an influence on the scattering data. But for the latter part a boundedness assumption is not to restrictive.

This introduction of regular perturbations of K_0 is analogous to the Kato's class potentials if K_0 is the Laplace operator and $p(t,x,y)$ is the Wiener density function. Therefore the formulation of this kind of stochastic spectral analysis only makes sense, if it enables us to include not only the canonical Gaussian case, but also some other physically interesting examples.

This is why we include some examples of unperturbed Hamiltonians K_0, that are included in our discussion of BASSA.

Example 1. Consider second order differential operators of the form

$$-\sum_{i,j=1}^{n}\frac{\partial}{\partial x_i}\left(a_{ij}(x)\frac{\partial}{\partial x_j}\right). \tag{9}$$

One may impose several conditions on the coefficients $a_{ij}(x)$ for which the hypotheses in BASSA are satisfied. Kochubeĭ [Ko] used the following:

$$\frac{\partial^2 a_{ij}}{\partial x_i \partial x_j}\in L^1_{\text{loc}}(\mathbf{R}^n)\cap L^{n/2}_{\text{loc}}(\mathbf{R}^n);\quad \sum_{i,j=1}^{n}\frac{\partial^2 a_{ij}}{\partial x_i\partial x_j}\le 0,$$

where the functions $\frac{\partial^2 a_{ij}}{\partial x_i \partial x_j}$ are locally Hölder continuous. Then he proved that the underlying process is a strong Markov process with the Feller property. Further conditions can be imposed, in order that the formally symmetric operator in (9) gives rise to an essentially self-adjoint operator K_0. In particular, unbounded coefficients $a_{ij}(x)$ are possible. For this reason the corresponding diffusion process cannot be canonical in the sense that $p(t,x,y)$ is dominated by the Wiener density.

In [Da] Davies considers the Feller property of semigroups on manifolds. For flat spaces his conditions allow coefficients which need not be differentiable.

Example 2. Let K_0 be the relativistic Hamiltonian $\sqrt{-\Delta+m^2}-m$. Then Carmona, Masters and Simon [Ca, Ma, Si] showed that this operator is the generator of a Feller process with transition density

$$p(t,x,y)=\frac{1}{(2\pi)^n}\frac{t}{\sqrt{|x-y|^2+t^2}}\int_{\mathbf{R}^n}e^{mt-\sqrt{(|x-y|^2+t^2)(u^2+m^2)}}du. \tag{10}$$

(In fact in [Ca, Ma, Si] the authors study more general situations.) The density in (10) is symmetric and it satisfies a boundedness assumption of the form

$$p(t,x,y)\le\frac{e^{mt}}{(2\pi t)^n}\int_{\mathbf{R}^n}e^{-|u|}du.$$

Example 3. It is also possible to include certain pseudo-differential operators. Jacob [Ja] considered

$$L(x,D)\varphi(x) = \sum_{j=1}^{n} b_j(x)\Lambda_j^r(D_j)\varphi(x), \quad \varphi \in C_0^\infty(\mathbf{R}^n), \tag{11}$$

with

$$\Lambda_j^r(D_j) = \frac{1}{(2\pi)^{n/2}} \int_{\mathbf{R}^n} e^{ix.\xi}\Lambda_j^r(\xi_j)\,\hat{\varphi}(\xi)d\xi$$

and with

$$\Lambda_j^r(\xi) = \left(1+|\xi|^2\right)^{r/2}, \quad 0 < r \le 2.$$

He proved that $L(x,D)$ is a Feller generator, provided that, for $1 \le j \le n$, the function b_j belongs to $C^\infty(\mathbf{R}^n)$, that $\partial^\alpha b_j$ is bounded and that $\dfrac{\partial b_j}{\partial x_j} = 0$. The underlying process is a strong Markov process, possessing the Feller property and having a symmetric density which can be estimated by

$$p(t,x,y) \le ct^{-n/r}e^{bt}, \tag{12}$$

where c and b are certain positive constants. This means that all conditions for BASSA are fulfilled.

Schrödinger semigroups are often used to obtain spectral theoretical results. A recent survey was given in [De]. However this theory is restricted to single free operators, most of the time the Laplacian.

In stochastic spectral analysis one may consider classes of generators K_0, i.e. one can give unified proofs for a variety of physically interesting situation. The most important tools are the stochastic representations of semigroups and resolvents. In this fashion one can employ operator theory, integral operators together with stochastic analysis.

3. PERTURBATIONS OF GENERALIZED SCHRÖDINGER OPERATORS

Here we want to compare situations where a Feller generator K_0 is perturbed by different Kato-Feller potentials V and W. Several spectral theoretical consequences are due to the behaviour of sandwiched differences of resolvents or semi-groups. In order to prove that the operator $e^{-(K_0+V)} - e^{-(K_0+W)}$ is a trace class operator, one can factorize the difference

$$\begin{aligned} &e^{-2(K_0+V)} - e^{-2(K_0+W)} \\ &= \left[e^{-(K_0+V)}\varphi^{-1}\right]\left[\varphi\left(e^{-(K_0+V)} - e^{-(K_0+W)}\right)\right] \\ &\quad + \left[\left(e^{-(K_0+V)} - e^{-(K_0+W)}\right)\varphi\right]\left[\varphi^{-1}e^{-(K_0+W)}\right], \end{aligned} \tag{13}$$

where φ denotes a multiplication operator with a non-vanishing function. Then one shows that every factor is a Hilbert Schmidt operator.

Another reason to consider sandwiched differences is an estimate for resolvent differences in the limit absorption case. By the Hilbert identity one may estimate

$$\begin{aligned}&\left\|\varphi^{-1}\left(R_V(\lambda \pm i0) - R_W(\lambda \pm i0)\right)\varphi^{-1}\right\| \\ &\leq \left(1 + |\lambda - a|\left\|\varphi^{-1}R_V(\lambda \pm i0)\,\varphi^{-1}\right\|\right) \\ &\quad \times \left(1 + |\lambda - a|\left\|\varphi^{-1}R_W(\lambda \pm i0)\,\varphi^{-1}\right\|\right) \\ &\quad \times \left\|\varphi^{-1}\left(R_V(a) - R_W(a)\right)\varphi^{-1}\right\|,\end{aligned} \tag{14}$$

with $R_V(a) = \left(aI + K_0 \dot{+} V\right)^{-1}$, where a is any resolvent value of R_V and of R_W and where λ is a real value. Inequality (14) holds if the right-hand side is finite. In what follows we consider such sandwiched differences.

Proposition 1. Suppose BASSA and let V and W be two Kato-Feller potentials. Let φ and ψ be non-vanishing Borel measurable functions on $\mathbf{R}^n$. Then the following inequality is valid for $0 \leq \gamma \leq 1$ and for arbitrary α, $\beta \geq 0$:

$$\begin{aligned}&\left\|\varphi\left[\left(K_0 \dot{+} V + aI\right)^{-1} - \left(K_0 \dot{+} W + aI\right)^{-1}\right]\psi\right\|_{2,2} \\ &\leq \left\||\psi|^{2\beta}\left(K_0 + aI\right)^{-1}|V - W|^{2\gamma}\left(K_0 \dot{+} 2V + aI\right)^{-1}|\varphi|^{2\alpha}\right\|_{\infty}^{1/2} \\ &\quad \times \left\||\varphi|^{2-2\alpha}\left(K_0 + aI\right)^{-1}|V - W|^{2-2\gamma}\left(K_0 \dot{+} 2W + aI\right)^{-1}|\psi|^{2-2\beta}\right\|_{\infty}^{1/2}.\end{aligned} \tag{15}$$

Proof. Let f and g be functions in $L^2(\mathbf{R}^n)$. Since resolvents can be written as Laplace transforms we have by Duhamel's formula

$$\begin{aligned}&\left|\left\langle \varphi\left[\left(K_0 \dot{+} V + aI\right)^{-1} - \left(K_0 \dot{+} V + aI\right)^{-1}\right]\psi f, g\right\rangle\right| \\ &\leq \int_0^\infty d\lambda e^{-a\lambda}\int_0^\lambda ds\left|\left\langle \varphi\left[e^{-s(K_0 \dot{+} V)}\left(W - V\right)e^{-(\lambda - s)(K_0 \dot{+} W)}\psi f\right], g\right\rangle\right| \\ &\leq \int_0^\infty d\lambda e^{-a\lambda}\int_0^\lambda ds\int dx \\ &\quad |W(x) - V(x)|^\gamma\left[e^{-(\lambda - s)(K_0 + W)}|\psi f|\right](x)\,|W(x) - V(x)|^{1-\gamma}\left[e^{-s(K_0 \dot{+} V)}|\varphi f|\right](x)\end{aligned} \tag{16}$$

$$\begin{aligned}&= \int_0^\infty d\lambda e^{-a\lambda}\int_0^\lambda ds\int dx \\ &\quad \mathsf{E}_x\left\{|W(x) - V(x)|^\gamma\,|\psi(\omega(\lambda - s))|\,|f(\omega(\lambda - s))|\right. \\ &\quad \left.\left(e^{-s(K_0 \dot{+} 2V)}|\varphi|^{2\alpha}\right)^{1/2}(x)\,|W(x) - V(x)|^{1-\gamma}e^{-\int_0^{\lambda - s}W(\omega(\sigma))d\sigma}\left(e^{-sK_0}|\varphi|^{2-2\alpha}|g|^2\right)^{1/2}\right\},\end{aligned} \tag{17}$$

with

$$\left(\left[e^{-s(K_0 \dot{+} V)}|\varphi g|\right](x)\right)^2 \leq \left[e^{-s(K_0 \dot{+} 2V)}|\varphi|^{2\alpha}\right](x)\left[e^{-s(K_0)}|\varphi|^{2-2\alpha}|g|^2\right](x).$$

The quantity in (17) can be estimated by a product of two factors:

$$\left[\int_0^\infty d\lambda e^{-a\lambda}\int_0^\lambda ds\int dx \right. \tag{18}$$

$$\left. \mathsf{E}_x\left\{|W(x)-V(x)|^{2\gamma}\,|\psi(\omega(\lambda-s))|^2\left[e^{-s(K_0\dot{+}2V)}\,|\varphi|^{2\alpha}\right](x)\right\}\right]^{1/2}$$

$$\times\left[\int_0^\infty d\lambda e^{-a\lambda}\int_0^\lambda ds\int dx \right. \tag{19}$$

$$\left. \mathsf{E}_x\left\{|\psi(\omega(\lambda-s))|^{2-2\beta}\,e^{-2\int_0^{\lambda-s}W(\omega(\sigma))d\sigma}\,|W(x)-V(x)|^{2-2\gamma}\left[e^{-sK_0}\,|\varphi|^{2-2\alpha}\,|g|^2\right](x)\right\}\right]^{1/2}.$$

Since

$$\int_0^\infty d\lambda e^{-a\lambda}\int_0^\lambda ds\,|g_1(s)|\,|g_2(\lambda-s)|=\int_0^\infty dse^{-as}\int_0^\infty due^{-au}\,|g_1(s)|\,|g_2(u)|\,,$$

the factor in (18) is dominated by

$$\left\|\int_0^\infty due^{-au}\int_0^\infty dse^{-as}e^{-s(K_0\dot{+}2V)}\,|\varphi|^{2\alpha}\,|W-V|^{2\gamma}\,e^{-sK_0}\,|\psi|^{2\beta}\,|f|^2\right\|_1^{1/2}$$

$$\leq\left\{\sup_{y\in\mathbf{R}^n}\int_{\mathbf{R}^n}dx\int_0^\infty due^{-au}\int_0^\infty dse^{-as}\right.$$

$$\left.\left[e^{-s(K_0\dot{+}2V)}\,|\varphi|^{2\alpha}\right](x)\,|W(x)-V(x)|^{2\gamma}\,p(u,x,y)\,|\psi(y)|^{2\beta}\right\}^{1/2}\|f\|_2$$

$$=\left\|\int_0^\infty due^{-au}\int_0^\infty dse^{-as}\,|\psi|^{2\beta}\,e^{-uK_0}\,|W-V|^{2\gamma}\,e^{-s(K_0\dot{+}2V)}\,|\varphi|^{2\alpha}\right\|_\infty^{1/2}\|f\|_2$$

$$=\left\||\psi|^{2\beta}\,(K_0+aI)^{-1}\,|W-V|^{2\gamma}\,(K_0\dot{+}2V+aI)^{-1}\,|\varphi|^{2\alpha}\right\|_\infty^{1/2}\|f\|_2\,. \tag{20}$$

For the factor (19) we get

$$=\left\|\int_0^\infty due^{-au}\int_0^\infty dse^{-as}\,|W-V|^{2-2\gamma}\,e^{-(\lambda-s)(K_0\dot{+}2W)}\,|\psi|^{2-2\beta}\,e^{-sK_0}\,|\varphi|^{2-2\alpha}\,|g|^2\right\|_1^{1/2}$$

$$\leq\left\||\varphi|^{2-2\alpha}\,(K_0+aI)^{-1}\,|W-V|^{2-2\gamma}\,(K_0\dot{+}2W+aI)^{-1}\,|\psi|^{2-2\beta}\right\|_\infty^{1/2}\|g\|_2\,. \tag{21}$$

The inequalities in (20) and (21) yield the desired estimate in (15).

q.e.d.

In this first proposition the only conditions imposed on α, β and γ are: $\alpha \geq 0$, $\beta \geq 0$ and $0 \leq \gamma \leq 1$. To some extent we will study such general α, β and γ in [De, Cas-2]. The main purpose of the present article is to estimate the resolvent difference in (14) for the limit absorption case and to study the continuity of the S-matrix. For this purpose it suffices to choose particular values for α, β and γ. For the choice $\alpha = 1$, $\beta = 0$ and $\gamma = \frac{1}{2}$, the estimates in (15) become simpler as is indicated in the following corollary.

Corollary 2. Suppose BASSA and let V and W be two Kato-Feller potentials. In addition let φ be a non-vanishing Borel measurable function on $\mathbf{R}^n$. Then the following inequality is valid:

$$\begin{aligned} &\left\|\varphi\left[\left(K_0\dotplus V + aI\right)^{-1} - \left(K_0\dotplus W + aI\right)^{-1}\right]\varphi\right\|_{2,2} \\ &\leq c\left\|\left(K_0 + aI\right)^{-1}|V - W|\left(K_0\dotplus 2V + aI\right)^{-1}|\varphi|^2\right\|_{\infty}^{1/2} \\ &\quad\left\|\left(K_0 + aI\right)^{-1}|V - W|\left(K_0\dotplus 2W + aI\right)^{-1}|\varphi|^2\right\|_{\infty}^{1/2}. \end{aligned} \tag{22}$$

In the presence of Kato-Feller potentials we may and do define so-called Kato-Feller norms.

Definition. Let $V = V_+ - V_-$ be a Borel measurable function for which V_+ as well as V_- belongs to KF. Then the Kato-Feller norm $\|V\|_{\mathrm{KF}}$ is defined by

$$\begin{aligned} \|V\|_{\mathrm{KF}} &:= \sup_{x\in\mathbf{R}^n}\int_0^1 ds\int_{\mathbf{R}^n} p(s,x,y)\,|V(y)|\,dy \\ &= \sup_{x\in\mathbf{R}^n}\int_0^1 ds\left[e^{-sK_0}|V|\right](x). \end{aligned} \tag{23}$$

In order to analyze (22) in terms of the Kato-Feller norm we need some properties of the function φ. The choice of φ depends on the estimates one has in mind. For many operators K_0 in (13) it suffices to take φ in such a way that φ^{-1} belongs to $L^2(\mathbf{R}^n)$. In (14) one has to choose φ in such a way that $\left\|\varphi^{-1}R_V(\lambda \pm i0)\varphi^{-1}\right\|$ is bounded. For $K_0 = -\frac{1}{2}\Delta$ and an Agmon-potential V one may utilize $\varphi(x) = \left(1 + |x|^2\right)^{\delta/2}$ with $\delta > \frac{1}{2}$.

Therefore it is sufficient to take for φ a "nice" function, e.g. a polynomially bounded one with the property that $\varphi(x) \geq 1$ and for which the following kind of pointwise estimates hold:

$$\left[e^{-tK_0}|\varphi|^\rho\right](x) \leq ce^{bt}|\varphi(x)|^\rho,$$

b, $c \geq 0$, ρ arbitrary, or

$$\left[\left(K_0 + aI\right)^{-1}|\varphi|^\rho\right](x) \leq c\,|\varphi(x)|^\rho,$$

for some $\rho \geq 0$.

Proposition 3. Suppose BASSA and let V and W be two Kato-Feller potentials. In addition let φ be a non-vanishing Borel measurable with the property that

$$\left[(K_0 + aI)^{-1} |\varphi|^{2\rho}\right](x) \leq c|\varphi(x)|^{2\rho} \tag{24}$$

for some $\rho > 1$. Then

$$\left\|\varphi\left[(K_0 \dot{+} V = aI)^{-1} - (K_0 \dot{+} W + aI)^{-1}\right]\varphi\right\|_{2,2} \leq C(a)\left\||V - W|\,|\varphi|^2\right\|_{\mathrm{KF}}, \tag{25}$$

where $C(a)$ tends to 0 if a tends to $+\infty$.

Proof. Because of (24) the following inequality is valid:

$$\begin{aligned}
&\left[(K_0 \dot{+} 2V + aI)^{-1} |\varphi|^2\right](x) \\
&\leq \left[\int_0^\infty d\lambda e^{-a\lambda} \mathsf{E}_x\left\{e^{-2\rho\int_0^\lambda V(\omega(\sigma))d\sigma}\right\}\right]^{1/\rho}\left[\int_0^\infty d\lambda e^{-a\lambda}\mathsf{E}_x\left\{|\varphi(\omega(\lambda))|^{2\rho'}\right\}\right]^{1/\rho'} \\
&\leq c|\varphi(x)|^2,
\end{aligned} \tag{26}$$

where $\frac{1}{\rho} + \frac{1}{\rho'} = 1$, $1 < \rho, \rho' < \infty$. We estimate the first factor on the right-hand side of (22):

$$\begin{aligned}
&\sup_{x\in\mathbf{R}^n}\left[\int_0^\infty d\lambda e^{-a\lambda}\mathsf{E}_x\left\{|V(\omega(\lambda)) - W(\omega(\lambda))|\left((K_0 \dot{+} 2V + aI)^{-1}|\varphi|^2\right)(\omega(\lambda))\right\}\right] \\
&\leq c\sup_{x\in\mathbf{R}^n}\left[\int_0^\infty d\lambda e^{-a\lambda}\mathsf{E}_x\left\{|V(\omega(\lambda)) - W(\omega(\lambda))|\,|\varphi(\omega(\lambda))|^2\right\}\right] \\
&= c\left\|(K_0 + aI)^{-1}|V - W|\,|\varphi|^2\right\|_\infty
\end{aligned}$$

and hence

$$\begin{aligned}
&\left\|\varphi\left[(K_0 \dot{+} V = aI)^{-1} - (K_0 \dot{+} W + aI)^{-1}\right]\varphi\right\|_{2,2} \leq c\left\|(K_0 + aI)^{-1}|V - W|\,|\varphi|^2\right\|_\infty \\
&\leq c\sup_{x\in\mathbf{R}^n}\sum_{k=0}^\infty\int_k^{k+1} d\lambda e^{-a\lambda}\left[e^{-\lambda K_0}|V - W|\,|\varphi|^2\right](x) \\
&\leq c\sup_{x\in\mathbf{R}^n}\sum_{k=0}^\infty e^{-ak}\int_0^1 dse^{-as}\left[e^{-kK_0}e^{-sK_0}|V - W|\,|\varphi|^2\right](x) \\
&\leq c\sum_{k=0}^\infty e^{-ak}\sup_{y\in\mathbf{R}^n}\int_0^1 dse^{-as}\left[e^{-sK_0}|V - W|\,|\varphi|^2\right](y) \leq C\left\||V - W\|\,|\varphi|^2\right\|_{\mathrm{KF}}.
\end{aligned}$$

q.e.d.

Next we consider a sequence of Kato-Feller potentials $\{V_\alpha\}$, tending to V_{α_0} in the Kato-Feller norm as $\alpha \to \alpha_0$. In this case one may use the same methods as above. To obtain a uniform upper bound for the factors in (22) we need a stability condition with respect to α. This stability condition guarantees that the first factor in (26) admits a uniform upper bound.

Stability condition. Let $\{V_\alpha\}$ be a sequence of Kato-Feller potentials with respect to K_0. Then, from Khas'minskii's lemma it follows that

$$\sup_{x\in\mathbf{R}^n} \mathsf{E}_x \left\{ e^{-\int_0^\lambda V_\alpha(\omega(s))ds} \right\} \leq B_\alpha e^{A_\alpha \lambda}. \tag{27}$$

The stability condition consists in a uniform bound on B_α and on A_α in the sense that, for some non-negative numbers B and A independent of α,

$$B_\alpha \leq B, \qquad A_\alpha \leq A, \tag{28}$$

for all admissible α.

Remark. A simple sufficient condition for (28) to hold is provided by Khas'mainskii's lemma. Write each V_α as a difference of its positive and negative part, i.e. $V_\alpha = V_{\alpha,+} - V_{\alpha,-}$. Assume that there exists a (sufficiently small) strictly positive number T with the property that for all α the following inequality is valid:

$$\sup_{x\in\mathbf{R}^n} \mathsf{E}_x \left\{ \int_0^T V_{\alpha,-}(\omega(\sigma))d\sigma \right\} \leq d < 1, \tag{29}$$

or, what is the same,

$$\sup_{x\in\mathbf{R}^n} \int_0^T \left[e^{-\sigma K_0} V_{\alpha,-}\right](x)d\sigma \leq d < 1. \tag{30}$$

(Compare this with the Kato-Feller norm in (23).) Then the stability condition is satisfied with

$$B = (1-d)^{-1}; \tag{31}$$

$$A = -T^{-1}\ln(1-d). \tag{32}$$

Corollary 4. Suppose BASSA and let $\{V_\alpha\}$ be sequence of Kato-Feller potentials with respect to K_0. Suppose that the sequence $\{V_\alpha\}$ verifies the stability condition (28) and let φ be a Borel function as in Corollary 2 or in Proposition 3, satisfying the pointwise estimate:

$$\left[(K_0 + aI)^{-1} |\varphi|^{2\rho}\right](x) \leq c\,|\varphi(x)|^{2\rho} \tag{33}$$

for some real number $\rho > 1$. Then

$$\left\|\varphi\left[(K_0\dot{+}V_\alpha + aI)^{-1} - (K_0\dot{+}V_{\alpha_0} + aI)^{-1}\right]\varphi\right\| \leq C\,\|(V_\alpha - V_{\alpha_0})\varphi^2\|_{\mathrm{KF}}\,.$$

4. CONVERGENCE OF SCATTERING MATRICES

The final aim of this article is to analyze the convergence of scattering matrices for sequences of Kato-Feller potentials. For the existence and completeness of the corresponding systems sufficient conditions can be given via trace class criteria or via the Cook criterion (see [De, Cas-2]). One possibility is to ensure that the operators in (13) are trace class operators. If the wave operators for the scattering systems $(K_0\dot{+}V_\alpha, K_0)$ and $(K_0\dot{+}V_{\alpha_0}, K_0)$ exist, i.e. if the operators

$$\Omega_{\pm,\alpha} = \text{s-}\lim_{t\to\pm\infty} e^{it(K_0\dot{+}V_\alpha)}e^{-itK_0}P_{ac}(K_0)$$

and

$$\Omega_{\pm,\alpha_0} = \text{s-}\lim_{t\to\pm\infty} e^{it(K_0\dot{+}V_{\alpha_0})}e^{-itK_0}P_{ac}(K_0)$$

exist, then the scattering operators

$$S_\alpha = \Omega^*_{+,\alpha}\Omega_{-,\alpha} \quad \text{and} \quad S_{\alpha_0} = \Omega^*_{+,\alpha_0}\Omega_{-,\alpha_0}$$

are available and these operators S_α and S_{α_0} commute with K_0. The use of the spectral transformation Φ_{K_0} of K_0 on $\sigma_{ac}(K_0)$ enables us to introduce the scattering matrices:

$$[\Phi_{K_0}S_\alpha g]_\lambda =: S_\alpha(\lambda)\,[\Phi_{K_0}g]_\lambda \tag{34}$$

and

$$[\Phi_{K_0}S_{\alpha_0} g]_\lambda =: S_{\alpha_0}(\lambda)\,[\Phi_{K_0}g]_\lambda\,, \tag{35}$$

where $[\Phi_{K_0}g]_\lambda$ denotes the representative of $g \in P_{ac}(K_0)\mathcal{H}$ in $\mathcal{H}_{K_0}(\lambda)$. Here $\mathcal{H}_{K_0}$ is a fiber of the spectral resolution of $P_{ac}(K_0)\mathcal{H}$ with respect to K_0 (see [Ba, Wo] p. 189).

Choose a non-negative function φ, such that its reciprocal φ^{-1} is bounded. Suppose that, for some λ_0, the inequality

$$\|V_\alpha\varphi^{-1}R_{\alpha_0}(\lambda_0 - i0)\varphi^{-1}\| < 1, \tag{36}$$

is valid. Then the scattering matrix defined in (34) is represented by

$$S_\alpha(\lambda_0) = I_{\mathcal{H}_{K_0}(\lambda_0)} - 2\pi i F(\lambda_0)\,(V_\alpha - V_\alpha R_\alpha(\lambda_0 - i0)\,V_\alpha)\,F^*(\lambda_0). \tag{37}$$

Here $F(\lambda_0)$ is a bounded operator mapping the weighted L^2-space $L^2_\varphi(\mathbb{R}^n)$ (given by functions f such that $\int dx\,|\varphi(x)|^2\,|f(x)|^2 < \infty$) onto $\mathcal{H}_{K_0}(\lambda_0)$.

Theorem 5. Suppose BASSA and let $\{V_\alpha\}$ be sequence of Kato-Feller potentials with respect to K_0. In addition suppose that the wave operators for $(K_0\dot{+}V_\alpha, K_0)$ and for $(K_0\dot{+}V_{\alpha_0}, K_0)$ exist. Moreover suppose that the sequence $\{V_\alpha\}$ satisfies the stability condition. Finally let $\varphi \geq 1$ be a Borel measurable function on $\mathbb{R}^n$ such that the following conditions are satisfied:

(1) Put

$$v_\varphi = \sup_{\alpha,\alpha_0} \sup_{x \in \mathbf{R}^n} |\varphi(x) V_\alpha(x)| \tag{38}$$

and suppose $v_\varphi < \infty$.

(2) Write $v = \sup_{\alpha,\alpha_0} \sup_{x\in\mathbf{R}^n} |V_\alpha(x)|$ and let $\lambda \in \mathbf{R}$ be such that

$$\left\|\varphi^{-1} R_\alpha(\lambda \pm i0)\varphi^{-1}\right\| \leq \frac{1}{v}. \tag{39}$$

(3) There exists a finite constant c such that

$$\left[(K_0 + aI)^{-1} \varphi^{2\rho}\right](x) \leq c\varphi(x)^{2\rho} \tag{40}$$

for some $\rho > 1$ and also for $\rho = 1$.

(4) For the limit potential V_{α_0} the quantity d_{α_0} defined by

$$d_{\alpha_0} = \left\|\varphi^{-1} R_{\alpha_0}(\lambda \pm i0)\varphi^{-1}\right\| \tag{41}$$

is finite.

Then the scattering matrices $S_\alpha(\lambda)$ and $S_{\alpha_0}(\lambda)$ exist and there exists a constant C such that for all α sufficiently close to α_0 (i.e. $|\alpha - \alpha_0| < \epsilon$), the following inequality is valid:

$$\|S_\alpha(\lambda) - S_{\alpha_0}(\lambda)\|^2 \leq C \left\|(V_\alpha - V_{\alpha_0})\varphi^2\right\|_{\mathrm{KF}}. \tag{42}$$

The constant C depends on ϵ, on λ, on φ and on v. It can be given in terms of the constants defined in the assumptions (see the proof).

Proof. From (38) and (39) it follows that the scattering matrices are given by the representation in (37). Therefore we have

$$\begin{aligned}
&\left[\frac{1}{2\pi}\langle [S_\alpha(\lambda) - S_{\alpha_0}(\lambda)] f, [S_\alpha(\lambda) - S_{\alpha_0}(\lambda)] f\rangle\right]^{1/2} && (43)\\
&\leq \langle F(\lambda)(V_\alpha - V_{\alpha_0}) F^*(\lambda) f, F(\lambda)(V_\alpha - V_{\alpha_0}) F^*(\lambda) f\rangle && \text{(I)}\\
&\quad + \langle F(\lambda) V_\alpha [R_\alpha(\lambda - i0) - R_{\alpha_0}(\lambda - i0)] V_\alpha F^*(\lambda) f, && \text{(II)}\\
&\quad F(\lambda) V_\alpha [R_\alpha(\lambda - i0) - R_{\alpha_0}(\lambda - i0)] V_\alpha F^*(\lambda) f\rangle &&\\
&\quad + \langle F(\lambda)(V_\alpha - V_{\alpha_0}) R_{\alpha_0}(\lambda - i0) V_\alpha F^*(\lambda) f, && \text{(III)}\\
&\quad F(\lambda)(V_\alpha - V_{\alpha_0}) R_{\alpha_0}(\lambda - i0) V_\alpha F^*(\lambda) f\rangle &&\\
&\quad + \langle F(\lambda) V_{\alpha_0} R_{\alpha_0}(\lambda - i0)(V_\alpha - V_{\alpha_0}) F^*(\lambda) f, && \text{(IV)}\\
&\quad F(\lambda) V_{\alpha_0} R_{\alpha_0}(\lambda - i0)(V_\alpha - V_{\alpha_0}) F^*(\lambda) f\rangle. &&
\end{aligned}$$

Next we estimate every term in (43). We set $W = V_\alpha - V_{\alpha_0}$. From (39) we infer, employing the notation

$$F(\lambda) =: D_\lambda \varphi^{-1}, \tag{44}$$

$$\begin{aligned}
\|F^*(\lambda)F(\lambda)\| &= \|\varphi^{-1} D_\lambda^* D_\lambda \varphi^{-1}\| = \left\|\varphi^{-1}\frac{dE_0(\lambda)}{d\lambda}\varphi^{-1}\right\| \\
&\leq \frac{1}{2\pi}\left\|\varphi^{-1}[R_{\alpha_0}(\lambda - i0) - R_{\alpha_0}(\lambda + i0)]\varphi^{-1}\right\| \leq \frac{1}{\pi v}.
\end{aligned} \tag{45}$$

We also use

$$\|F(\lambda)\|^2_{L^2_\varphi \to \mathcal{H}_{K_0}(\lambda)} = \|F^*(\lambda)F(\lambda)\|_{L^2_\varphi \to L^2_\varphi} \,. \tag{46}$$

For the first term (I) in (43) we get:

$$\begin{aligned}
&\langle F(\lambda)WF^*(\lambda)f, F(\lambda)WF^*(\lambda)f\rangle^{1/2} \\
&= \langle W\varphi^{-1}D^*_\lambda D_\lambda \varphi^{-1} W\varphi^{-1}D^*_\lambda f, \varphi^{-1}D^*_\lambda f\rangle^{1/2} \\
&= \Big\langle W\varphi^{-1}(K_0+aI)^{-1}\varphi\varphi^{-1}(K_0+aI)\frac{dE_0(\lambda)}{d\lambda}(K_0+aI)\varphi^{-1} \\
&\qquad \varphi(K_0+aI)^{-1}\varphi W\varphi^{-1}D^*_\lambda f, \varphi^{-1}D^*_\lambda f\Big\rangle^{1/2} \\
&\le \left\|W\varphi^{-1}(K_0+aI)^{-1}\varphi\right\| \left\|\varphi^{-1}(K_0+aI)\frac{dE_0(\lambda)}{d\lambda}(K_0+aI)\varphi^{-1}\right\|^{1/2} \|\varphi^{-1}D^*_\lambda\|\,\|f\| \\
&\le (\lambda+a)\frac{1}{\sqrt{\pi v}}\|f\|\left\|W\varphi^{-1}(K_0+aI)^{-1}\varphi\right\| .
\end{aligned} \tag{47}$$

In order to estimate term (II) in (43) we write

$$R = R_\alpha(\lambda - i0) - R_{\alpha_0}(\lambda - i0).$$

Then we see

$$\begin{aligned}
&\langle F(\lambda)V_\alpha R V_\alpha F^*(\lambda)f, F(\lambda)V_\alpha R V_\alpha F^*(\lambda)f\rangle^{1/2} \\
&= \Big\langle V_\alpha\varphi^{-1}\frac{dE_0(\lambda)}{d\lambda}\varphi^{-1}V_\alpha R V_\alpha \varphi^{-1}D^*_\lambda f, RV_\alpha\varphi^{-1}D^*_\lambda f\Big\rangle^{1/2} \\
&= \Big\langle \varphi V_\alpha\varphi^{-1}\frac{dE_0(\lambda)}{d\lambda}\varphi^{-1}V_\alpha\varphi\varphi^{-1}R\varphi^{-1}\varphi V_\alpha\varphi^{-1}D^*_\lambda f, \varphi^{-1}R\varphi^{-1}\varphi V_\alpha\varphi^{-1}D^*_\lambda f\Big\rangle^{1/2} \\
&\le \frac{v_\varphi^2}{\sqrt{\pi v}}\|\varphi^{-1}R\varphi^{-1}\|\,\|\varphi^{-1}D^*_\lambda\|\,\|f\| \\
&\le \frac{v_\varphi^2}{\pi v}\|f\|\,\|\varphi^{-1}[R_\alpha(\lambda-i0)-R_{\alpha_0}(\lambda-i0)]\varphi^{-1}\| .
\end{aligned} \tag{48}$$

Now we have (see also (14)):

$$\begin{aligned}
&\|\varphi^{-1}[R_\alpha(\lambda-i0)-R_{\alpha_0}(\lambda-i0)]\varphi^{-1}\| \\
&\le (1+|\lambda-a|\,\|\varphi^{-1}R_{\alpha_0}(\lambda-i0)\varphi^{-1}\|) \\
&\qquad \times \|\varphi[R_\alpha(a)-R_{\alpha_0}(a)]\varphi\| \\
&\qquad \times [2+|\lambda-a|\,\|\varphi^{-1}[R_\alpha(\lambda-i0)-R_{\alpha_0}(\lambda-i0)]\varphi^{-1}\|]
\end{aligned}$$

(use Corollary 2.2 and Proposition 3)

$$\le c\left(1+|\lambda-a|\,d_{\alpha_0}\right)\left\|(V_\alpha-V_{\alpha_0})\varphi^2\right\|_{\mathrm{KF}}$$
$$+\,c\,|\lambda-a|\left(1+|\lambda-a|\,d_{\alpha_0}\right)\left\|\varphi^{-1}\left[R_\alpha(\lambda-i0)-R_{\alpha_0}(\lambda-i0)\right]\varphi^{-1}\right\|$$
$$\times\left\|(V_\alpha-V_{\alpha_0})\varphi^2\right\|_{\mathrm{KF}}\,.$$

Therefore, if a is large enough or if $|\alpha-\alpha_0|$ is small enough we obtain:

$$\left\|\varphi^{-1}\left[R_\alpha(\lambda-i0)-R_{\alpha_0}(\lambda-i0)\right]\varphi^{-1}\right\|\le C(\alpha,\alpha_0,\lambda,a)\left\|(V_\alpha-V_{\alpha_0})\varphi^2\right\|_{\mathrm{KF}}\,. \tag{49}$$

Inequalities (48) and (49) provide us with the necessary estimates to find an upper bound for the term (II):

$$\text{term (II)}\le\frac{v_\varphi^2}{\pi v}\,\|f\|\,c(\alpha,\alpha_0,\lambda,a)\left\|(V_\alpha-V_{\alpha_0})\varphi^2\right\|_{\mathrm{KF}}\,. \tag{50}$$

The estimates for the terms (III) and (IV) are similar. Thus we consider here only term number (III). Again we write $W=V_\alpha-V_{\alpha_0}$. An upper bound for (III) is obtained as follows:

$$\langle F(\lambda)WR_{\alpha_0}(\lambda-i0)V_\alpha F^*(\lambda)f,F(\lambda)WR_{\alpha_0}(\lambda-i0)V_\alpha F^*(\lambda)f\rangle^{1/2}$$
$$=\Big\langle\varphi W\varphi^{-1}(K_0+aI)^{-1}\varphi\varphi^{-1}(K_0+aI)\frac{dE_0(\lambda)}{d\lambda}(K_0+aI)\varphi^{-1}$$
$$\varphi(K_0+aI)^{-1}\varphi^{-1}W\varphi\varphi^{-1}R_{\alpha_0}(\lambda-i0)\varphi^{-1}\varphi V_\alpha F^*(\lambda)f,$$
$$\varphi^{-1}R_{\alpha_0}(\lambda-i0)\varphi^{-1}\varphi V_\alpha F^*(\lambda)f\Big\rangle^{1/2}$$
$$\le\left\|W(K_0+aI)^{-1}\varphi\right\|(\lambda+a)\frac{1}{\pi v}d_{\alpha_0}v_\varphi\,\|f\|\,. \tag{51}$$

A combination of (51), (50) and (47) yields an estimate for the norm of the scattering matrices:

$$\|S_\alpha(\lambda)-S_{\alpha_0}(\lambda)\|\le c\Big[\left\|(V_\alpha-V_{\alpha_0})\varphi^{-1}(K_0+aI)^{-1}\varphi\right\|_{2,2} \tag{52}$$
$$+\left\|(V_\alpha-V_{\alpha_0})(K_0+aI)^{-1}\varphi\right\|_{2,2}+\left\|(V_\alpha-V_{\alpha_0})\varphi^2\right\|_{\mathrm{KF}}\Big]\,.$$

Since $\varphi^{-1}\le 1$, we see

$$\left\|(V_\alpha-V_{\alpha_0})\varphi^{-1}(K_0+aI)^{-1}\varphi\right\|\le\left\|(V_\alpha-V_{\alpha_0})(K_0+aI)^{-1}\varphi\right\|\,. \tag{53}$$

So, finally, we estimate the right-hand side of (53) in terms of the Kato-Feller norm:

$$\left|\left\langle W(K_0+aI)^{-1}\varphi f,g\right\rangle\right|^2=\left|\int_{\mathbf{R}^n}dx\,W(x)\int_0^\infty d\lambda e^{-a\lambda}\mathsf{E}_x\left\{\varphi(\omega(\lambda))f(\omega(\lambda))\overline{g(x)}\right\}\right|^2$$
$$\le\int_{\mathbf{R}^n}dx\int_0^\infty d\lambda e^{-a\lambda}\mathsf{E}_x\left\{|W(x)|^2\varphi^2(x)\,|f(\omega(\lambda))|^2\right\}$$
$$\int_{\mathbf{R}^n}dy\int_0^\infty d\mu e^{-a\mu}\mathsf{E}_y\left\{\varphi(\omega(\lambda))^2\varphi(y)^{-2}\,|g(y)|^2\right\}$$
$$\le\left\|(K_0+aI)^{-1}|W|^2|\varphi|^2\right\|_\infty\|f\|\int dx\,|g(x)|\left[(K_0+aI)^{-1}\varphi^2\right](x)\varphi^{-2}(x).$$

From assumption (3) together with Proposition 3 we conclude

$$\left\|(V_\alpha - V_{\alpha_0})(K_0+aI)^{-1}\varphi\right\|_{2,2} \le C\left\|(K_0+aI)^{-1}|V_\alpha - V_{\alpha_0}|^2\varphi^2\right\|_\infty^{1/2}$$
$$\le C'\left\||V_\alpha - V_{\alpha_0}|^2\varphi^2\right\|_{\mathrm{KF}}^{1/2}.$$

Since the $\max(\sup_\alpha \|V_\alpha\|_\infty, \|V_{\alpha_0}\|_\infty)$ is supposed to be finite, it follows from (52) that the norm of $S_\alpha(\lambda) - S_{\alpha_0}(\lambda)$ is dominated by

$$\text{constant } \times \left\|(V_\alpha - V_{\alpha_0})\varphi^2\right\|_{\mathrm{KF}}^{1/2},$$

which is the dominating term in (52) compared to $\|(V_\alpha - V_{\alpha_0})\varphi^2\|_{\mathrm{KF}}$ if $|\alpha-\alpha_0| < \epsilon$.
q.e.d.

Acknowledgement. The first author wants to thank Erik Balslev for the kind hospitality and for the scientifically stimulating atmosphere during his stay in Aarhus. Moreover he is grateful to the DFG for supporting the project on "Schrödinger Operators" in collaboration with Werner Kirsch from Bochum. The second author is indebted to the University of Antwerp (UIA) and to the Belgian Fund for Scientific Research (NFWO) for their material support. He also thanks the European Community for its support of the Science project "Evolutionary systems: deterministic and stochastic evolution equations, control theory and mathematical biology".

REFERENCES

[Cas-1] J.A. van Casteren, Generators of strongly continuous semigroups, Research Notes in Mathematics 115, Pitman, London 1985.

[Cas-2] J.A. van Casteren, A pointwise inequality for generalized Schrödinger semigroups, in Symposium "Partial differential equations" Holzhau 1988, Teubner Texte zur Mathematik **112**, 298-312, Leipzig (1989).

[Ca, Ma, Si] R. Carmona, W.C. Masters and B. Simon, Relativistic Schrödinger operators, J. Funct. Analysis **51**, 117-142, (1990).

[Da] E.B. Davies, Heat kernel bounds, conservation of probability and the Feller property, Preprint King's College London, 1991.

[De] M. Demuth, On topics in spectral and stochastic analysis for Schrödinger operators, in "Recent Developments in Quantum Mechanics", 223-234, Kluwer, Dordrecht (1990).

[De, Cas-1] M. Demuth and J.A. van Casteren, On spectral theory for selfadjoint Feller generators, Reviews in Math. Physics **1**, 325-414 (1989).

[De, Cas-2] M. Demuth and J.A. van Casteren, On spectral theory for selfadjoint Feller generators II, Preprint University of Antwerp (UIA), to be published 1992.

[Ja] N. Jacob, A class of elliptic differential operators generating symmetric Dirichlet forms, Preprint University of Erlangen-Nürnberg, 1990.

[Ko] A.N. Kochubeĭ, Singular parabolic equations, hypersingular integrals and Markov processes, Math. Izvestiya USSR, **33**, 233-259 (1989).

Some Transport and Spectral Properties of Disordered Media

J.-M. Combes*
P.D. Hislop*†‡

Départment de Mathématiques
Université de Toulon et du Var
83130 La Garde
France

1 Introduction

In this paper, we study some relations between the local geometry of a potential V on $\mathbf{R}^n$ and the spectrum of the corresponding self-adjoint Hamiltonian $H = -\Delta + V$ on $L^2(\mathbf{R}^n)$. More specifically, for a fixed energy interval I, we want to examine the effects of strong local fluctuations of V near energies in I on the spectrum of $H, \sigma(H)$, in the interval I. We consider, firstly, the dynamics generated by such a potential as expressed by the diffusion constant. We give sufficient conditions on V in terms of the potential barriers of V at energies in I so that the system evolves in a sub-diffusive manner. Secondly, we study the nature of the spectrum of H in I. We show that with sufficiently strong local fluctuations of V the absolutely continuous spectrum of H in I is empty. Our motivation for these preliminary studies is to explore these questions in a simple disordered but deterministic situation. We are primarily interested in one-electron models of random disordered media. Here, we give an application of these ideas to the one-dimensional Poisson model and prove the absence of absolutely continuous spectrum at all energies (when the single-site potential $u \geq 0$) and the absence of diffusion at low energies. We will discuss the Anderson-type and Poisson-type models for potentials in $\mathbf{R}^n$ in another paper.

Our results on the absence of diffusion rely heavily on the quantum tunneling estimates for the localized resolvent proved in [BCD]. Traditionally, the absence diffusion has been demonstrated as a consequence of *a priori*, almost sure exponential decay estimates on the Green's function. This approach was used by Fröhlich and Spencer [FS] in their study of the Anderson model on $\mathbf{Z}^n$ (see

*also: Centre de Physique Théorique, CNRS, Luminy Case 907, 13288 Marseille, Cédex 9, France
†permanent address: Mathematics Department, University of Kentucky, Lexington KY 40506-0027 USA
‡supported in part by NSF grants DMS91-06479 and INT 90-15895

also Martinelli and Holden [MH]). In a recent article, Combes and Sigal [CS] demonstrated that sub-diffusive behavior occurs for deterministic, disordered systems under very general conditions on the size and distribution of potential wells resonant at energies in I. These conditions are verified, for example, in the heirarchial model of Jona-Lasinio, Martinelli, and Scoppola [JMS]. In a sense, our results are complimentary to the results of Combes and Sigal in that we study the distribution of potential barriers at energies in I.

Recently, Simon and Spencer [SS] introduced a new idea into the study of the absolutely continuous spectrum for disordered systems. They use the techniques of trace-class scattering theory to prove the absence of absolutely continuous spectrum for lattice models by constructing wave operators for a pair (H, H_D), where H_D is a self-adjoint operator with pure point spectrum. Roughly, this approximate operator H_D is obtained from H by imposing Dirichlet boundary conditions in the regions where V becomes arbitrarily large. This work was extended to unbounded potentials on $\mathbf{R}$ by Hislop and Nakamura [HN]. They studied the effect of an external electric field on some random Schrödinger operators on $L^2(\mathbf{R})$ of Anderson-type. Our results are generalizations of these ideas to Schrödinger operators on $L^2(\mathbf{R}^n)$.

We complete this introduction with a discussion of the meaning of diffusion in one-electron models. The notion of diffusion for a single-electron model can be formulated in several ways. Perhaps the most basic definition is in terms of the average square distance traveled by a particle in time t given an initial state ψ_0 at $t = 0$. We say that the evolution of ψ_0 under the Hamiltonian H is *diffusive* if there exists a finite constant D_{ψ_0}, $0 < D_{\psi_0} < \infty$, such that

$$\langle \psi_0, e^{iHt} x^2 e^{-iHt} \psi_0 \rangle \sim D_{\psi_0} t \text{ as } t \to \infty \,. \tag{1.1}$$

(We assume that $U(t)\psi_0 \in D(x)\ \forall\, t \geq 0$).

Definition 1.1. (1) A state ψ_0 is diffusive (resp. sub-diffusive) if $U(t)\psi_0 \in D(x)\ \forall\ t \geq 0$ and $\exists\, 0 < D_{\psi_0} < \infty$ such that

$$\lim_{t\to\infty} t^{-1} \|\, |x| U(t)\psi_0 \|^2 = D_{\psi_0} \text{ (resp. } 0) \,. \tag{1.2}$$

(2) A Hamiltonian H has diffusive behavior (resp. sub-diffusive behavior) for energies in $I \subset \mathbf{R}$ if $\exists$ dense set $\mathcal{D}_I \subset E_I(H)\mathcal{H}$ such that each $\psi \in \mathcal{D}_I$ is diffusive (resp. sub-diffusive). Here, $E_I(H)$ is the spectral projection for H and I.

We have the following results showing some of the relations between the spectral properties of H and diffusive behavior. First, we study how the spectral properties of H affect diffusion.

Proposition 1.2. (1) Suppose $H\psi_i = E_i\psi_i$, $\psi_i \in D(x) \subset \mathcal{H}$ and let $\psi = \sum_{i=1}^N c_i\psi_i$, $N < \infty$, then $D_\psi = 0$.

(2) Suppose $I \subset \sigma_c(H)$ (the continuous spectrum of H) and H obeys a strict Mourre estimate on I with conjugate operator $A \equiv x \cdot \nabla + \nabla \cdot x$, i.e., $\exists\, \alpha > 0$ such that $E_I(H)[H,A]E_I(H) \geq \alpha E_I(H)$. Then for any ψ such that $U_t\psi \in E_I(H)\mathcal{H} \cap D(x) \cap H^1(\mathbf{R}^n)$ $\forall t \geq 0$, $D_\psi = \infty$.

PROOF: The proof of (1) is trivial. To prove (2), we write,

$$\begin{aligned} \| \, |x| U(t)\psi\|^2 &= i\int_0^t ds \left\langle U_s\psi, [H, x^2]U_s\psi\right\rangle + \| \, |x|\psi\|^2 \\ &= -2i\int_0^t ds \, \langle U_s\psi, AU_s\psi\rangle + \| \, |x|\psi\|^2 \, . \end{aligned} \tag{1.3}$$

Clearly the constant term in (1.3) doesn't contribute to the limit in (1.2). Repeating the differentiation, the last integral on the right in (1.3) becomes

$$2i\int_0^t ds \int_0^s dr \, \langle U_r\psi, [H,A]U_r\psi\rangle - 2it\langle\psi, A\psi\rangle \, . \tag{1.4}$$

The first term in (1.4) is strictly positive and

$$\lim_{t\to\infty} t^{-1}\int_0^t ds \int_0^s dr \, \langle U_r\psi, E_I(H)[H,A]E_I(H)U_r\psi\rangle \geq \lim_{t\to\infty} \frac{\alpha t}{2}\|\psi\|^2 = +\infty \, . \tag{1.5}$$

As for the second term, $\langle\psi, (-iA)\psi\rangle$ is real and finite by hypothesis so $\exists c_0$, $|c_0| < \infty$, such that

$$\lim_{t\to\infty} t^{-1}\left(2t\,\langle\psi, (-iA)\psi\rangle\right) = c_0 \, . \tag{1.6}$$

It follows from (1.3)—(1.6) that $D_\psi = +\infty$. ∎

This simple proposition supports the idea that localized states (here, finite linear combinations of eigenfunctions) and extended states associated with the absolutely continuous spectrum do not diffuse. Recall that the strict Mourre estimate rules out the possibility of eigenvalues and singular continuous spectrum in I. However, the above proof shows that for absolutely continuous spectrum as described in part (2), $\exists$ constants c_0, c_1, $0 < c_0 \leq c_1 < \infty$ such that

$$0 < c_0 \leq \lim_{t\to\infty} t^{-2}\| \, |x|U(t)\psi\|^2 \leq c_1 < \infty \, ,$$

i.e., we have roughly $\langle x(t)^2\rangle \sim v_0^2 t^2$, which is ballistic motion. It is not known whether states in the singular spectral subspace diffuse. It would be interesting to study this in Pearson's examples [P]. With regard to point (1) of Proposition 1.2, it is not known whether a state ψ, which is an infinite linear combination of eigenfunctions, is diffusive. It appears as if this may depend on the statistics of the corresponding eigenvalue spacings. There is, however, one result, due to Simon [S], concerning the dynamical properties of a Hamiltonian H with pure point spectrum. This result states that the evolution is sub-ballistic.

Proposition 1.3 ([S]). Let V be a multiplication operator on $L^2(\mathbf{R}^n)$ and let H be the self-adjoint operator associated with the form closure of $-\Delta + V$ on $Q(-\Delta) \cap Q(V)$, the intersection of the form domains of $-\Delta$ and V, respectively. Suppose $Q(H) \subset Q(-\Delta)$. Let $\phi \in D(x) \cap Q(H)$. If H has only point spectrum, then

$$\lim_{t \to \infty} t^{-2} \left\langle U_t \phi, x^2 U_t \phi \right\rangle = 0 .$$

We now consider the converse question: if H is diffusive in I, what can we say about $\sigma(H) \cap I$? It is clear that if H is diffusive in I, then $\sigma_{pp}(H) \cap I = \phi$ for eigenstates do not diffuse. Otherwise, not much is known. One weak result, observed by Spencer [Sp], is the following.

Proposition 1.4 ([Sp]). Suppose $\forall\, \psi \in \mathcal{D}_I \subset E_I(H)\mathcal{H}$ and $\mathcal{D}_I$ dense, $\| \, |x| U(t) \psi \| < c_1 < \infty \; \forall\, t \geq 0$. Then $\sigma_c(H) \cap I = \phi$.

Proof: Let $P_c(I)$ be the projection onto the continuous spectral subspace for H in I. Let $B_R(0) \subset \mathbf{R}^n$ be the ball of radius R about the origin and let F_R be the projection onto $\mathbf{R}^n | B_R(0)$. By the RAGE Theorem [RS3], for any $\psi \in \mathcal{H}$,

$$\|P_c(I)\psi\|^2 = \lim_{R \to \infty} \overline{\lim_{T \to \infty}} \frac{1}{T} \int_0^T \|F_R U_s \psi\|^2 \, ds .$$

We can replace F_R by $|x| R^{-1} F_R$ and obtain

$$\|P_c(I)\psi\|^2 \leq \lim_{R \to \infty} \frac{1}{R^2} \overline{\lim_{T \to \infty}} \frac{1}{T} \int_0^T \| \, |x| U_s \psi\|^2 \, ds = 0 , \tag{1.7}$$

where we used the bound on $\| \, |x| U_s \psi \|$. ∎

We remark that formula (1.7) does suggest a method of proving the absence of continuous spectrum in I.

We now consider how the geometry of potential wells and barriers for energies in I affect the diffusion constant for initial states ψ_0 with energy localized in I. One might conjecture that if the distance in the Agmon metric for energy I_+, where $I = [I_-, I_+]$, is infinite as $|x| \to \infty$ (suppose, for example, that V is finite a.e.), then $D_{\psi_0} < \infty$. That this is not the case can be seen by studying a periodic potential in $\mathbf{R}$.

Proposition 1.5. Suppose V is a smooth periodic potential on $\mathbf{R}$ with maximum V_0 and that E_n is a band lying below V_0. Let $u_k^n(x)$ be a Bloch function, $k \in B$, the Brillouin zone, for the band E_n. Let $f \in C_0^\infty(B)$ such that (1.14) below holds and form the wave packet

$$\psi_n(x) \equiv \int_B f(k) e^{ikx} u_k^n(x) \,. \tag{1.8}$$

Then $D_{\psi_n} = \infty$.

Proof: Let $v_k^n(x) \equiv e^{ikx} u_k^n(x)$ be a Bloch wave function for the nth band. Let $\epsilon_n(k)$ be the band energy function. Then for $H = \frac{-d^2}{dx^2} + V$, we have

$$H v_k^n = \epsilon_n(k) v_k^n \,.$$

Consequently, we can compute

$$\begin{aligned} xe^{-iHt}\psi_n(x) &= -i \int_B f(k) u_k^n(x) e^{-i\epsilon_n(k)t} \left(\frac{\partial}{\partial k} e^{ikx} \right) dx \\ &= +i \int_B \left[\frac{\partial f}{\partial k} u_k^n e^{-i\epsilon_n(k)t+ikx} + \frac{\partial u_k^n}{\partial k} f(k) e^{-i\epsilon_n(k)t+ikx} \right. \\ &\quad \left. + \left(-it \frac{\partial \epsilon_n}{\partial k} \right) f(k) u_k^n e^{-i\epsilon_n(k)t+ikx} \right] dk \end{aligned} \tag{1.9}$$

since f vanishes at the boundary of B. Using the orthogonality of the Bloch wave functions, we obtain from (1.9),

$$\begin{aligned} \|xe^{-iHt}\psi\|^2 &= \int_B \left| \frac{\partial f}{\partial k} - it \frac{\partial \epsilon_n}{\partial k} f \right|^2 dk + \int_{\mathbf{R}} dx \, |W_t(x)|^2 \\ &\quad + 2Re \left\{ \int_{\mathbf{R}} dx \left[\int_B dk \left(\frac{\partial f}{\partial k} - it \frac{\partial \epsilon_n}{\partial k} \right) u_k^n e^{-\epsilon_n(k)t+ikx} \right]^* W_t(x) \right\} \end{aligned} \tag{1.10}$$

where

$$W_t(x) \equiv \int_B dk f(k) \left(\frac{\partial u_k^n}{\partial k} \right) e^{-\epsilon_n(k)t+ikx} \,. \tag{1.11}$$

Consider the first term on the right in (1.9). It is

$$\int_B \left| \frac{\partial f}{\partial k} \right|^2 + t^2 \int_B \left| \frac{\partial \epsilon_n}{\partial k} f \right|^2 - 2Re \left\{ it \int_B \left(\frac{\partial f^*}{\partial k} \frac{\partial \epsilon_n}{\partial k} f \right) \right\} , \tag{1.12}$$

so that the limit as $t \to \infty$ of t^{-1} times (1.12) is $+\infty$. As for the second term, we apply the method of stationary phase [RS3] to analyze the integral. We note that due to the continuous differentiability of $\epsilon_n(k)$ on B, $\exists \, \delta > 0$ such that

$$\left\{ \frac{\partial \epsilon_n(k)}{\partial k} | k \in \text{supp} \left(\frac{\partial u_k^n}{\partial k} f(k) \right) \right\} \subset (-\delta, \delta) \,.$$

Consequently, for any $m \in \mathbb{Z}^+$, $\exists\, c_m > 0$ such that

$$|W_t(x)| \leq c_m (1 + |x| + t)^{-m} ,$$

provided $x/t \notin (-\delta, \delta)$. This gives the estimate

$$\int_{\mathbb{R}\setminus(-t\delta,t\delta)} dx\, |W_t(x)|^2 < k_0 , \tag{1.13}$$

for some $k_0 > 0$. As for the other interval, we use the hypothesis that

$$\text{supp}\left(f(k)\frac{\partial u_k^n}{\partial k}\right) \cap \left\{k \in B | \frac{\partial^2 \epsilon_n}{\partial k^2} = 0\right\} = \phi \tag{1.14}$$

so that $|W_t(x)| \leq c_0 t^{-\frac{1}{2}}$, for some $c_0 > 0$. Combining (1.13) with this result, we find $\exists\, c_0 > 0$ such that $\forall\, t \geq 0$,

$$\int_{\mathbb{R}} dx\, |W_t(x)|^2 < c_0 \tag{1.15}$$

and, hence, the second term on the right in (1.9), when multiplied by t^{-1}, vanishes in the limit. Similarly, the last integral in (1.9) can be bounded by $(c_0 + |t|c_1)$ by the Schwarz inequality, (1.15) and (1.12). Thus if $g(t)$ represents the third term on the right in (1.9), $t^{-1}g(t)$ is bounded. Combining this with (1.12) and (1.15), we find that $D_\psi = +\infty$. ■

As we will see in Section 2, we expect sub-diffusive behavior when both the distance in the Agmon metric is infinite as $|x| \to \infty$ and the location of the potential barriers is disordered.

There is another definition of the diffusion constant (for one-electron models) in the literature. It is called Kubo's formula for the diffusion constant for a state ψ at energy E. It is given by

$$D_\psi^K(E) \equiv \lim_{\epsilon \to 0} \frac{2\epsilon^2}{\pi} \| \, |x| R(E + i\epsilon)\psi \|^2 , \tag{1.16}$$

where $R(z) \equiv (H - z)^{-1}$ is the resolvent. (We include the density of states factor in (1.16), see [FS]). We may also consider an integrated diffusion constant

$$D_\psi^K \equiv \int_{-\infty}^{\infty} dE\, D_\psi(E) . \tag{1.17}$$

The relation between D_ψ and D_ψ^K is given in the following proposition.

Proposition 1.6. Suppose $D_\psi < \infty$, then $D_\psi^K = D_\psi$.

PROOF: We note that

$$D_\psi \equiv \lim_{\epsilon \to 0} \epsilon^2 \int_0^\infty e^{-\epsilon s} \| \, |x| U_s \psi \|^2 \, ds , \tag{1.18}$$

which follows from the hypothesis on the existence of the limit (1.2) and the Lebesgue dominated convergence theorem. We apply the Plancherel theorem to (1.18) to obtain

$$\begin{aligned} D_\psi &= \lim_{\epsilon\to 0} \epsilon^2 \int_0^\infty dt \int_{\mathbb{R}} dx \left| |x| e^{-i(H-\epsilon/2)t} \psi(x) \right|^2 \\ &= \lim_{\epsilon\to 0} \frac{\epsilon^2}{2\pi} \int_{\mathbb{R}} dE \, \| \, |x| R(E+i\epsilon/2)\psi|^2 = D_\psi^K . \end{aligned}$$

■

This result is not directly useful, for if we do not know $D_\psi < \infty$ *a priori*, we cannot use formulas (1.16)–(1.17) to compute D_ψ. There is a weaker version, however, indicated by Fröhlich and Spencer [FS] and actually used by Martinelli and Holden [MH]. For a state ψ, we define a mean diffusion coefficient by

$$\bar{D}_\psi \equiv \lim_{T\to\infty} \int_1^T s^{-1} \| \, |x| U_s \psi \|^2 \, ds .$$

We relate this to D_ψ^K as follows. We set $T = \epsilon^{-1}$ and, taking $\eta \in (0,1)$, we obtain

$$\begin{aligned} T^{-1} \int_{T\eta}^T s^{-1} \| \, |x| U_s \psi \|^2 \, ds &\le \epsilon^2 e \eta^{-1} \int_0^\infty e^{-\epsilon s} \left\| |x| e^{-iH_s} \psi \right\|^2 \\ &= \epsilon^2 e (2\pi\eta)^{-1} \int_{-\infty}^\infty dE \, \| \, |x| R(E+i\epsilon/2)\psi \|^2 . \end{aligned}$$

In [HN], it is proved that if

$$\lim_{\epsilon\to 0} \epsilon^2 \int_{\mathbb{R}} dE \, \| \, |x| R(E+i\epsilon/2)\psi \|^2 = 0 ,$$

then $\bar{D}_\psi = 0$. Consequently, we obtain

Proposition 1.7. If $D_\psi^K = 0$, then $\bar{D}_\psi = 0$.

We now turn to the case when we have a random family of self-adjoint operators H_ω, $\omega \in \Omega$ with (Ω, P) a probability space (see, for example, the review article of Kirsch [K]). We define a deterministic diffusion constant by averaging over the potential. For example, one usually refers to the average of (1.16) as the fixed energy Kubo diffusion coefficient

$$\left\langle D_\psi^K(E) \right\rangle = \lim_{\epsilon\to 0} \frac{2\epsilon^2}{\pi} \int_\Omega dP(\omega) \, \| \, |x| R_\Omega(E+i\epsilon/2)\psi \|^2 .$$

The advantage of working with this quantity is that it requires almost sure estimates at one energy only. By contrast, if we consider D_ψ averaged over the potential, we would need a probability estimate at all energies (see (2.16) below). We will take a slightly different approach below. In Section 2, we will prove a deterministic result giving sufficient conditions on the configuration of the classically forbidden region for energy I_+ which imply $D_\psi = 0$ for $\psi \in \mathcal{D}_I$. We will then in Section 4 show for the Poisson model in one-dimension that there exists a set of potentials $\Omega_0 \subset \Omega$ of full measure $P(\Omega_0) = 1$ for which this configuration is achieved. Consequently, thinking of D_ψ, $\psi \in \mathcal{D}_I$, as a random variable on ω (note that $\mathcal{D}_I$ and hence ψ depend on ω), we have $\langle D_\psi \rangle = 0$. This implies $\left\langle D_\psi^K(E) \right\rangle = 0$ for almost any $E \in I$.

2 Absence of Diffusion

In this section we prove that the diffusion constant D_ψ, defined in (1.2), vanishes for a family of initial states ψ which are decaying in x and have compact energy support, provided the potential V has sufficiently many barriers at the energies in the spectral support of ψ. We fix an interval $I \equiv [I_-, I_+] \subset \mathbf{R}$. Let $G(I_+)$ denote the classically forbidden region for energy I_+, i.e.,

$$G(I_+) \equiv \{x | V(x) > I_+\} \ . \tag{2.1}$$

The potential V must satisfy the following conditions relative to the energy I_+.

$(V1)$ $H \equiv -\Delta + V$ is self-adjoint on $D(H)$;

$(V2)$ $\exists \ N_0$ such that for all $n > N_0$ $\exists$ regions $I_n \equiv \{x | e^n < |x| < e^{n+1}\}$ containing sub-regions $\{x | a_n \le |x| \le b_n\} \subset G(I_+)$ and $|b_n - a_n| \ge \mathcal{O}(n)$;

$(V3)$ $\exists \ R_0 > 0$ such that $(I_+ - V(x))_+ \mid \{x \mid |x| > R_0\} < c_0$..

Next, we describe the family of initial states ψ to which our results apply. Take an interval $J \subset I$, $J = [J_-, J_+]$ such that $|J_\pm - I_\pm| > 2\delta_0 > 0$ for some $\delta_0 > 0$. We require ψ satisfy

$(W1)$ $\psi \in D(H)$, $\|\psi\| = 1$, and $E_J(H)\psi = \psi$, where $E_J(H)$ is the spectral projector for H and interval J;

$(W2)$ if J_n is a smooth function with supp $J_n \subset \{x \mid |x| > a_{n-1}\}$, $\exists \ c_0 > 0$, and a positive function $g(n) \ge \mathcal{O}(n)$ such that

$$\|J_n \psi\| \le c_0 e^{-g(n)} \, .$$

Under the assumptions $(V1)$—$(V3)$, we will construct states satisfying $(W1)$—$(W2)$ in the appendix. The function g is a linear combination of distances occuring in $(S1)$—$(S4)$ below and given explicitly in (A.15). The hypotheses of Theorem 2.1 and the methods of the appendix imply that such ψ and $e^{-iHt}\psi \in D\left(|x|^{2^{p-1}}\right)$, for some $p \ge 2$.

For a potential V, we define the Agmon pseudo-metric at energy I_+ by $ds^2 \equiv (V - I_+)_+ dx^2$ and let $\rho_{I_+}(x,y)$ be the corresponding geodesic distance from x to y. For any subset $\Omega \subset G(I_+)$, let $d_{I_+}(x,\Omega) \equiv \inf\{\rho_{I_+}(x,y) | y \in \Omega\}$. This is a Lipschitz function and satisfies

$$|\nabla d_{I_+}(x,\Omega)|^2 \leq (V(x) - I_+)_+ \quad \text{a.e.}. \tag{2.2}$$

Let $G(I_+)^c \equiv \mathbf{R}^n \backslash G(I_+)$ and, for any $\Omega \in G(I_+)$, define $d_{I_+}(\Omega) \equiv \inf\{\rho_{I_+}(x,\Omega) | x \in G(I_+)^c\}$. For any two sets $\Omega_1, \Omega_2 \subset G(I_+)$, (and, more generally, any two subsets at $\mathbf{R}^n$), define

$$d_{I_+}(\Omega_1,\Omega_2) \equiv \inf\{\rho_{I_+}(x,y) | x \in \Omega_1, y \in \Omega_2\} .$$

We now state our main theorem.

Theorem 2.1. Let the potential V satisfy $(V1)$—$(V3)$ and let the initial state ψ satisfy $(W1)$—$(W2)$. Assume there exist disjoint closed subsets $\kappa_n, \alpha_n, \omega_n, \Omega_n$ of $B_n \equiv \{x \mid a_n \leq |x| \leq b_n\}$ with non-empty interiors and C^2-boundaries such that κ_n is a neighborhood of a_n, $\kappa_n \cap \alpha_n = \phi$, and for each $\omega \in S^{n-1}$, $\sup\{x \mid (|x|,\omega) \in \alpha_n\} < \inf\{x \mid (|x|,\omega) \in \omega_n\}$ and $\sup\{x \mid (|x|,\omega) \in \omega_n\} < \inf\{x \mid (|x|,\omega) \in \Omega_n\}$. For any $\delta > 0$, define

$$W_{2\delta} \equiv (V - I_+)_+^{1/2} | \{x \in G(I_+) | d_{I_+}(x, G(I_+)^c) < 2\delta\} .$$

Let $\tilde{\chi}_n$ be the characteristic function for $\{(|x|,\omega) |$ for each $\omega \in S^{n-1}$, $|x| \geq \inf\{|x| \mid (|x|,\omega) \in \alpha_{n-1}\}\}$. Let $e(p,n) \equiv n^{-1}e^{2^p n}$. Assume $\exists\, p \geq 2$, $\delta > 0$, and $N_0' \geq N_0$ such that the following series converge:

(S1) $\displaystyle\sum_{n \geq N_0'} e(p,n) e^{-d_{I_+}(\kappa_{n-1},\alpha_{n-1})}$

(S2) $\displaystyle\sum_{n \geq N_0'} e(p,n) e^{-2d_{I_+}(\alpha_{n-1},\omega_{n-1})}$

(S3) $\displaystyle\sum_{n \geq N_0'} e(p,n) e^{-2d_{I_+}(\alpha_{n-1})} \|W_{2\delta}\tilde{\chi}_n\|^2$

(S4) $\displaystyle\sum_{n \geq N_0'} e(p,n) e^{-2d_{I_+}(\omega_{n-1})} \|W_{2\delta}\tilde{\chi}_n\|^2 .$

Then the diffusion constant vanishes, i.e., $D_\psi = 0$.

Proof: We define $D_\psi^{(p)}$, for any $p \in \mathbf{Z}^+$, in analogy with D_ψ:

$$D_\psi^{(p)} \equiv \lim_{t \to \infty} t^{-1} \langle \psi_t, x^{2^p} \psi_t \rangle , \tag{2.3}$$

where $\psi_t \equiv e^{-iHt}\psi$ and $\psi \in E_J(H)\mathcal{H}$ as described in the theorem. Of course, when $p = 1$, $D_\psi^{(1)} = D_\psi$, the diffusion constant. We will prove that for some $p \geq 2$, as in the theorem, $D_\psi^{(p)} < c_0 < \infty$. From this, it follows that $D_\psi = 0$, since

$$D_\psi = \lim_{t\to\infty} t^{-1}\langle\psi_t, x^2\psi_t\rangle \leq= \lim_{t\to\infty} t^{-1+2^{1-p}} \left[t^{-1}\langle\psi_t, x^{2^p}\psi_t\rangle\right]^{\frac{1}{2^{1-p}}} = 0\,, \tag{2.4}$$

where we used the Schwarz inequality. We divide the proof of the theorem into several steps.

(1) **Partition of Unity.** As in $(V2)$, we consider potential barriers to lie in the regions $B_n = \{x \mid a_n < |x| < b_n\}$ so that $|b_n - a_n| \geq \mathcal{O}(n)$ and $e^n < a_n < b_n < e^{n+1}$. (In the one-dimensional case, the barriers lie in $[a_n, b_n]$ for $n \in \mathbf{Z}$ and $|n|$ sufficiently large). Let $\Omega_n \subset B_n$ as in the theorem. We choose a partition of unity $\{\tilde{J}_n\}$ such that $\sum_n \tilde{J}_n = 1$ and

$$\operatorname{supp}\left|\nabla\tilde{J}_n\right| \subset \Omega_{n-1} \cup \Omega_n \tag{2.5}$$

for each n sufficiently large (see Figure 2.1). Note that $|x| \left|\left(\operatorname{supp}\tilde{J}_n\right)\right. \leq e^{n+1}$ so for $p \geq 2$, as in the theorem,

$$\langle\psi_t, x^{2^p}\psi_t\rangle \leq \sum_n e^{2^p(n+1)}\langle\psi_t, \tilde{J}_n\psi_t\rangle\,. \tag{2.6}$$

We introduce a second family of functions $\{J_n\}$ such that $J_n \in C^2$, $\operatorname{supp} J_n \subset [e^{n-1}, \infty)$, and $J_n\tilde{J}_k = \tilde{J}_k\;\forall\; k = n, n+1, \ldots$. We choose J_n such that $\operatorname{supp}|\nabla J_n|$ is contained in a set $\omega_{n-1} \subset B_{n-1}$ satisfying the conditions of the theorem. Since $\tilde{J}_n J_k \neq 0$ for $k = 1, \ldots, n+1$, we can replace $\tilde{J}_n$ in (2.6) and obtain

$$\langle\psi_t, x^{2^p}\psi_t\rangle \leq \sum_n \frac{e^{2^p(n+1)}}{n}\langle\psi_t, J_n\psi_t\rangle\,. \tag{2.7}$$

(2) **First Reduction.** As usual, we write

$$\begin{aligned}\langle\psi_t, J_n\psi_t\rangle &= \int_0^t \frac{d}{ds}\langle\psi, e^{iHs}J_n e^{-iHs}\psi\rangle ds + \langle\psi, J_n\psi\rangle \\ &= i\int_0^t \langle\psi, W(J_n)\psi\rangle ds + \langle\psi, J_n\psi\rangle\,,\end{aligned} \tag{2.8}$$

where $W(J_n) \equiv [-\Delta, J_n]$. Formula (2.3) becomes

$$D_\psi^{(p)} \leq \lim_{t\to\infty}\frac{i}{t}\sum_n \frac{e^{2^p(n+1)}}{n}\int_0^t \langle\psi_s, W(J_n)\psi_s\rangle ds + \lim_{t\to\infty}\frac{i}{t}\sum_n \frac{e^{2^p(n+1)}}{n}\langle\psi, J_n\psi\rangle\,. \tag{2.9}$$

The second term on the right in (2.9) can be evaluated due to the exponential decay of ψ, $(W2)$,

$$\|J_n\psi\| \leq c_0 e^{-g(n)}\,, \tag{2.10}$$

since $\tilde{\tilde{J}}_n J_n = J_n$. Then we have for some $p \geq 2$,

$$\sum_n \frac{e^{2p(n+1)}}{n} e^{-g(n)} < \infty, \tag{2.11}$$

by $(S1)$, $(S3)$ and $(S4)$ and, consequently, the contribution to (2.9) from the second term vanishes. We now write the first term in terms of the resolvent. We use the following fact:

$$\text{if } f(t) \geq 0 \text{ on } [0,\infty), \text{ then } \frac{1}{t}\int_0^t f(s)ds \leq e\epsilon \int_0^\infty e^{-\epsilon s} f(s)ds, \text{ for any } \epsilon > 0. \tag{2.12}$$

This follows by setting $t = 1/\epsilon$ in the integral on the left in (2.12) and using the positivity of f to replace $\int_0^{1/\epsilon}$ by $\int_0^\infty$. Then, using (2.12) to write the integral in (2.9), we obtain

$$D_\psi^{(p)} \leq \overline{\lim_{\epsilon\to 0}}\, \epsilon e \sum_n \frac{e^{2p(n+1)}}{n} \int_0^\infty e^{-\epsilon s} |\langle \psi_s, W(J_n)\psi_s\rangle|\, ds. \tag{2.13}$$

Let χ_n be a (smoothed) characteristic function on $\omega_{n-1} \subset \operatorname{supp} W(J_n)$. Then

$$|\langle \psi_s, W(J_n)\psi_s\rangle| \leq \|\chi_n \psi_s\|^2 + \|W(J_n)\psi_s\|^2,$$

and (2.13) becomes

$$D_\psi^{(p)} \leq \overline{\lim_{\epsilon\to 0}}\, \epsilon e \sum_n \frac{e^{2p(n+1)}}{n} \left\{ \int_0^\infty e^{-\epsilon s} \|W(J_n)\psi_s\|^2 + \int_0^\infty e^{-\epsilon s} \|\chi_n \psi_s\|\, ds \right\}. \tag{2.14}$$

We now apply the vector-valued Plancherel theorem [RS3] to each integral. If X_n denotes either χ_n or $W(J_n)$, then each integral in (2.14) contributes a term of the form:

$$\overline{\lim_{\epsilon\to 0}}\, \frac{\epsilon e}{\sqrt{2\pi}} \sum_n \frac{e^{2p(n+1)}}{n} \int_{-\infty}^\infty e^{-\epsilon s} \|X_n R(E + i\epsilon/2)\psi\|^2\, dE. \tag{2.15}$$

Recall that $\operatorname{supp} X_n = \omega_{n-1} \subset G(I_+)$, the classically forbidden region for energy I_+. We will prove that for some $p \geq 2$ as in the theorem,

$$\overline{\lim_{\epsilon\to 0}}\, \epsilon \sum_n \frac{e^{2pn}}{n} \int_{-\infty}^\infty \|X_n R(E + i\epsilon/2)\psi\|^2\, dE < \infty. \tag{2.16}$$

From this, it follows that $D_\psi^{(p)} < \infty$ for some $p \geq 2$.

(3) **Second Reduction.** We introduce a third family of functions, $\{\tilde{\tilde{J}}_n\}$, $\tilde{\tilde{J}}_n \in C^2$ and such that $\operatorname{supp}|\nabla \tilde{\tilde{J}}_n| \equiv \alpha_{n-1} \subset [a_{n-1}, b_{n-1}]$, satisfying the conditions of the theorem, and $\tilde{\tilde{J}}_n J_n = J_n$. Let H_n be defined to be the self-adjoint Hamiltonian on $L^2(\operatorname{supp} \tilde{\tilde{J}}_n)$ given by $H_n = -\Delta + V_n$, where $V_n \equiv \sup(V, I_+)$ on $\operatorname{supp} \tilde{\tilde{J}}_n$, with Dirichlet boundary conditions on $\partial(\operatorname{supp} \tilde{\tilde{J}}_n)$. Note that $\sigma(H_n) \subset [I_+, \infty)$. Let $R_n(z) = (H_n - z)^{-1}$. We have the geometric resolvent equation

$$\begin{aligned}\tilde{\tilde{J}}_n R(E + i\epsilon/2) = R_n(E + i\epsilon/2)\tilde{\tilde{J}}_n &+ R_n(E + i\epsilon/2) W(\tilde{\tilde{J}}_n) R(E + i\epsilon/2) \\ &+ R_n(E + i\epsilon/2)(I_+ - V)_+ \tilde{\tilde{J}}_n R(E + i\epsilon/2),\end{aligned} \tag{2.17}$$

so consequently,

$$X_n R(E+i\epsilon/2)\psi = X_n R(E+i\epsilon/2)\,\tilde{\tilde{J}}_n\,\psi + X_n R_n(E+i\epsilon/2) Z_n R(E+i\epsilon/2)\psi\,, \tag{2.18}$$

where we write $Z_n \equiv W(\tilde{\tilde{J}}_n) + (I_+ - V)_+\,\tilde{\tilde{J}}_n$. Each term of (2.18) contributes an integral of the type (2.16). We first consider the low-energy contribution ($\leq I_+$) of each integral coming from (2.18). We evaluate each separately, beginning with the first term on the right in (2.18). We then consider the high-energy part of both integrals.

(4) **Single Resolvent Integral.** Consider the first term on the right in (2.18). We use exponential decay of localized resolvents to treat the low-energy ($\leq I_+$) part of the integral.

low energy:

$$\varlimsup_{\epsilon\to 0} \sum_n \frac{e^{2p_n}}{n}\epsilon \int_{-\infty}^{I_+} \left\| X_n R_n(E+i\epsilon/2)\,\tilde{\tilde{J}}_n\,\psi\right\|^2 dE\,. \tag{2.19}$$

The results of [BCD] are applicable to estimate $\left\|X_n R_n\,\tilde{\tilde{J}}_n\,\psi\right\|$. Following the same steps leading to equation (II.19) of [BCD], we find

$$\begin{aligned}\left\|X_n R_n(E+i\epsilon/2)\,\tilde{\tilde{J}}_n\,\psi\right\|^2 &\leq Re\left\langle \chi_n^2 e^{-\rho_n}\,\tilde{\tilde{J}}_n\,\psi, R_n(E+i\epsilon/2)\,\tilde{\tilde{J}}_n\,\psi\right\rangle + e^{-2d_{I_+}(\alpha_{n-1})}\left\|R_n(E+i\epsilon/2)\,\tilde{\tilde{J}}_n\,\psi\right\| \\ &\leq c_0\epsilon^{-1}e^{-d_{I_+}(\Omega_{n-1},\alpha_{n-1})}\left\|\tilde{\tilde{J}}_n\,\psi\right\|^2 + e^{-2d_{I_+}(\alpha_{n-1})}\left\|R_n(E+i\epsilon/2)\,\tilde{\tilde{J}}_n\,\psi\right\|^2 \quad (2.2\end{aligned}$$

where, as defined above, $\rho_n(x) = d_{I_+}(\Omega_{n-1}, x)$, and $d_{I_+}(K) = d_{I_+}(K, G(I_+)^c)$ for any $K \subset G(I_+)$. Evaluation of the energy integral of (2.20) is facilitated by Stone's formula for the spectral projections:

$$\lim_{\epsilon\to 0} \epsilon \int_a^b \|R(E+i\epsilon/2)\psi\|^2\, dE \leq c\,\|E([a,b])\,\psi\|^2 \leq c\,\|\psi\|^2\,, \tag{2.21}$$

where E is the spectral projection for H. Furthermore, by assumption ($V3$), we take $n > N_0$ so the low-energy integral is finite. Integration of (2.20) over $E \leq I_+$ yields

$$\begin{aligned}\int_{-\infty}^{I_+} \left\|X_n R_n(E+i\epsilon/2)\,\tilde{\tilde{J}}_n\,\psi\right\|^2 dE \quad &\leq \quad c_0\epsilon^{-1}e^{-d_{I_+}(\Omega_{n-1},\alpha_{n-1})}e^{-2g(n)} \\ &\quad + e^{-2d_{I_+}(\alpha_{n-1})}e^{-2g(n)}\,. \end{aligned} \tag{2.22}$$

Consequently, by assumption ($S1$), and ($S3$)—($S4$) the low-energy contribution (2.19) is finite.

(5) **Double Resolvent Integral.** We now consider the low-energy part of the second term on the right in (2.18). We consider only the term involving $W(\tilde{\tilde{J}}_n)$. The term involving $(I_+ - V)_+\,\tilde{\tilde{J}}_n$ is treated in the same way using ($V3$).

low energy:

$$\varlimsup_{\epsilon\to 0} \sum_n \frac{e^{2p_n}}{n}\epsilon \int_{-\infty}^{I_+} \left\|X_n R_n(E+i\epsilon/2) W(\tilde{\tilde{J}}_n) R(E+i\epsilon/2)\psi\right\|^2 dE\,. \tag{2.23}$$

To estimate the integrand, we use the analog of Proposition III.2 of [BCD] which yields

$$\int_{-\infty}^{I_+} dE \left\| X_n R_n(E+i\epsilon/2) W(\tilde{\tilde{J}}_n) R(E+i\epsilon/2)\psi \right\|^2 \le e^{-2d_{I_+}(\alpha_{n-1})} e^{2\delta} b(\delta)$$
$$\times \left\{ \left\| W_{2\delta} \tilde{\tilde{J}}_n \right\|^2 \int_{-\infty}^{I_+} \|R(E+i\epsilon/2)\psi\|^2 \, dE + \int_{-\infty}^{I_+} \left\| W_{2\delta} R_n(E+i\epsilon/2) \tilde{\tilde{J}}_n \psi \right\|^2 dE \right\}, \quad (2.24)$$

where, as above, $b(\delta) \equiv \delta^{-2}(1+2\delta)e^{2\delta}$ and $W_{2\delta} \equiv (V - I_+)_+^{1/2} | \Omega_{2\delta}$, with $\Omega_{2\delta}$ defined to be the set on the right in the theorem. The first integral in (2.24) can be evaluated using Stone's formula, (2.21), and its contribution to (2.23) is

$$\sum_n \frac{e^{2P(n+1)}}{n} e^{-2d_{I_+}(\alpha_{n-1})} e^{2\delta} b(\delta) \left\| W_{2\delta} \tilde{\tilde{J}}_n \right\|^2, \quad (2.25)$$

which is finite by $(S3)$. The second integral in (2.24) is evaluated in a similar manner using Stone's formula for the resolvent of H_n. As a result, its contribution to (2.23) is bounded above by

$$\sum_n \frac{e^{2P(n+1)}}{n} e^{-2d_{I_+}(\alpha_{n-1})} e^{2\delta} b(\delta) \left\| W_{2\delta} \tilde{\tilde{J}}_n \right\|^2 \left\| \tilde{\tilde{J}}_n \psi \right\|^2$$
$$\le c_0 \sum_n \frac{e^{2P(n+1)}}{n} e^{-2d_{I_+}(\alpha_{n-1})} e^{2\delta} b(\delta) \left\| W_{2\delta} \tilde{\tilde{J}}_n \right\|^2, \quad (2.26)$$

using the exponential decay of ψ. This sum (2.26) is also finite by assumption $(S3)$.

(6) **High Energy.** We turn to the estimation of

$$\overline{\lim_{\epsilon \to 0}} \sum_n \frac{e^{2Pn}}{n} \epsilon \int_{I_+}^{\infty} \|X_n R(E+i\epsilon/2)\psi\|^2 \, dE. \quad (2.27)$$

It will be convenient to express $R(E+i\epsilon/2)\psi$ in terms of a contour integral over Γ_J, a rectangular contour around J and passing midway through the intervals $[I_-, J_-]$ and $[J_+, I_+]$. Let $\delta_0 \equiv \frac{1}{2}\min\{|J_- - I_-|, |I_+ - J_+|\}$. The contour representation is

$$R(E+i\epsilon/2)\psi = (2\pi i)^{-1} \oint_{\Gamma_J} (z - E - i\epsilon/2)^{-1} R(z)\psi dz, \quad (2.28)$$

which follows from the spectral representation of H and the fact that $E \ge I_+ > J_+$ (see Figure 2.2). Recalling the construction of ψ in the appendix, we can write

$$\Psi \equiv R(E+i\epsilon/2)\psi = -(2\pi i)^{-1} \oint_{\Gamma} \tilde{f}(z;\, E;\, \epsilon) R(z)\phi dz,$$

where

$$\tilde{f}(z;\, E;\, \epsilon) \equiv f(z)(z - E - i\epsilon/2)^{-1}.$$

It is clear that $\tilde{f}$ satisfies the same conditions as f in a neighborhood of J. Consequently $X_n\Psi$ satisfies the same estimate as in (A.15), where the coefficients depend on $\tilde{f}$ directly through the norm

$$\sup_{z\in\Gamma}\left|\tilde{f}(z;E;\epsilon)\right| \leq \|f\|_\infty |E-J_+|^{-1},$$

independent of ϵ. This shows that the energy integral in (2.27) converges. As for the sum, estimate (A.15) shows that it converges due to ($S1$) and ($S3$). Since these estimates are independent of ϵ, the limit in (2.27) is zero.

(7) It follows from (2.19) and (2.22), and from (2.23), and estimates (2.25)—(2.26) that the low-energy contribution to (2.16) is finite. The high-energy part of (2.16) is zero. This then proves that $D_\psi^{(p)} < \infty$, for some $p \geq 2$. By the argument given in (2.4), we conclude $D_\psi = 0$. ■

3 Absence of Absolutely Continuous Spectrum

In this section we prove a theorem about the absence of absolutely continuous spectrum for a Hamiltonian $H = -\Delta + V$ on an interval I provided the resonant wells for V at energies in I are sufficiently scarce. The idea is motivated by a result of Simon and Spencer [SS] who provide the following theorem for lattice Hamiltonians.

Theorem 3.1 ([SS]). Let $\{C_k\}_{k=1}^\infty$ be a family of disjoint hypercubes in $\mathbb{Z}^n$ of size l_k. Define the distance between these cubes as follows:

$$d_k \equiv \min\left\{\|\alpha-\beta\| \mid \alpha\in C_k,\ \beta\in C_m,\ k\neq m\right\}.$$

Suppose that for any $\epsilon > 0$,

$$\sum_k l_k^{n-1}e^{-\epsilon d_k} < \infty. \tag{3.1}$$

Let V be a function on $\mathbb{Z}^n$ such that

$$V(\alpha) \geq 0 \text{ if } \alpha \notin \bigcup_n C_n.$$

Then $\sigma_{ac}(-\Delta+V)\cap(-\infty,0) = \phi$, where $-\Delta$ is the lattice Laplacian.

Condition (3.1) is a geometric condition on the size (l_k) and the distribution (d_k) of resonant wells which insures that tunneling between these wells is sufficiently suppressed to prevent the existence of extended states. The main idea behind the proof is to introduce an operator H_D which is a direct sum of local operators with discrete spectrum in $(-\infty,0)$ and a strictly positive operator. The local operators h_a, one for each cube C_a, are obtained by restricting $H = -\Delta + V$ to a cube A_a of side $l_a + d_a$, centered at the same point as C_a, and imposing Dirichlet boundary conditions. Considering $\mathrm{tr}\left[(H - H_D)E_{(a,b)}(H)\right]$ for $a < b < 0$, we see that we obtain a term from each ∂A_a. But at any point of ∂A_a, the wave packet with energy localized in (a,b) is exponentially small. Consequently, the contribution of ∂A_a to the total trace is $c_0 l_a^{n-1} e^{-\epsilon d_a/2}$. Condition (3.1) guarantees that the sum is finite. We can apply Pearson's Theorem [RS3] in scattering theory to conclude that the absolutely continuous subspaces of H and H_D for $(-\infty, 0)$ are unitarily equivalent. Hence, H has no absolutely continuous spectrum in $(-\infty, 0)$. Note that the perturbation by Dirichlet boundary conditions on a finite discrete boundary is trace class.

This idea was used to prove an analogous result for one-dimensional Schrödinger operators on $L^2(\mathbf{R})$ in [HN] (see Section 4). Their theorem states that if there are potential barriers of increasing height and not too rapidly decreasing width which extend out to infinity in both directions, then the Hamiltonian has no absolutely continuous spectrum. We remark that for both the lattice and continuous one-dimensional cases, the widths of the resonant potential wells do not matter. In [HN], this result was applied to study Anderson-type random potentials with a Stark term.

We now state the conditions for a potential V on $\mathbf{R}^n$ and a compact interval $I = [I_-, I_+]$. We assume that V is a real-valued function such that $H = -\Delta + V$ is self-adjoint on a domain $D(H) \subset L^2(\mathbf{R}^n)$. We make the following geometric assumptions on V with respect to I_+. Recall from (2.1) that $G(I_+)$ is the classically forbidden region for V and energy I_+.

$(G1)$ $R \equiv \{x \mid V(x) \leq I_+\} = \cup_a U_a$, is a countable union of disjoint regions U_a which are the potential wells for V and I_+.

We note that $G(I_+) = \mathbf{R}^n \backslash R$. Recall from Section 2 the notation concerning distances in the Agmon metric ds^2.

$(G2)$ For each a in $(G1)$, define $d_a \equiv \inf_{a' \neq a} d_{I_+}(U_a, U_{a'})$. We assume $\exists\ \delta_0 > 0$ such that $d_a \geq \delta_0 > 0\ \forall\ a$.

As a consequence of $(G2)$, we can define regions $W_a \supset U_a$ by

$$W_a \equiv \left\{x \mid d_{I_+}(x, U_a) < \frac{1}{2} d_a\right\}. \tag{3.2}$$

These regions have the following property:

$$W_a \cap W_b = \phi, \quad b \neq a. \tag{3.3}$$

We define a covering of $\{U_a\}$ as follows. Let

$$W'_a \equiv \left\{x \mid d_{I_+}(x, U_a) < \frac{1}{4}d_a\right\}, \tag{3.4}$$

and set

$$J_a(x) \equiv \begin{cases} \left(\frac{1}{4}d_a\right)^{-1}\left(\frac{1}{4}d_a - d_{I_+}(x, W'_a)\right) & x \in W_a\,, \\ 0 & x \in \mathbf{R}^n \backslash W_a \end{cases} \tag{3.5}$$

so that $J_a|W'_a = 1$, J_a is Lipschitz continuous, and

$$|\nabla J_a(x)|^2 \leq \left(\frac{1}{4}d_a\right)^{-1}(V - I_+) \text{ on } W_a\backslash W'_a, \tag{3.6}$$

and zero elsewhere, as follows from (3.5) and (2.2). Finally, let $N \equiv \mathbf{R}^n \backslash \cup_a W_a$, so $N \subset G(I_+)$ and

$$d_{I_+}(N, R) \geq \inf_a(d_a/2) \geq \delta_0/2\,. \tag{3.7}$$

We need one other assumption on H. Let χ_a be the characteristic function of $W_a\backslash W'_a$. For some $m > n/2$, define constants $\tilde{c}_a$ by

$$\tilde{c}_a \equiv \|(-\Delta + 1)^m \chi_a E_{ac}(I)\| \tag{3.8}$$

where $E_{ac}(I)$ is the spectral projector for the absolutely continuous part of the spectrum of H in the interval I. As in Theorem 2.1, for any δ_a, $0 < 2\delta_a < (d_a/4)$, define

$$W_{\delta_a} \equiv (V(x) - I_+)_+^{\frac{1}{2}} \mid \{x \in G(I_+) \mid d_{I_+}(x, U_a) < \delta_a\}\,. \tag{3.9}$$

Then we require

$(G3)$ $\sum_a \tilde{c}_a \|W_{2\delta_a}\| e^{-(d_a/4 - 2\delta_a)} \left\{\int_{W_a\backslash W'_a} \left(1 + |x|^2\right)^\delta\right\}^{\frac{1}{2}} < \infty$ for some $\delta > \frac{n}{2}$.

We now turn to assumptions on the local Hamiltonians. For each well U_a, we define a local Hamiltonian H_a on $L^2(\mathbf{R}^n)$ by $H_a = -\Delta + V_a$, where

$$V(x) = \begin{cases} V(x) & x \in W_a\,, \\ \sup(I_+, V(x)) & x \in \mathbf{R}^n \backslash W_a\,. \end{cases} \tag{3.10}$$

The local Hamiltonian has only one well for interval I. We make the following assumption on the spectrum of H_a in $(-\infty, I_+]$,

$(S1)$ $\forall\, a$, $\sigma(H_a) \cap (-\infty, I_+]$ is discrete.

Of course, this is not a very restrictive hypothesis. We will also need the following. Let $N_a(\lambda) \equiv$number of eigenvalues of H_a less than or equal to λ, counting multiplicity.

$(S2)$ $N_a(I_+) < \infty \; \forall \, a$.

An alternate hypothesis can be given as follows. Recall from (3.8) that χ_a is the characteristic function of $W_a \backslash W_a'$. We write $E_a(J)$ for the spectral projector for H_a and set J.

$(S3)$ $\exists \, m > n/2$ and c_a, $0 < c_a < \infty$, such that $\forall \, a$,

$$\|(-\Delta + 1)^m \chi_a E_a((-\infty, I_+])\| < c_a \, .$$

This is a local regularity condition on V_a in N. Note that if V is regular so that the eigenfunctions of H_a are C^∞ and exponentially decaying in $W_a \backslash W_a'$, then $(S2)$ implies $(S3)$.

We can now state our main theorem.

Theorem 3.2. Assume $(G1)$—$(G3)$ and $(S1)$. Suppose either

(1) $(S2)$ holds and $\exists \, \delta_a$, $0 < 2\delta_a < d_a/4 \; \forall \, a$, such that for some $\epsilon > 0$,

$$\sum_a N_a(I_+) e^{-(1-\epsilon)(3d_a/4)+4\delta_a} \|W_{2\delta_a}\|^2 < \infty \, , \tag{3.11}$$

where $W_{2\delta_a}$ is defined in (3.9); or

(2) $(S3)$ holds and $\exists \, \delta > n/2$ and δ_a, $0 < 2\delta_a < d_a/4$, $\forall \, a$, such that

$$\sum_a c_a \left\{ \int_{W_a \backslash W_a'} \left(1 + |x|^2\right)^\delta dx \right\}^{\frac{1}{2}} e^{-(d_a/2 - 4\delta_a)} \|W_{2\delta_a}\|^2 < \infty \, , \tag{3.12}$$

where c_a is defined in $(S3)$. Then under condition (1) or (2), $\sigma_{ac}(H) \cap I = \phi$.

Remark 3.3. (1) Using the Cwikel-Lieb-Rosenbljun bound on $N_a(I_+)$ for $n \geq 3$ [RS4], we see that we can replace $N_a(I_+)$ in (3.11) by

$$\left(\sup_{U_a} |V - I_+|_-^{n/2} \right) \text{vol}(U_a) \, , \tag{3.13}$$

provided V is bounded below on U_a. Hence we obtain a formula similar to (3.1) although we lose one factor of the radius in this estimate since (3.1) involves the surface area of the well.

(2) Roughly, the integral in (3.12) is the effective size of the boundary of the well and corresponds to the l_a^{n-1} term in (3.1). We can estimate c_a if $N_a(I_+) < \infty$ and $\psi_{a,k} \in H^m(\mathbf{R}^n)$ where $H_a \psi_{a,k} = e_k \psi_{a,k}$ and $e_k \leq I_+$. In this case, Agmon estimates [A] on eigenfunction decay give

$$\begin{aligned} \|(-\Delta + 1)^m \chi_a E_a((-\infty, I_+])\| &\leq \sum_k \|(-\Delta + 1)^m \chi_a \psi_{a,k}\| \\ &\leq c_0 N_a(I_+) e^{-(1-\epsilon)d_a/4} \, . \end{aligned}$$

Using this bound in (3.12) we recover an expression similar to (3.11).

(3) To compare this theorem with the one-dimensional result of [HN], it is important to note that their result establishes the absence of absolutely continuous spectrum at all energies. In that case, the width of the potential barriers need not increase (in fact, they may decrease), but the height must increase in both directions (see Section 4).

To proof Theorem 3.2, we need a slight variation of Pearson's Theorem [RS3] on trace class scattering theory in a two-Hilbert space context. We do not give the proof of this proposition since it is a slight modification of that in [RS3].

Proposition 3.4. Let H_0 and H be two self-adjoint operators on Hilbert spaces $\mathcal{H}_0$ and $\mathcal{H}$, respectively. Let $J\colon \mathcal{H}_0 \to \mathcal{H} \in B(\mathcal{H}_0, \mathcal{H})$. Let $I, I' \subset \mathbf{R}$ be two intervals and define $\tilde{J} \equiv E_0(I')J^* E_{ac}(I)$: $\mathcal{H} \to \mathcal{H}_0$, where E_0 is the spectral projection for H_0 and E_{ac} is the spectral projection for the absolutely continuous part of H. Suppose there exists an operator $C\colon \mathcal{H} \to \mathcal{H}_0$ such that $\forall\, \phi \in D(H_0)$ and $\forall\, \psi \in D(H)$,

$$(\phi, C\psi) = (H_0\phi, \tilde{J}\psi) - (\tilde{J}^*\phi, H\psi) \tag{3.14}$$

and $C^*C \in \mathcal{I}_1(\mathcal{H})$, the trace ideal over $\mathcal{H}$. Then the wave operators defined by

$$s - \lim_{t\to\pm\infty} e^{iH_0t}E_0(I')J^*e^{-iHt}E_{ac}(I) \equiv \Omega_\pm(H_0, H, J, I', I) \tag{3.15}$$

exist as bounded operators from $E_{ac}(I)\mathcal{H}$ to $E_0(I')\mathcal{H}_0 \cap \mathcal{H}_{ac}(H_0)$, where $\mathcal{H}_{ac}(H_0)$ is the absolutely continuous subspace of H_0.

Proof of Theorem 3.2:

Let H_a be the local Hamiltonian for the well U_a as defined in (3.10) above. Define $\mathcal{H}_0 = \oplus_a L^2(\mathbf{R}^n)$ and the Hamiltonian $H_0 = \oplus_a H_a$ on its natural domain

$$\left\{ \psi = (\psi_a) \mid \psi_a \in D(H_a) \quad \text{and} \quad \|H_0\psi\| = \sum_a \|H_a\psi_a\| < \infty \right\}. \tag{3.16}$$

We apply Proposition 3.4 to the pairs $(H_0, \mathcal{H}_0)$ and $(H, \mathcal{H})$, $\mathcal{H} = L^2(\mathbf{R}^n)$. The operator $J\colon \mathcal{H}_0 \to \mathcal{H}$ is defined using the functions J_a introduced in (3.5) above. For $\psi = (\psi_a) \in \mathcal{H}_0$, define

$$J\psi = \sum_a J_a\psi_a \in \mathcal{H}, \tag{3.17}$$

and $\|J\| \leq 1$ since $|J_a| \leq 1$. We note that $JJ^*\colon \mathcal{H} \to \mathcal{H}$ and $J^*J\colon \mathcal{H}_0 \to \mathcal{H}_0$ are given by

$$\begin{aligned} JJ^*w &= \left(\sum J_a^2\right) w \neq w \\ J^*J\psi &= \bigoplus_a J_a^2\psi_a \neq \psi, \end{aligned} \tag{3.18}$$

but that $JJ^* - 1_{\mathcal{H}}$ and $J^*J - 1_{\mathcal{H}_0}$ are localized in N. Since (3.18) holds, the operators $\Omega_\pm$ are not isometries. We will deal with this below.

We choose the interval $I' = (-\infty, I_+]$ and we take I to be $I_\Delta \equiv [I_-, I_+ - \Delta]$ for any $\Delta > 0$ small. We will show below that this is no restriction. We define modified wave operators as in (3.15)

$$\Omega_\pm(H_0, H, J, (-\infty, I_+], I_\Delta) \equiv s - \lim_{t \to \pm\infty} e^{iH_0 t} \tilde{J} e^{-iHt}, \tag{3.19a}$$

where

$$\tilde{J} \equiv E_0((-\infty, I_+]) J^* E_{ac}(I_\Delta). \tag{3.19b}$$

To prove the existence of $\Omega_\pm$, we must show that C^*C is trace class on $\mathcal{H}$, where C: $\mathcal{H} \to \mathcal{H}_0$ is defined (as in (3.14)) by

$$\begin{aligned} C &\equiv H_0\tilde{J} - \tilde{J}H \\ &= E_0((-\infty, I_+])(H_0 J^* - J^* H) E_{ac}(I_\Delta). \end{aligned} \tag{3.20}$$

Using the definition of J in (3.17) and of H_0, we find for $w \in \mathcal{H}$,

$$Cw = \bigoplus E_a(I') W(J_a) E_{ac}(I_\Delta) w, \tag{3.21}$$

where the interaction term

$$W(J_a) \equiv [-\Delta, J_a], \tag{3.22}$$

is defined almost everywhere (by (3.6)) and is supported in $W_a \backslash W_a' \subset\subset G(I_+)$. From (3.21) we obtain

$$C^*Cw = \sum_a E_{ac}(I_\Delta) W(J_a)^* E_a(I') W(J_a) E_{ac}(I_\Delta) w. \tag{3.23}$$

We now proceed to estimate (3.23). Let χ_a be a smooth cut-off function such that $\chi_a W(J_a) = W(J_a)$ and the support of χ_a lies within $W_a \backslash W_a'$ (up to a negligible amount). We first compute

$$\|\chi_a E_a((\infty, I_+]) \chi_a\|_1 \tag{3.24}$$

under $(S2)$. Let $\{\phi_i\}$ be an orthonormal basis of eigenfunctions of h_a for $E_a((-\infty, I_+])\mathcal{H}$. Then

$$\operatorname{Tr}\left(\chi_a^2 E_a(I')\right) = \sum_{i=1}^{N_a(I_+)} \|\chi_a \phi_i\|^2. \tag{3.25}$$

Since $\operatorname{supp} \chi_a \subset G(I_+)$, standard eigenfunction decay estimates [A] yield for any $\epsilon > 0$,

$$\|\chi_a \phi_i\| \le c_0 e^{-(1-\epsilon)d_a/4}. \tag{3.26}$$

Consequently, our estimate for (3.24) is

$$\|\chi_a E_a(I')\chi_a\|_1 \leq c_0 N_a(I_+)e^{-(1-\epsilon)d_a/4}. \tag{3.27}$$

Next, we estimate (3.24) under assumption ($S3$). For $m \in \mathbf{Z}^+$ and $m > n/2$, write

$$\begin{aligned}\mathrm{Tr}\left(\chi_a^2 E_a(I')\right) &\leq \|(1-\Delta)^m \chi_a E_a(I')\| \; \|\chi_a(1-\Delta)^{-m}\|_1 \\ &\leq c_a \|\chi_a(1-\Delta)^{-m}\|_1 . \end{aligned} \tag{3.28}$$

To estimate the trace in (3.28), we use a standard result (see for example [RS3]). For any $\delta > n/2$,

$$\|f(\nabla)g(x)\|_1 \leq \|f\|_{2,\delta}\|g\|_{2,\delta}, \tag{3.29}$$

where

$$\|f\|_{2,\delta}^2 = \int \left(1+|x|^2\right)^\delta |f(x)|^2 .$$

Applying (3.29) to (3.28) we get

$$\|\chi_a(1-\Delta)^{-m}\|_1 \leq c_0(m)\left\{\int_{W_a\setminus W_a'} \left(1+|x|^2\right)^\delta\right\}^{\frac{1}{2}}, \tag{3.30}$$

where

$$c_0(m) \equiv \int \left(1+|x|^2\right)^{\delta-2m} dx ,$$

which is finite. Hence from (3.28) and (3.30),

$$\|\chi_a E_a(I')\chi_a\|_1 \leq c_a \left\{\int_{W_a\setminus W_a'} \left(1+|x|^2\right)^\delta\right\}^{\frac{1}{2}} . \tag{3.31}$$

Finally, we turn to the estimate of

$$\|W(J_a)E_{ac}(I_\Delta)\| , \tag{3.32}$$

which follows from a result of [BCD] on the decay of wave packets in the classically forbidden region $G(I_+)$. It is at this point that we need $\Delta > 0$.

Proposition 3.5 ([BCD]). Let $\phi \in E([I_-, I_+ - \Delta])\mathcal{H}$, $\Delta > 0$. Let $\Omega \subset G(I_+)$ be open with $d_{I_+}(\Omega) > 0$ and χ_Ω the characteristic function on Ω. Then for any $0 < 2\delta < d_{I_+}(\Omega)$ and for any $\beta > 0$,

$$\|\chi_\Omega\phi\|_{H^1(\Omega)} \leq e^{-d_{I_+}(\Omega)}c(\beta,\Delta)b(\delta)^{\frac{1}{2}}e^\delta\|W_{2\delta}\| \; \|\phi\| , \tag{3.33}$$

where $W_\delta \equiv (I_+ - V) \mid \{x \mid d_{I_+}(x, G(I_+)^c) < \delta\}$, $b(\delta) \equiv (\delta^2)^{-1}(1+2\delta)e^{2\delta}$, and $c(\beta,\Delta)$ satisfies $c(\beta,\Delta) \to +\infty$ as $\beta \to 0$ or $\Delta \to 0$, $c(\beta,\Delta) \to c(\Delta) < \infty$ as $\beta \to \infty$.

Applying (3.33) to (3.32), we obtain

$$\|W(J_a)E_{ac}(I_\Delta)\| \leq c(\beta,\Delta)e^{-d_a/4}b(\delta_a)^{\frac{1}{2}}e^{\delta_a}\|W_{2\delta_a}\|, \tag{3.34}$$

where we take β independent of a. Combining (3.27) and (3.34) we see that if $(S2)$ holds, we obtain the estimate

$$\mathrm{Tr}(C^*C) \leq c_0 \sum_a N_a(I_+)e^{-(1-\epsilon)(3d_a/4)+4\delta_a}\|W_{2\delta_a}\|^2, \tag{3.35}$$

where c_0 depends on (Δ,β). Similarly, we combine (3.31) and (3.34), to find an estimate in the case $(S3)$ holds:

$$\mathrm{Tr}(C^*C) \leq c_0 \sum_a c_a \left\{ \int_{W_a\setminus W_a'} (1+|x|^2)^\delta dx \right\}^{\frac{1}{2}} e^{-(d_a/2-4\delta_a)}\|W_{2\delta_a}\|^2, \tag{3.36}$$

Under either assumption $(S2)$ or $(S3)$ of Theorem 3.2, either (3.11) or (3.12), respectively, imply that $\mathrm{Tr}(C^*C) < \infty$. Now, by Proposition 3.4, this guarantees the existence of the modified wave operators (3.18). By construction, $\sigma_{ac}(H_0) \cap I' = \phi$. From this we wish to conclude $\sigma_{ac}(H) \cap [I_-, I_+ - \Delta] = \phi$ for any $\Delta > 0$. Since $\Omega_\pm$ are not partial isometrics, this requires a separate argument. It is sufficient to show that if $\psi \in E_{ac}(I_\Delta)\mathcal{H}$, then $\psi = 0$. Consider the strictly positive quantity

$$0 < \Delta\|\psi\|^2 \leq \langle \psi, (I_+ - H)\psi \rangle = \langle \psi_t, (I_+ - H)\psi_t \rangle \ , \tag{3.37}$$

where $\psi_t \equiv e^{-iHt}\psi$, as usual. We introduce a partition of unity for $\mathbf{R}^n$ based on J_a constructed above. Define $\tilde{J}_i$, $i = 0, \ldots,$ by

$$\begin{aligned} \tilde{J}_i &= J_{a(i)}, \ \ i \neq 0 \\ \tilde{J}_0 &= \sqrt{1 - \sum_a J_a^2}, \end{aligned}$$

so that $\sum \tilde{J}_i^2 = 1$. Note that $\tilde{J}_0$ is localized in N. With this choice, we can apply the localization formula (the IMS formula) [CFKS] to H:

$$H = \sum_{i=0} \tilde{J}_i H \tilde{J}_i - W, \tag{3.38}$$

where

$$W = \sum_{i=0} |\nabla \tilde{J}|^2,$$

and W is defined a.e. by the Lipschitz property of J_a. Furthermore, we note that

$$\sum_{i\neq 0} \tilde{J}_i H \tilde{J}_i = J H_0 J^* . \tag{3.39}$$

Now consider the right side of (3.37). Using (3.38)—(3.39), it is

$$\langle J^*\psi_t,(I_+ - H_0)J^*\psi_t\rangle + \langle J_0\psi_t,(I_+ - H)J_0\psi_t\rangle - \langle\psi_t, W\psi_t\rangle\,. \tag{3.40}$$

The second term of (3.40) is negative since $H|N > I_+$.

We also have

$$\lim_{t\to\infty}\langle\psi_t, W\psi_t\rangle = 0\,, \tag{3.41}$$

which can be seen as follows. First note that $|\nabla J_0|^2 = \sum_a \eta_a^2$ where $\operatorname{supp}\eta_a = \operatorname{supp}|\nabla J_a|^2 \subset W_a\backslash W_a'$ and $\eta_a^2 < c_0$. Next, it follows as above that

$$\tilde{W} \equiv \sum_{i=0} E_{ac}(I_\Delta)|\nabla J_i|^2 E_{ac}(I_\Delta) = \sum_a E_{ac}(I_\Delta)\left(\eta_a^2 + |\nabla J_a|^2\right)E_{ac}(I_\Delta) \tag{3.42}$$

is a trace class operator on $\mathcal{H}$ with trace norm bounded as in (3.30). This follows from estimate (3.29) since, for $m > \frac{n}{2}$,

$$(-\Delta+1)^{-m}\left(|\nabla J_a|^2 + \eta_a^2\right) \in \mathcal{I}_1(\mathcal{H})\,,$$

and by local regularity of V in $W_a\backslash W_a'$, $(-\Delta+1)^m\chi_a E_{ac}(I_\Delta)$ is bounded by $\tilde{c}_a$ as in (3.8). Finally, summability of (3.42) follows from estimate (3.34)

$$\|\chi_a E_{ac}(I_\Delta)\| \le c_I e^{-(d_a/4-2\delta_a)}\|W_{2\delta_a}\|\,, \tag{3.43a}$$

where $\operatorname{supp}\chi_a \supset \operatorname{supp}|\nabla J_a|$, and δ_a, $0 < 2\delta_a < d_a/4$, are defined above. Whence, we see from condition $(G3)$ that

$$\|\tilde{W}\|_1 \le 2\sum_a \tilde{c}_a\left\|(-\Delta+1)^{-m}\left(|\nabla J_a|^2+\eta_a^2\right)\right\|_1 e^{-(d_a/4-2\delta_a)}\|W_{2\delta_a}\| < \infty\,. \tag{3.43b}$$

The result (3.41) now follows as $\psi \in E_{ac}(I)\mathcal{H}$ and $\tilde{W}$ is compact. Consequently, from (3.37) and (3.40), we obtain

$$0 < \lim_{t\to\infty}\langle J^*\psi_t,(I_+ - H_0)E_0((-\infty,I_+])\,J^*\psi_t\rangle = \lim_{t\to\infty}\left\|(I_+ - H_0)_+^{\frac{1}{2}}\Omega(t)\psi\right\|^2 \tag{3.44}$$

since the term with $E_0((I_+,\infty))$ is negative.

We now prove that

$$\lim_{t\to\infty}\langle\Omega(t)\psi,(I_+ - H_0)\Omega(t)\psi\rangle = 0\,. \tag{3.45}$$

¿From the definition of $J^*\colon \mathcal{H}\to\mathcal{H}_0$, we obtain

$$H_0J^* = J^*H + M\,,$$

where $M\colon \mathcal{H} \to \mathcal{H}_0$ is defined by

$$Mu = \oplus_a W(J_a) E_{ac}(I_\Delta) u \,. \tag{3.46}$$

Consequently, we can write

$$(I_+ - H_0)\tilde{J}\psi_t = \tilde{J}(I_+ - H)\psi_t + E_0(I')M\psi_t \,,$$

where $\tilde{J} \equiv E_0(I')J^* E_{ac}(I_\Delta)$ as above. Let us write $U_t^0 \equiv \exp(-iH_0 t)$ and $U_t \equiv \exp(-iHt)$. Then we obtain

$$\lim_{t\to\infty} \langle \Omega(t)\psi, (I_+ - H_0)\Omega(t)\psi\rangle = \lim_{t\to\infty} \left\langle \Omega(t)\psi, U_t^0 \tilde{J}(I_+ - H)\psi_t \right\rangle + \left\langle \Omega(t)\psi, U_t^0 E_0(I')M\psi_t \right\rangle . \tag{3.47}$$

The first term on the right in (3.47) can be easily evaluated as $(I_+ - H)\psi \in E_{ac}(I)\mathcal{H}$. ¿From the existence of the wave operators, we obtain

$$\lim_{t\to\infty} \left\langle \Omega(t)\psi, U_t^0 \tilde{J} U_t (I_+ - H)\psi \right\rangle = \langle \Omega_+\psi, \Omega_+(I_+ - H)\psi\rangle = 0 \,. \tag{3.48}$$

As for the second term in (3.47), note that

$$\left| \left\langle \Omega(t)\psi, U_t^0 E_0(I')M\psi_t \right\rangle \right| \le \|\Omega(t)\psi\| \; \|E_0(I')M\psi_t\| \,. \tag{3.49}$$

The second factor on the right in (3.49) has the form

$$\begin{aligned} \|E_0(I')M\psi_t\|^2 &= \sum_a \langle \psi_t, E_{ac}(I_\Delta) W(J_a)^* E_a(I') W(J_a) E_{ac}(I_\Delta)\psi_t\rangle \\ &= \langle \psi_t, C^* C \psi_t\rangle \,. \end{aligned}$$

Consequently, both terms on the right in (3.49) vanish in the limit $t \to \infty$. These results prove that the limit on the right in (3.47) and hence in (3.44) vanishes. This fact contradicts the left side of (3.44) so $\psi = 0$. This proves that $E_{ac}(I_\Delta)\mathcal{H} = \{0\}$ for any $\Delta > 0$. Finally, we note that $\sigma_{ac}(H) \cap [I_-, I_+ - \Delta] = \phi$ $\forall\ \Delta > 0$ implies that $\sigma_{ac}(H) \cap [I_-, I_+] = \phi$ by the absolute continuity of the spectral measure of H with respect to Lebesgue measure. ∎

4 An Application to Poisson Potentials on R

We want to apply the ideas of the previous two sections to a family of positive Poisson potentials on **R**. We assume a non-negative bounded function u satisfies

$(U1)$ $u \ge 0$, supp $u \supset [-1, 1]$, $u \in L^1(\mathbf{R})$ and $u \,|\, [-1, 1] \ge 1$.

Let $\{X_m(\omega)\}_{m\in\mathbf{Z}}$, for each ω, be a set of Poisson distributed points in $\mathbf{R}$, i.e., for any Borel set $S \subset \mathbf{R}$,

$$P\{\exists\ k \text{ points } X_m(\omega) \text{ in } S\} = \frac{(\lambda|S|)^k}{k!}e^{-\lambda|S|}. \tag{4.1}$$

The constant λ is called the density of the Poisson distribution. We form a Poisson potential with a single-site potential u satisfying $(U1)$ as follows:

$$V_\omega(x) = \sum_m u\left(x - X_m(\omega)\right) \geq 0. \tag{4.2}$$

Under the hypothesis $(U1)$, Howland and Herbst [HH] proved that $V_\omega(x)$ is an ergodic, stationary process and, as a function of a.e. ω, it is finite measurable function of x. Under the additional hypothesis that

$(U2)$ $\sum_n \sup_{|x+n|<1} |u(x)| < \infty,$

they show that $V_\omega(x) = \mathcal{O}(\log|x|)$ a.s. as $|x| \to \infty$. Consequently, we can define $H_\omega = \frac{-d^2}{dx^2} + V_\omega$ on $L^2(\mathbf{R})$ on its natural domain. This is an ergodic, measurable family of self-adjoint operators which is a.s. essentially self-adjoint on $C_0^\infty(\mathbf{R})$. Pastur's theorem [K] states that $\exists \sum \subset \bar{\mathbf{R}}_+$ such that $\sigma(H_\omega) = \sum$ a.s..

We first turn to the question of a.s. absence of diffusion at low energy. First we prove with probability one that V_ω has barriers of height at least 1 and width $c|n|$, for any fixed $c > 0$, in each interval $[e^{|n|}, e^{|n|+1}]$ and $[-e^{|n|+1}, -e^{|n|}]$, $n \in \mathbf{N}$, for all but finitely many n.

Proposition 4.1. Let u satisfy $(U1)$—$(U2)$ and let $V_\omega(x)$ be as defined in (4.2). Let $c > 0$ be fixed and suppose the density λ (in (4.1)) satisfies $e(1-e^{-\lambda})^c > 1$. Then $\exists\ \Omega_0 \subset \Omega$, $P(\Omega_0) = 1$ such that for $\omega \in \Omega_0$, V_ω is such that $\exists$ an interval I_n of length $c|n|$ in $[e^{|n|}, e^{|n|+1}]$ and in $[-e^{|n|+1}, -e^{|n|}]$, for all but finitely many n, and $V_\omega | I_n \geq 1$.

PROOF: For simplicity, we consider $x \geq 0$; the argument for $x < 0$ is the same after small notational changes. It suffices to show that a.s. for all n large, we can find an interval $I_n \subset [e^n, e^{n+1}]$, $|I_n| = cn$, with at least one $X_j(\omega)$ in each of the nc non-overlapping subintervals of length 1. It follows from (4.1) that

$$P\{\ \exists \geq 1 \text{ point } X_j(\omega) \text{ in each non-overlapping unit sub-interval of } I_n\} = (1-e^{-\lambda})^{nc}$$

(where, by nc, we mean the greatest integer $\leq nc$). Then, if A_n is the event "$V \,|\, I_n \geq 1$", we find

$$P(A_n) \geq (1-e^{-\lambda})^{cn} \tag{4.3}$$

since other configurations of the $X_j(\omega)$'s contribute to the event. Next, the interval $I_n \subset [e^n, e^{n+1}]$ can be located in any of the $(nc)^{-1}e^n(e-1)$ non-overlapping intervals in $[e^n, e^{n+1}]$. Let K_n be the event "$\exists$ no $I_n \subset [e^n, e^{n+1}]$ such that $V \mid I_n \geq 1$". Then, we get the estimate from (4.3),

$$P(K_n) \leq \left[1-(1-e^{-\lambda})^{nc}\right]^{\frac{2e^n}{nc}} . \tag{4.4}$$

We apply the Borel-Cantelli lemma to the events K_n. It is easy to check that the condition on the density λ implies $\sum_n P(K_n) < \infty$. Consequently, for all but finitely many n, $K_n^c = \{\exists\, I_n \subset [e^n, e^{n+1}]$ such that $V \mid I_n \geq 1\}$ occurs with probability 1. ■

Theorem 4.2. Let u satisfy $(U1)$—$(U2)$ and let $V_\omega(x)$ be as defined in (4.2). Suppose we are given $\epsilon_0 > 0$ small. If the density satisfies $\lambda > \left|\log(1-e^{-1/c})\right|$, with $c \geq 3\epsilon_0^{-\frac{1}{2}} c_0$, for some $c_0 > 4$, then for $\psi \in E([0, 1-\epsilon_0))\, L^2(\mathbf{R})$ satisfying $(W1)$-$(W2)$, $D_\psi = 0$. If, in addition, supp u is compact, then $\sigma(H_\omega) = [0, \infty)$ P–a.s..

Proof: If supp u is compact, it is easy to show that the set of potentials such that V_ω vanishes on a sequence of open sets with increasing diameter has positive measure. For such a V_ω, $\sigma(H_\omega) = [0, \infty)$, as can be proven by constructing Weyl sequences. Since $\sigma(H_\omega)$ is deterministic, $\sigma(H_\omega) = [0, \infty)$ P–a.s.. As for diffusion, we verify the hypotheses of Theorem 2.1 with $p = 2$ for any $\omega \in \Omega_0$ constructed in Proposition 4.1 for c as given in the theorem. We choose $\kappa_n, \alpha_n, \omega_n \subset I_n$ such that $|\kappa_n|$, $|\alpha_n|$, and $|\omega_n|$ are independent of n and

$$d_{I_+}(\alpha_{n-1}), d_{I_+}(\omega_{n-1}), d_{I_+}(\alpha_{n-1}, \omega_{n-1}) \geq \epsilon_0^{\frac{1}{2}} \cdot \frac{1}{3} \cdot |I_n| . \tag{4.5}$$

We obtain from Proposition 4.1 that $|I_n| = c|n|$ and, from (4.5) and the bound on c,

$$d_{I_+}(\alpha_{n-1}), d_{I_+}(\omega_{n-1}), d_{I_+}(\alpha_{n-1}, \omega_{n-1}) \geq \frac{1}{3}\epsilon_0^{\frac{1}{2}} \left(\frac{3c_0}{\epsilon_0^{\frac{1}{2}}}\right) |n| = c_0|n| . \tag{4.6}$$

We can now directly verify $(S1)$ and $(S2)$ provided $c_0 > 4$.

Next, we must estimate $\|W_{2\delta}\tilde{\chi}_n\|^2$. Let us define

$$\chi_n \equiv \min\{x \mid (V(x) - (1-\epsilon_0)) \leq 0\} .$$

We recall from [HH] that $\exists\, \Omega_1 \subset \Omega$, $P(\Omega_1) = 1$ such that $\omega \in \Omega_1$ implies

$$V_\omega(x) = \mathcal{O}(\log|x|) .$$

We take $\Omega_0 \cap \Omega_1$ and obtain a set of measure 1 which we call again Ω_0. If $\chi_n = \mathcal{O}(e^n)$, then

$$\|W_{2\delta}\tilde{\chi}_n\|^2 \leq d|n|, \text{ a.s.,}$$

and by the choice of α_{n-1} and ω_{n-1} made above $(S3)$ and $(S4)$ are finite. In general, suppose $\chi_n = \mathcal{O}(f(n))$, where $f(x) \geq \mathcal{O}(e^{|x|})$, then $\exists\, d > 0$ such that

$$\|W_{2\delta}\tilde{\chi}_n\|^2 \leq d \log|f(n)|\,, \text{ a.s..}$$

We can now redefine α_{n-1} and ω_{n-1} so that

$$d_{1-\epsilon_0}(\alpha_{n-1}, \omega_{n-1}) \geq \epsilon_0^{\frac{1}{2}}\left(f(n) - a_n)\right)/3\,.$$

Since the exponential terms in $(S3)$—$(S4)$ will be dominated by $f(n)$, it is seen as above that they converge. Consequently, for any $\psi \in E([0, 1-\epsilon_0))\, L^2(\mathbf{R})$ satisfying $(W1)$-$(W2)$, $D_\psi = 0$ P-a.s.. ■

Next we turn to the question of absence of absolutely continuous spectrum. This can be proved at low energy using Theorem 3.1. However, one can show $\sigma_{ac}(H_\omega) = \phi$ P–a.s. directly using the results of [HN]. Let us recall their main theorem on absence of absolutely continuous spectrum.

Theorem 4.3 ([HN]). Suppose $\exists$ sequences $\{x_n\}$, $\{\delta_n\}$, $\{E_n\}$, $n \in \mathbf{Z}$ and $c, \epsilon > 0$ such that

(i) $E_n \to \infty$ as $|n| \to \infty$;

(ii) $x_n \to \pm\infty$ as $n \to \pm\infty$;

(iii) $0 \leq E_n \leq \inf_{|x-x_n|<\delta_n} V(x) \leq \sup_{|x-x_n|<\delta_n} V(x) \leq cE_n$;

(iv) $E_n\delta_n^2 \geq E_n^\epsilon$;

(v) $H = -\frac{d^2}{dx^2} + V$ is essentially self-adjoint on $C_0^\infty(\mathbf{R})$.

Then $\sigma_{ac}(H) = \phi$.

We note that the upper bound constraint in Condition (iii) can be eliminated. To use this result, we show $\exists\, \Omega_0 \subset \Omega$, $P(\Omega_0) = 1$, such that for each $\omega \in \Omega_0$, V_ω has a sequence of barriers of increasing height and constant width one, extending to $\pm\infty$. We can take $\delta_n = 1$ and $\epsilon = 1$ in Theorem 4.3. The important point here is that there is no restriction on the separation between these barriers as there is in higher dimensions.

Theorem 4.4. Let u satisfy $(U1)$—$(U2)$ and V_ω be as in (4.2), then $\sigma_{ac}(H_\omega) = \phi$ P–a.s..

PROOF: For each $p \in \mathbf{N}$, define the event

$$K_n \equiv \exists\, p \text{ points of a Poisson ensemble in } (n, n+1)\,.$$

Since the intervals for each n are disjoint, the events K_n are independent. Then by (4.1),

$$P(K_n) \geq \frac{\lambda^p}{p!} e^{-\lambda}\,,$$

and hence $\sum_n P(K_n) = \infty$. By the Borel-Cantelli lemma, the events K_n occur infinitely often. As a consequence of $(U1)$, there exist infinitely many intervals of length one on which $V_\omega \geq p$ P–a.s.. Call this set of potentials Ω_p. Then $\Omega_0 \equiv \bigcap_{p \geq 1} \Omega_p$ has full P-measure. If $\omega \in \Omega_0$, there exists a sequence B_n of unit length intervals extending to $\pm\infty$ as $n \to \pm\infty$ such that $V_\omega \,|\, B_n \geq |n|$. We now apply Theorem 4.3, with the modification of removing the upper bound on condition (iii), to conclude that $\sigma_{ac}(H_\omega) = \phi$. ∎

We note that a similar result, with more restrictive decay conditions on u and using other techniques, was proved in Carmona and Lacroix [CL]. We point out that the a.s. nature of the spectrum is unknown. It is expected to be dense pure point.

Appendix: Decaying States with Compact Energy Support

The purpose of this appendix is to show how to construct states ψ satisfying the condition $(W1)$—$(W2)$ of Section 2.

Proposition A.1. Let $I = [I_-, I_+] \subset \mathbf{R}$ and $J = [J_-, J_+] \subset I$ with $|J_\pm - I_\pm| \geq \Delta > 0$. Let Γ be a rectangular contour about J passing through $J_\pm$. Let f be a function analytic in a neighborhood of J which includes Γ and such that $f^{(k)}(J_\pm) = 0$ for $k = 0, 1$. Let $H = -\Delta + V$ be the self-adjoint Hamiltonian of Section 2 with V satisfying $(V1)$—$(V3)$. Then for any $\phi \in C_0^\infty(\mathbf{R})$, the integral

$$\psi = (2\pi i)^{-1} \oint_\Gamma f(z) R(z) \phi \, dz \tag{A.1}$$

converges strongly and ψ satisfies $(W1)$—$(W2)$ with the specific decay estimate given in (A.15).

PROOF:

(1) **Existence and Spectral Support.** Let $\Gamma = \Gamma_v \cup \Gamma_h$, where $\Gamma_v = \Gamma_v^+ \cup \Gamma_v^-$, with

$$\Gamma_v^\pm = \{J_\pm \pm is \,|\, s \in [-a, a]\}\,, \tag{A.2}$$

and $\Gamma_h = \Gamma_h^+ \cup \Gamma_h^-$, with

$$\Gamma_h^- \;=\; \{s - ia \,|\, s \in [J_-, J_+]\} \tag{A.3a}$$

$$\Gamma_h^+ \;=\; \{s - ia \,|\, s \in [J_-, J_+]\}\,. \tag{A.3b}$$

Here, a is chosen so $J_\pm \pm ia$ lies within the radius of convergence of f at each point of J. It is clear that the integrals

$$\oint_{\Gamma_h^\pm} f(z)R(z)\phi dz \tag{A.4}$$

exist. Consider the integral over Γ_v^+. Since $f^{(k)}(J_+) = 0$ for $k = 0, 1$, $\exists$ an analytic function g such that for z near I_+,

$$f(z) = (z - J_+)^2 f^{(2)}(J_+) + (z - J_+)^3 g(z)\,. \tag{A.5}$$

Consequently, we make the definition

$$\int_{-a}^{a} ds\, f(J_+ + is)R(z + is)\phi \equiv \lim_{\epsilon\to 0}\left\{\int_{-a}^{-\epsilon} + \int_{\epsilon}^{a}\right\} ds\, f(J_+ + is)R(J_+ + is)\phi\,. \tag{A.6}$$

It is easy to show using (A.5) that the strong limit in (A.6) exists. Hence, ψ exists. Now we have from (A.6)

$$\lim_{a\to 0}\int_{-a}^{a} ds\, f(J_\pm + is)R(J_\pm + is)\phi = 0\,.$$

Consequently, we can write

$$\psi = \lim_{a\to 0}\int_{\Gamma_h^+\cup\Gamma_h^-} f(z)R(z)\phi dz \tag{A.7}$$

where $\Gamma_h^\pm$ are defined in (A.3a)—(A.3b). To evaluate (A.7), we have

$$\begin{aligned}
&\lim_{a\to 0}\int_{J_-}^{J_+} ds\,[f(s - ia)R(s - ia) - f(s + ia)R(s + ia)]\,\phi \\
&= -\lim_{a\to 0}\int_{J_-}^{J_+} ds f(s + ia)\,[R(s + ia) - R(s - ia)]\,\phi \\
&\quad + \lim_{a\to 0}\int_{J_-}^{J_+} ds\,[f(s + ia) - f(s - ia)]\,R(s - ia)\phi\,.
\end{aligned}$$

The first integral can be evaluated using Stone's formula. The result is

$$-\frac{1}{2}\left(E_J(H) + E_{\overset{\circ}{J}}(H)\right) f(H)\phi, \tag{A.8}$$

where $\overset{\circ}{J} \equiv \mathrm{Int}(J)$. As for the second, we expand f about s so that

$$f(s + ia) - f(s - ia) = 2iaf'(s) + \mathcal{O}(a^2)\,,$$

uniformly on J. The $\mathcal{O}(a^2)$ contribution vanishes in the limit. For the first term, we write

$$\begin{aligned}\left\|2ia\int_{J_-}^{J_+} f'(s)R(s-ia)ds\right\|^2 &= 4a^2\int_J ds\int_J dt\, f'(s)\overline{f'(t)}\,\langle R(t-ia)\phi, R(s-ia)\phi\rangle \\ &= 4a^2\int_J\int_J ds\,ds\,[(s-t)+2ia]^{-1} f'(s)\overline{f'(t)}\ [\langle\phi, R(t+ia)\phi\rangle \\ &\qquad - \langle\phi, R(s-ia)\phi\rangle]\ .\end{aligned}$$

Since $|\langle\phi, R(s+ia)\phi\rangle| \leq a^{-1}$ and f' vanishes at $J_\pm$, we can evaluate each integral using integration-by-parts and show that the term vanishes in the limit. It is then clear from (A.7) that $E_J(H)\psi = \psi$.

(2) **Decay**. Recall the functions $\tilde{\tilde{J}}_n$ from Section 2 with $\mathrm{supp}|\nabla\tilde{\tilde{J}}_n| \subset \alpha_{n-1}$. Let $H_n = -\Delta + V_n$ be the operator on $L^2(\mathrm{supp}\ \tilde{\tilde{J}}_n)$ with Dirichlet boundary conditions on $\partial(\mathrm{supp}\ \tilde{\tilde{J}}_n)$ and $V_n \equiv \sup(V, I_+)$. Let $R_n(z)$ be the resolvent. Applying the geometric resolvent formula (2.17) in this case to $\tilde{\tilde{J}}_n\,\psi$, we obtain

$$\tilde{\tilde{J}}_n\,\psi = (2\pi i)^{-1}\oint_\Gamma f(z)R_n(z)\left[W(\tilde{\tilde{J}}_n) + (I_+ - V)_+\,\tilde{\tilde{J}}_n\right]R(z)\phi dz \tag{A.9}$$

where we have chosen n so large that $\tilde{\tilde{J}}_n\,\phi = 0$. We estimate the first term in (A.9) using the factorization of $W(\tilde{\tilde{J}}_n)$ as described in [BCD]. The second term is similar because of $(V3)$. We write

$$W(\tilde{\tilde{J}}_n) = M_2^{(n)^*}M_1^{(n)}\ , \tag{A.10}$$

and for the first integrand in (A.9),

$$\begin{aligned}\left\|R_n(z)W(\tilde{\tilde{J}}_n)R(z)\phi\right\| &= \sup_{\eta\neq 0}\left\langle R_n(z)^*\eta,\ M_2^{(n)^*}M_1^{(n)}R(z)\phi\right\rangle \\ &\leq \sup_{\eta\neq 0}\left\|M_2^{(n)}R_n(z)\eta\right\|\ \left\|M_1^{(n)}R_n(z)\phi\right\|\ .\end{aligned} \tag{A.11}$$

To estimate the first factor on the right in (A.11), we note that $M_2^{(n)}$ is localized in α_{n-1}. We can apply Lemma II.1 of [BCD] and obtain

$$\left\|M_2^{(n)}R_n(z)\right\| \leq \|R_n(z)\|^{\frac{1}{2}} + e^{-d_{I_+}(\alpha_{n-1})}b^{\frac{1}{2}}(\delta)\,\|W_\delta\tilde{\chi}_n R_n(z)\|\ , \tag{A.12}$$

for some $\delta > 0$ and W_δ as defined in Theorem 2.1. As for the second factor on the right in (A.11), let ξ_n be a function with $\kappa_{n-1} \equiv \mathrm{supp}\,|\nabla\xi_n| \subset [a_{n-1}, b_{n-1}]$, $\xi_n\phi = \phi$, and $d_{I_+}(\kappa_{n-1}, \alpha_{n-1}) > 0$. We take κ_{n-1} as close to a_{n-1} as possible. Using result (II.19) of [BCD], we find the estimate

$$\begin{aligned}\left\|M_1^{(n)}R(z)\xi_n\phi\right\|^2 &\leq e^{-2d_{I_+}(\kappa_{n-1},\alpha_{n-1})}\,\|R(z)\phi\| \\ &\quad + e^{-2d_{I_+}(\alpha_{n-1})}b^{\frac{1}{2}}(\delta)\,\|W_\delta\tilde{\chi}_n R(z)\|^2\ .\end{aligned} \tag{A.13}$$

Because $z \in \Gamma$, $\|R_n(z)\phi\| \leq \Delta^{-1}$. Combining (A.12) and (A.13), we obtain upper bound for (A.11),

$$\begin{aligned} \left\| R_n(z) W(\tilde{\tilde{J}}_n) R(z)\phi \right\| \leq\ & c_0 e^{-d_{I_+}(\kappa_{n-1},\alpha_{n-1})} \|R(z)\| \\ &+ c_1 e^{-d_{I_+}(\kappa_{n-1},\alpha_{n-1}) - d_{I_+}(\alpha_{n-1})} \|W_\delta \tilde{\chi}_n\| \, \|R(z)\| \\ &+ c_2 e^{-d_{I_+}(\alpha_{n-1})} \|W_\delta \tilde{\chi}_n\| \, \|R(z)\| \\ &+ c_3 e^{-2d_{I_+}(\alpha_{n-1})} \|W_\delta \tilde{\chi}_n\|^2 \|R(z)\| , \end{aligned} \qquad (A.14)$$

where c_i depends on Δ and δ. Returning to (A.9), we see that the integral converges as in part (1) above. Moreover, we obtain:

$$\begin{aligned} \| \tilde{\tilde{J}}_n \psi \| \leq\ & c_0 e^{-d_{I_+}(\kappa_{n-1},\alpha_{n-1})} + c_1 e^{-d_{I_+}(\kappa_{n-1},\alpha_{n-1}) - d_{I_+}(\alpha_{n-1})} \\ & c_2 e^{-d_{I_+}(\alpha_{n-1})} \|W_\delta \tilde{\chi}_n\| + c_3 e^{-2d_{I_+}(\alpha_{n-1})} \|W_\delta \tilde{\chi}_n\|^2 , \end{aligned} \qquad (A.15)$$

proving the necessary decay estimate. ■

Acknowledgement. P.H. would like to thank Erik Balslev for his kind invitation to participate in the Workshop on Schrödinger Operators held in Aarhus, Denmark in the spring of 1991. He enjoyed the warm hospitality of Erik Balslev, Arne Jensen, and Erik Skibsted (and the friendly cafés of Aarhus) during his stay in Aarhus. He would also like to thank Shu Nakamura for many discussions on random Schrödinger operators.

References

[A] Agmon, S.: Lectures on exponential decay of solutions of second order elliptic operators. Princeton, N.J.: Princeton University Press (1982).

[BCD] Briet, Ph., Combes, J.-M, Duclos, P.: Spectral Stability under Tunneling. Commun. Math. Phys. **126**, 133–156 (1989).

[CFKS] Cyclon, H.L., Froese, R.G., Kirsch, W., Simon, B.: *Schrödinger Operators with Application to Quantum Mechanics and Global Geometry.* Berlin: Springer-Verlag (1987).

[CL] Carmona, R., Lacroix, J.: *Spectral Theory of Random Schrödinger Operators.* Boston: Birkhauser Boston (1990).

[CS] Combes, J.-M., Sigal, I.M.: Dynamical Behaviour for some Multiple Well Schrödinger Operators, CPT preprint, Marseille, France (1990).

[FS] Fröhlich, J., Spencer, T.: Absence of diffusion in the Anderson tight-binding model for large disorder or low energy. Commun. Math. Phys. **88**, 151–184 (1983).

[HH] Herbst, I., Howland, J.: The Stark ladder and other one-dimensional external field problems. Commun. Math. Phys. **80**, 23–42 (1981).

[HN] Hislop, P.D., Nakamura, S.: Stark Hamiltonians with Unbounded Random Potentials. Rev. Math. Phys. **2**, 479–494 (1990).

[JMS] Jona-Lasinio, G., Martinelli, F., Scoppola, E.: Multiple tunnelings in d-dimensions: A quantum particle in a hierarchical potential. Ann. Inst. H. Poincaré **42**, 73–108 (1985).

[K] Kirsch, W.: Random Schrödinger Operators, in Schrödinger Operators: Proceedings, Sønderborg, Denmark 1988. H. Holden and A. Jensen (Eds.). Berlin: Springer-Verlag (1989).

[MH] Martinelli, F., Holden, H.: On Absence of Diffusion near the Bottom of the Spectrum for a Random Schrödinger Operator on $L^2(\mathbf{R}^\nu)$. Commun. Math. Phys. **93**, 197–217 (1984).

[P] Pearson, D.: Singular continuous measures in scattering theory. Commun. Math. Phys. **60**, 13 (1978).

[RS3] Reed, M., Simon, B.: Methods of Modern Mathematical Physics, III. Scattering Theory. New York: Academic Press (1979).

[RS4] Reed, M., Simon, B.: Methods of Modern Mathematical Physics, IV. Analysis of Operators. New York: Academic Press (1978).

[S] Simon, B.: Absence of Ballistic Motion. Commun. Math. Phys. **134**, 209–212 (1990).

[SS] Simon, B., Spencer T.: Trace class perturbations and the absence of absolutely continuous spectra. Commun. Math. Phys. **125**, 113–125 (1989).

[Sp] Spencer, T.: The Schrödinger equation with a random potential, in Critical Phenomena, Random Systems, Gauge Theories, Les Houches Summer School XLIII (1984), K. Osterwalder and R. Stora (Eds.).

Figure 2.1. Various families of functions for the proof of Theorem 2.1.

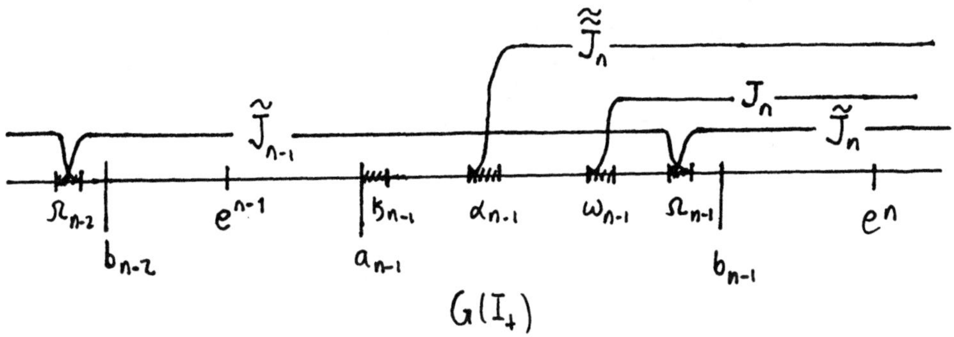

Figure 2.2. Contour Γ_J for the proof of Theorem 2.1 and the contour Γ for the proof of Proposition A.1.

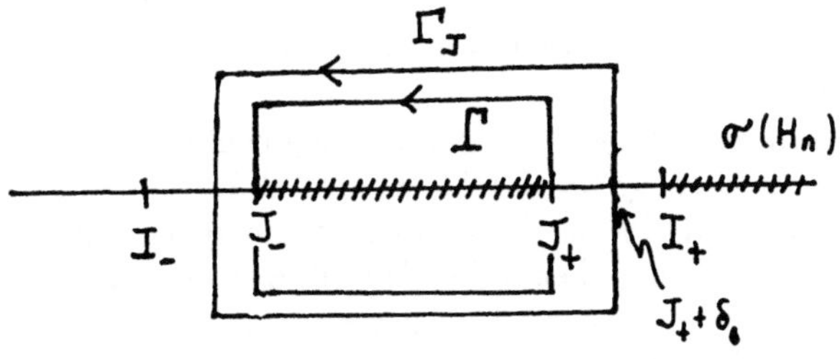

Spectral Theory of Schrödinger Operators with Very Long Range Potentials

Jaime Cruz
Universidad de las Américas Puebla, México

1. Introduction

In this work we describe some results on the spectral theory of the Schrödinger operator $H = -\Delta + V$ acting on $L^2(\mathbf{R}^n)$, where V is a bounded real-valued function of class C^l on $\mathbf{R}^n \setminus \{0\}$ with $n \geq 2$ that satisfies

$$\partial_x^\alpha V(x) = O(|x|^{-|\alpha|}), \qquad \text{as} \quad |x| \to \infty, |\alpha| \leq l. \tag{1.1}$$

We obtain at high energies the limiting absorption principle in the framework of Besov spaces, existence and uniqueness of the generalized eigenfunctions, and an eigenfunction expansion for H.

Since (1.1) can be thought of as the long-range-potential condition with $\epsilon = 0$, we refer to potentials satisfying (1.1) as *very long-range potentials*. We note however that very long-range potentials are not necessarily one-body potentials. In fact, the class of potentials satisfying (1.1) contains all homogeneous potentials of degree zero, essentially all one-body potentials, and potentials that may not have a limit in any direction at infinity. In this sense, we may view the very long-range potentials as being more complicated than N-body potentials.

A class of Schrödinger operators which contains those with very long-range potentials has been considered by Saitō [17], who establishes a radiation condition and the limiting absorption principle in the framework of the spaces $L^{2,s}$ for that class; in addition, the spectral and scattering theory for Schrödinger operators with homogeneous potentials of degree zero has been discussed by Herbst [10].

In Section 2 we introduce notation and present some heuristic ideas and in Section 3 we state our results. Complete proofs of these results will be published elsewhere.

Most of the results reported here were obtained in the author's thesis [8] at the University of Virginia under the supervision of Professors S. Agmon and I. Herbst to whom he expresses his deep gratitude. The hospitality of Professors E. Balslev and E. Skibsted during the Summer Activities on Schrödinger Operators at the Mathematisk Institute of the Aarhus Universitet, where part of this work was completed, should also be greatly appreciated.

2. Notation and Main Ideas

Let $H = -\Delta + V$ be a Schrödinger operator acting on $L^2(\mathbf{R}^n)$, where V is a very long-range potential. The self-adjoint realization of H exists by the Kato-Rellich theorem [9]. Although we know little about the structure of the essential spectrum $\sigma(H)$ for low energies it can easily be seen, for example using the Mourre estimate [16], that there exists $\lambda_0 > 0$ such that the absolutely continuous spectrum $\sigma_{ac}(H)$ contains the interval (λ_0, ∞).

For $\lambda_0 > 0$ sufficiently large we derive an eigenfunction expansion for H (Theorem 3); that is to say, we construct a partial isometry $\mathcal{F}$ on $L^2(\mathbf{R}^n)$ such that

$$\mathcal{F}(Hf)(\xi) = S(\xi)^2(\mathcal{F}f)(\xi), \qquad \xi\text{-a.e.},$$

for all $f \in L^2_{\lambda_0}(\mathbf{R}^n_x)$ for which $S(\cdot)(\mathcal{F}f)(\cdot) \in L^2_{\lambda_0}(\mathbf{R}^n_\xi)$, where S is a solution to the eikonal equation (2.2) satisfying (2.4). We define

$$L^2_{\lambda_0}(\mathbf{R}^n_x) = \{f \in L^2(\mathbf{R}^n) : f = E(\lambda_0, \infty)f\}$$

and

$$L^2_{\lambda_0}(\mathbf{R}^n_\xi) = \{g \in L^2(\mathbf{R}^n) : \ g = 0 \quad \text{for a.e. } \xi \text{ with } S(\xi)^2 \leq \lambda_0\} \ ,$$

where $E(\cdot)$ denotes the right-continuous spectral measure of H. The initial and final subspaces of $\mathcal{F}$ are $L^2_{\lambda_0}(\mathbf{R}^n_x)$ and $L^2_{\lambda_0}(\mathbf{R}^n_\xi)$.

The main role in the construction of $\mathcal{F}$ is played by the generalized eigenfunctions which we extract from the resolvent kernel of $(H - \lambda - i0)^{-1}$ for $\lambda > \lambda_0$ using a method developed by *Agmon* [1], *Ikebe* [11], and *Isozaki* [12].

Following *Agmon* [2] and *Constantine* [7], we first establish at high energies the limiting absorption principle for H in the framework of Besov spaces $B(\mathbf{R}^n)$ and $B^*(\mathbf{R}^n)$ (Theorem 1). The generalized eigenfunctions are indirectly provided by the ***boundary values at infinity*** of $(H - \lambda - i0)^{-1}f$, for $f \in B(\mathbf{R}^n)$ (Theorem 2). We obtain these boundary values by introducing appropriate coordinates on $\mathbf{R}^n$, which allow us to apply an abstract result of *Jäger* [13].

We derive the limiting absorption principle for H by studying the asymptotic behavior of solutions to the Schrödinger-like equation

$$-\Delta u - k^2 W u = f \qquad \text{on} \quad \mathbf{R}^n, \tag{2.1}$$

where f satisfies certain decay assumptions at infinity and $k \in \mathbf{C} \setminus \{0\}$ with $\operatorname{Im} k > 0$. For $|\lambda| > \|V\|_\infty$ we define

$$W(x) = 1 - \frac{V(x)}{\lambda}.$$

The reason we investigate the asymptotic behavior at infinity of solutions to (2.1) is that the study of solutions to the more natural P.D.E., namely, $-\Delta u + Vu - z = f$, where $z = \lambda + i\epsilon$, $\epsilon \neq 0$, leads to seeking solutions to the *complex* eikonal equation $\nabla S \cdot \nabla S + V = z$ on $\mathbf{R}^n$.

It is a well known fact in differential geometry [15] that the energy function

$$E(\xi) = \int W(\xi(t))|\dot{\xi}(t)|^2 dt$$

on $E_x \equiv \{\xi \in (H(0,1))^n : \xi(0) = 0, \xi(1) = x\}$, attains its minimum precisely on the set of minimal geodesics from 0 to x for the Riemannian metric $(ds)^2 = W(x)|dx|^2$. Hence the function $S_\lambda(x)$ defined by

$$(S_\lambda(x))^2 = \inf\{\int_0^1 W(\xi(t))| \cdot \xi(t)|^2 dt : \xi \in E_x\}$$

is the Agmon metric [3] and thus it plays an important role in the study of the asymptotic behavior of solutions to (2.1).

For V satisfying (1.1) *Lions* [14] proves that S_λ is the maximum solution in $W^{1,\infty}(\mathbf{R}^n)$ to the eikonal equation

$$|\nabla S_\lambda(x)| = W(x), \qquad x \in \mathbf{R}^n, \tag{2.2}$$

that satisfies $S_\lambda(0) = 0$.

Let P be a posistive solution of class C^2 to

$$\nabla \cdot (P(x)^2 \nabla S_\lambda(x)) = 0 \qquad \text{on} \quad \mathbf{R}^n \setminus \{0\}. \tag{2.3}$$

The basic idea in our study of the asymptotic behavior of solutions to (2.1), lies in observing that $\omega(x) \equiv P(x)^{-1}u(x)$ satisfies a Schrödinger-like equation with a potential that is found to be short-range for $|\lambda|$ large and that introducing appropriate coordinates on $\mathbf{R}^n$ makes the abstract approach of *Agmon* [4] applicable.

In what follows we will write S instead of S_λ.

We introduce coordinates on $\mathbf{R}^n$ as follows. Fix $x \in \mathbf{R}^n \setminus \{0\}$ and for $t > 0$ denote by $\eta_x(t)$ the integral curve of the vector field

$$X = \frac{\nabla S(x)}{|\nabla S(x)|^2}, \qquad x \in \mathbf{R}^n \setminus \{0\},$$

that satisfies $\eta_x(r) = x$, where $r = S(x)$. Since $S(\eta_x(t)) = t$ for $t > 0$ and η_x meets the manifold

$$S_1 \equiv \{x \in \mathbf{R}^n \,:\, S(x) = 1\}$$

only at $\sigma \equiv \eta_x(1)$, we may view $r = S(x)$ and $\sigma = \eta_x(1)$ as coordinates for x. We call these *eikonal coordinates on* $\mathbf{R}^n$. To show that r and σ constitute a legitimate coordinate system on $\mathbf{R}^n$ we use the facts, proved in [6] and [8], that for $|\lambda|$ sufficiently large, S is of class C^{l-1} on $\mathbf{R}^n \setminus \{0\}$ and

$$|x|^{|\alpha|-1} \left| \partial_x^\alpha \Big(S(x) - |x| \Big) \right| \le \frac{C}{|\lambda|}, \qquad x \in \mathbf{R}^n \setminus \{0\},\ |\alpha| \le l - 1, \tag{2.4}$$

where C is a constant depending only on V and its derivatives up to order $|\alpha|$.

Note that (2.4) implies that for $|\lambda|$ sufficiently large, S_1 is radially diffeomorphic to the unit sphere S^{n-1}; in addition, as $|\lambda| \to \infty$ the eikonal coordinates become the usal polar coordinates on $\mathbf{R}^n$.

We denote by $L^2(S_1)$ the space of functions on S_1 which are square-integrable with respect to the measure ν on S_1 that satisfies $d\nu = P(\sigma)^2 (W(\sigma))^{1/2} d\mu$, where μ is the usual Riemannian measure on S_1. We denote the norm and the inner product in $L^2(S_1)$ by $\|\cdot\|$ and $(\cdot,\cdot)$.

The relevance of S and P to the solutions of (2.1) and the role of eikonal coordinates in our work can be seen from the following heuristic arguments. Suppose that k is real and that f is of compact support. Then in a neighborhood of infinity (2.1) becomes

$$-\Delta u - k^2 W u = 0. \tag{2.5}$$

The standard procedure of separating the modulus A and the phase Φ of u as $u = A\exp(ik\Phi)$ and substituting in (2.5) yields

$$|\nabla\Phi|^2 - W = \frac{1}{k^2}\frac{\Delta A}{A},$$
$$\nabla\cdot(A^2\nabla\Phi) = 0.$$

These equations, under the assumption that $\Delta A/A$ is negligible, suggest that we seek a solution S to (2.2) and a positive solution P to (2.3) for which $\Delta P(x)/P(x)$ is small at infinity. We would then have, in some sense,

$$u(x) \sim P(x)\exp(ikS(x)) \qquad \text{as } |x| \to \infty.$$

This last motivates setting $\omega(x) = P^{-1}(x)u(x)$ and finding the differential equation that ω satisfies. A short calculation using (2.5) shows that

$$-\frac{1}{P^2W}\nabla\cdot(P^2(x)\nabla\omega(x)) - k^2\omega(x) = \frac{\Delta P}{PW}\omega(x). \tag{2.6}$$

It turns out that for large $|\lambda|$ the function $\Delta P/P$ is short range and that in eikonal coordinates (2.6) becomes

$$-\partial_{rr}\omega + \Lambda_r\omega - k^2\omega = \frac{\Delta P}{PW}\omega, \qquad r > 0,$$

where Λ_r is a family of positive operators on the Hilbert space $L^2(S_1)$. Furthermore, for β satisfying $0 < \beta < 2$ and v a smooth function on S_1 we have

$$-\frac{d}{dt}(\Lambda_r v, v) \geq \frac{\beta}{r}(\Lambda_r v, v), \qquad r > 0.$$

The above suggests that we view (2.6) as a Schrödinger equation with a short range potential and pursue a generalization of the abstract results in [4] and [13]. Thus we derive abstract results and applying those to the study of the asymptotic behavior of solutions to (2.1) we obtain the following:

Theorem 0. *Let V be as in (1.1) with $l \geq 5$, let $|\lambda|$ be sufficiently large, and suppose S is a C^4 solution to the eikonal equation satisfying (2.2). Introduce eikonal coordinates (σ, r) on $\mathbf{R}^n$. Let $P > 0$ be a C^2 solution to (2.3) and u be a solution to (2.1) with Im $k > 0$ and f of compact support. Write $P(x)$ as $P(\sigma, r)$ and $u(x)$ as $u(\sigma, r)$. Consider u as a function $u(\cdot\,, r)$ of r with values in $L^2(S_1)$. Assume further that*

$$\int_{\mathbf{R}^n} |u(x)|^2 dx < \infty.$$

Then there exists a function $\psi \in L^2(S_1)$ for which

$$\lim_{r\to\infty} P^{-1}(\cdot\,, r)\exp(-ikr)u(\cdot\,, r) = \psi \qquad \textit{strongly in } L^2(S_1).$$

If in addition $\psi = 0$ as an element of $L^2(S_1)$, then u vanishes on the unbounded component of $\mathbf{R}^n \setminus \operatorname{supp} f$.

A similar result is derived using eikonal coordinates and the work of Jäger [13] for $k \in \mathbf{R} \setminus \{0\}$.

It follows from Theorem 0 that, for all sufficiently large $|\lambda|$, the square integrable solutions u to (2.1) must satisfy the optimal upper bound

$$\|P^{-1}(\cdot\,,r)u(\cdot\,,r)\| = O(\exp(-mr)) \qquad \text{as } r \to \infty.$$

We refer to ψ in Theorem 0 as the *boundary value at infinity in $L^2(S_1)$* of u.

Using Theorem 0 and the strategy of *Agmon* [2] and *Constantin* [7] we establish the limiting absorption principle for $-\Delta - k^2 W$, from which the limiting absorption principle for H is derived.

To conclude this section we provide definitions of the Besov spaces B and B^* introduced in [5].

We denote by $B(\mathbf{R}^n)$ the space of all $u \in L^2(\mathbf{R}^n)$ satisfying

$$\|u\|_B = \sum_{j=1}^{\infty} R_j^{1/2} \left(\int_{\Omega_j} |u(x)|^2 dx \right)^{1/2} < \infty,$$

where $R_0 = 0$, $R_j = 2^{j-1}$ for $j > 0$, and $\Omega_j = \{x : R_{j-1} < |x| < R_j\}$. With this norm $B(\mathbf{R}^n)$ is a Banach space and its dual $B^*(\mathbf{R}^n)$ consists of all $u \in L^2_{\text{loc}}(\mathbf{R}^n)$ with

$$\|u\|_{B^*} = \left(\sup_{R>1} R^{-1} \int_{|x|<R} |u(x)|^2 dx \right)^{1/2} < \infty.$$

We denote by $B_0^*(\mathbf{R}^n)$ the closed subspace of $B^*(\mathbf{R}^n)$ consisting of all u satisfying

$$\lim_{R\to\infty} R^{-1} \int_{1 \le |x| < R} |u(x)|^2 dx = 0.$$

For $s \in \mathbf{R}$ we denote by $L^{2,s}(\mathbf{R}^n)$ the Hilbert space

$$L^{2,s}(\mathbf{R}^n) = \{u \in L^2_{\text{loc}}(\mathbf{R}^n) : (1 + |x|^2)^{s/2} u \in L^2(\mathbf{R}^n)\}$$

with norm

$$\|u\|_s^2 = \int_{\mathbf{R}^n} (1 + |x|^2)^s |u(x)|^2 dx.$$

Given two Banach spaces X and Y we denote by $\mathcal{L}(X, Y)$ the Banach space of all continuous linear operators $T : X \to Y$.

3. Results

In this Section we state our main results. In what follows, the functions S and P and the eikonal coordinates r and σ_x are as introduced in the previous section.

Theorem 1. *Let V be as in (1.1) with $l \geq 5$ and let $H = H_0 + V$, where $H_0 = -\Delta$. Set $R(z) = (H - z)^{-1}$. Then there exists $\lambda_0 > 0$ such that the following hold.*

i) For every $\lambda > \lambda_0$ there exists a linear operator $R(\lambda + i0) \in \mathcal{L}(B(\mathbf{R}^n), B^*(\mathbf{R}^n))$ for which

$$\lim_{\epsilon \downarrow 0} \langle R(\lambda + i\epsilon)f, g\rangle = \langle R(\lambda + i0)f, g\rangle$$

for every f, $g \in B(\mathbf{R}^n)$.

ii) Let $K = \{\lambda + i\epsilon : a \leq \lambda \leq b,\ 0 \leq \epsilon \leq 1\}$ with $b > a \geq \lambda_0$. Then there exists a constant C depending only on K and V for which

$$\|R(z)f\|_{B^*} \leq C\|f\|_B$$

for every $z \in K \cap \mathbf{C}_+$ and $f \in B(\mathbf{R}^n)$. Therefore this last estimate is valid for $R(\lambda + i0)f$ with $(\lambda, f) \in [a, b] \times B(\mathbf{R}^n)$.

iii) For any $f \in B(\mathbf{R}^n)$ and $s > 1/2$ the limit

$$\lim_{\epsilon \downarrow 0} R(\lambda + i\epsilon)f = R(\lambda + i0)f$$

exists in the norm of $L^{2,-s}(\mathbf{R}^n)$.

iv) For $f \in B(\mathbf{R}^n)$ we set $u \equiv R(\lambda + i0)f$. Then $u \in B^*(\mathbf{R}^n) \cap H^1_{\text{loc}}(\mathbf{R}^n)$ and, for $1/2 < s < 3/2$, u is the unique solution in $L^{2,-s}(\mathbf{R}^n) \cap H^1_{\text{loc}}(\mathbf{R}^n)$ to

$$-\Delta u + Vu - \lambda u = f$$

that satisfies the radiation condition

$$\nabla S \cdot (\nabla - i\sqrt{\lambda}\nabla S)u \in B^*_0(\mathbf{R}^n).$$

v) For f, $g \in B(\mathbf{R}^n)$, the maps $\lambda \mapsto \int (R(\lambda + i0)f)\bar{g}dx$ from $(0, \infty)$ to $\mathbf{C}$ and $\lambda \mapsto R(\lambda + i0)f$ from $(0, \infty)$ to $L^2(\mathbf{R}^n)$, $s > 1/2$, are continuous.

We derive Theorem 1 from a similar result for $-\Delta - k^2 W$ and a Mourre-estimate argument [16].

Theorem 2. *Let $\lambda_0 > 0$ be as in Theorem 1. Then for $k = \sqrt{\lambda} > \sqrt{\lambda_0}$ there exists a map $\Psi_k : B(\mathbf{R}^n) \mapsto L^2(S_1)$ satisfying:*

i) $\|\Psi_k(f)\| = \text{Im} \int_{\mathbf{R}^n} \left(R(k^2 + i0)f\right)\bar{f}dx$.

ii) The map $(k, f) \mapsto \Psi_k(f)$ is continuos from $(\lambda_0, \infty) \times B(\mathbf{R}^n)$ to $L^2(S_1)$.

iii) The range of Ψ_k contains $C_0^\infty(S_1)$.

iv) For every $f \in B(\mathbf{R}^n)$, $\Psi_k(f)$ is the unique function in $L^2(S_1)$ for which

$$u(x) = P(x)\exp(iS(x))\Psi_k(f)(\sigma_x) \qquad \text{mod } B^*_0(\mathbf{R}^n).$$

The functions $\Psi_k(f)$ are the boundary values at infinity of $R(k^2 + i0)f$. We obtain $\Psi_k(f)$ using eikonal coordinates and a result of *Jäger* [13].

For $f \in B(\mathbf{R}^n)$ and $\lambda > \lambda_0$ we define

$$\mathcal{F}(f)(\xi) = \begin{cases} \sqrt{\frac{2}{\pi}} W^{1/2}(\xi) S(\xi) P(\xi) \Psi_{S(\xi)}(f)(\sigma_\xi), & S(\xi)^2 > \lambda_0; \\ 0, & \text{otherwise.} \end{cases}$$

In the following theorem the spectral projections $E(\cdot)$ and the spaces $L^2_{\lambda_0}(\mathbf{R}^n_x)$ and $L^2_{\lambda_0}(\mathbf{R}^n_\xi)$ are as introduced in section 2. We will write E_{λ_0} instead of $E(\lambda_0, \infty)$.

Theorem 3. *Let* $\lambda_0 > 0$ *be as in* Theorem 1. *Then the following assertions hold.*

i) For any Borel set $A \subset (\lambda_0, \infty)$ *and* f, $g \in B(\mathbf{R}^n)$ *we have*

$$(E(A)f, g) = \int_{S(\xi)^2 \in A} (\mathcal{F}f)(\xi)\overline{(\mathcal{F}f)(\xi)} d\xi.$$

In particular we have

$$(E_{\lambda_0} f, g) = \langle \mathcal{F}f, \mathcal{F}g \rangle$$

and

$$\|E_{\lambda_0} f\|^2_{L^2(\mathbf{R}^n)} = \|\mathcal{F}f\|^2_{L^2(\mathbf{R}^n)}.$$

Therefore $\mathcal{F}$ *possesses a unique extension to a partial isometry also denoted by* $\mathcal{F}$, *with initial set* $L^2_{\lambda_0}(\mathbf{R}^n_x)$, *from* $L^2(\mathbf{R}^n)$ *onto* $L^2_{\lambda_0}(\mathbf{R}^n_\xi)$. *ii) For every bounded Borel function* χ *we have*

$$\begin{aligned}(E_{\lambda_0}\chi(H)f, g) &= \langle \chi(S(\cdot)^2)\mathcal{F}f, \mathcal{F}g \rangle \\ &= \int_{S(\xi)^2 > \lambda_0} \chi(S(\xi)^2)(\mathcal{F}f)(\xi)\overline{(\mathcal{F}g)(\xi)} d\xi\end{aligned}$$

for f *and* $g \in L^2_{\lambda_0}(\mathbf{R}^n_x)$. *In addition, for* $f \in L^2_{\lambda_0}(\mathbf{R}^n_x)$ *we have*

$$\mathcal{F}(\chi(H)f)(\xi) = \chi(S(\xi)^2)(\mathcal{F}f)(\xi) \qquad \xi\text{-a.e.}$$

iii) The domain of HE_{λ_0} *is* $\{f \in L^2_{\lambda_0}(\mathbf{R}^n_x) : S(\cdot)\mathcal{F}f \in L^2_{\lambda_0}(\mathbf{R}^n_\xi)\}$ *and*

$$\mathcal{F}(HE_{\lambda_0} f)(\xi) = S(\xi)^2(\mathcal{F}f)(\xi) \qquad \xi\text{-a.e.}$$

Similar results are obtained when we consider $R(k^2 - i0)f$ and when we add to V a short range perturbation $Q \in L^p_{\text{loc}}(\mathbf{R}^n)$ with $p \geq n/2$.

4. References

[1] Agmon, S. *Some new results in the spectral and scattering theory of differential operators on* $\mathbf{R}^n$. Seminaire Gaulaouic-Schwarz 1978-1979. Ècole Polytechnique Centre Mathèmatiques.

[2] Agmon, S. *A representation theorem for solutions of the Helmholtz equation and resolvent estimates for the Laplacian.* Preprint, University of Virginia, 1989.

[3] Agmon, S. *Lectures on exponential decay of solutions of second-order elliptic equations. Bounds on eigenfunctions of N-body Schrödinger operators.* Mathematical Notes 29, Princeton University Press, 1982.

[4] Agmon, S. *On the asymptotic behavior of solutions of Schrödinger type equations in unbounded domains.* Analyse de Mathématique et Applications, Gautier-Villars, Paris, 1988.

[5] Agmon, S. and Hörmander L. *Asymptotic properties of solutions of differential equations with simple characteristics.* Journal D'analyse Mathématique, vol 30, 1976.

[6] Barles, G. *On eikonal equations associated with Schrödinger operators with nonspherical radiation conditions.* Commun. in Partial Differential Equations 12, 263-283 (1987).

[7] Constantin, P. *Scattering theory for Schrödinger operators in a class of domains with noncompact boundaries.* J. Func. Anal., 44, 1981,pp.87-119.

[8] Cruz, J. *Ph.D. Thesis* University of Virginia, May 1991.

[9] Cycon, et. al. *Schrödinger operators* Springer Verlag 1987.

[10] Herbst, I.*Spectral and scattering theory for Schrödinger operators with potentials independent of* $|x|$. Preprint, University of Virginia, 1989.

[11] Ikebe, T. *Eigenfunction expansions associated with Schródinger operators and their applications to scattering theory.* Arch. Rational Mech. Anal. 5, 1960, pp 1-34.

[12] Isozaki, H. *Eikonal equations and spectral representations for long-range Schrödinger Hamiltonians.* Journal of Mathematics of Kyoto University. Vol. 20, No. 2, 1980.

[13] Jäger, W. *Das Asymptotische Verhalten von Lösungen eines Typs von Differential-Gleichungen.* Math. Z., 112, 1969, pp. 26-36.

[14] P. Lions. *Generalized Solutions of Hamilton-Jacobi Equations.* Pitman, London, (1982).

[15] Milnor, J. *Morse Theory.* Princeton University Press 1963.

[16] Mourre, E. *Absence of singular continuous spectrum for certain self-adjoint operators.* Commun. Math. Phys. **78**, 391–408 (1981).

[17] Saitō, Y. *Schrödinger operators with a nonspherical radiation condition.* Pacific Journal of Mathematics, Vol. 126, No. 2, 331-359, 1987.

Jaime Cruz-Sampedro
Departamento de Ciencias y Matemáticas
Universidad de las Américas Pue.
Ex-Hacienda de Santa Catarina Mártir
Cholula Puebla, México

Asymptotic completeness of long range N-body systems. Main ideas of a proof.

Jan Dereziński *
Centre de Mathematiques, Ecole Polytechnique
91128 Palaiseau cedex, France
on leave from
Department of Mathematical Methods in Physics
Warsaw University
Hoża 74, PL-00-682 Warszawa, Poland

December 19, 1991

Abstract

We sketch a proof of the asymptotic completeness of N-body quantum systems with potentials that decay like $x^{-\mu}$ where $\mu > \sqrt{3} - 1$.

*Supported in part by a grant from the Ministry of Education of Poland.

1 Introduction

The problem of the asymptotic completeness of N-body systems has a long and varied history. To our knowledge, the first result on this subject is due to Faddeev [Fa] who proved the asymptotic completeness of a certain class of three-body systems. His proof used the stationary method, which means that it it was based on the study of the resolvent of H. It contained an implicit assumption on the potentials, which was certainly a serious drawback of Faddeev's result. After Faddeev there were a number of other attempts to show the asymptotic completeness by the stationary method. All of them contained implicit assumptions except for [LoSig], where the three-body case was treated with rather restrictive assumptions on the potentials.

An important breakthrough was made by V.Enss who introduced time dependent method and phase space analysis to the problem of the asymptotic completeness [E1,2]. He was able to prove the asymptotic completeness and the existence of all wave operators of 3-body systems if the potentials decay like $x^{-\mu}$ with $\mu > \sqrt{3} - 1$.

In the general N-body short range case (which means $\mu > 1$) the poof of the asymptotic complete was first given by I.Sigal and A.Soffer [SigSof1]. Their proof was also based on the time dependent approach and on a detailed analysis of the propagation in phase space. After obtaining their result I.Sigal and A.Soffer tried to prove the long range case too. They obtained a number of interesting partial results [SigSof2,3] and were able to prove the asymptotic completeness in the 4-body case if $\mu \geq 1$ [SigSof4]. An important progress in the understanding of the N-body scattering was made by G.M.Graf who presented a new proof of the asymptotic completeness in the N-body short range case [Graf]. Some of the properties of propagation of N-body systems were studied in [De1,2]. In [De3] we give a proof of the asymptotic completeness for an arbitrary number of particles if $\mu > \sqrt{3} - 1$. There exist also a number of results about scattering in N-body classical systems [De4]. Some of these results are closely related to the proof of the asymptotic completeness of quantum systems.

In this article we would like to present some of the basic ideas of our proof of the asymptotic completeness in the long range N-body case. First we would like to state precisely our main result. It is convenient to formulate it in terms of the so-called generalized N-body Schrödinger operators first introduced by S.Agmon [A]. After a description of our result in a somewhat abstact language we will give an example of a physical system described by our result. We will also translate some of the abstract concepts into better known physical terms.

Let us begin with a description of the class of Hamiltonians that we are going to study. Let X be a Euclidean space. Let $\{X_a : a \in \mathcal{A}\}$ be a family of its subspaces closed with respect to intersection and containing $X_{a_{min}} := X$. The orthogonal complement of X_a will be denoted X^a. For any $x \in X$ we will write x_a and x^a to denote projections of x onto X_a and X^a respectively. We will write $a \subset b$ iff $X_a \supset X_b$. Note that $\{X_a : a \in \mathcal{A}\}$ contains $\bigcap_{a \in \mathcal{A}} X_a$ which will be denoted $X_{a_{max}}$.

If $a \in \mathcal{A}$ then $\#a$ denotes the maximal number of distinct $a_i \in \mathcal{A}$ such that $a = a_n \subset \ldots \subset a_1 = a_{max}$. We set $N := max\{\#a : a \in \mathcal{A}\}$.

We also define

$$Z_a := X_a \backslash \bigcup_{b \not\subset a} X_b.$$

Note that X is the disjoint sum of $\{Z_a : a \in \mathcal{A}\}$.

D, D_a and D^a denote the operators $\frac{1}{i}\nabla$, $\frac{1}{i}\nabla_a$ and $\frac{1}{i}\nabla^a$ on $L^2(X)$ respectively. Likewise, Δ, Δ_a and Δ^a denote the Laplacians corresponding to the variables x, x_a and x^a respectively.

Let C be a self-adjoint operator on a Hilbert space $\mathcal{H}$. If I is a Borel subset of $\mathbf{R}$ then $E_I(C)$ will denote the spectral projection of C onto I. $E^{pp}(C)$ denotes the spectral projection of C onto its pure point spectrum.

We assume that for every $a \in \mathcal{A}$ we are given a real function $X^a \ni x^a \mapsto V^a(x^a)$. Let $\mu_s > 0$ and $1 \geq \mu > 0$. We will assume that $V^a(x^a) = V^{sa}(x^a) + V^{la}(x^a)$ and the following conditions are satisfied:

$$\|(1-\Delta^a)^{-1} V^{sa}(x^a) \langle x^a \rangle^{1+\mu_s}\| \leq c; \tag{1.1}$$

$$|V^{la}(x^a)| \leq c \langle x^a \rangle^{-\mu}; \tag{1.2}$$

and

$$|\nabla V^{la}(x^a)| \leq c \langle x^a \rangle^{-1-\mu}. \tag{1.3}$$

Note that the above hypotheses imply that $V^a(x^a)$ is bounded relatively to $-\frac{1}{2}\Delta$ with the infinitesimal bound.

We set

$$V(x) := \sum_{a \in \mathcal{A}} V^a(x^a)$$

and

$$V_a(x) := \sum_{b \subset a} V^b(x^b).$$

Our basic Hamiltonian will be the self adjoint operator H defined by

$$H := -\frac{1}{2}\Delta + V(x). \tag{1.4}$$

It will be also useful to introduce subsystem Hamiltonians

$$H_a := -\frac{1}{2}\Delta + V_a(x).$$

Clearly, $H = H_{a_{max}}$.

We may identify $L^2(X)$ with $L^2(X_a) \otimes L^2(X^a)$. Then we can write:

$$H_a = -\frac{1}{2}\Delta_a \otimes 1 + 1 \otimes H^a$$

where

$$H^a := -\frac{1}{2}\Delta^a + V_a(x^a)$$

is the internal Hamiltonian of the subsystem a.

We set $I_a(x) := V(x) - V_a(x)$. We need also V_a^l, V_a^s, I_a^l and I_a^s which have obvious definitions.

We set

$$S_{at}(\xi_a) := \frac{t}{2}\xi_a^2 + \int_0^t I_a^l(s\xi_a)ds.$$

Let us state our main result. It roughly says that if potentials decay like $|x|^{-\mu}$ with $\mu > \sqrt{3} - 1$ then all the Dollard modified wave operators exist and the asymptotic completeness holds.

Theorem 1.1 *Assume that $\mu > \sqrt{3} - 1$ and for some $\mu_1 > 0$*

$$|\Delta^a V^{al}| \leq c\langle x^a\rangle^{-1-\mu_1}.$$

Then there exists

$$s - \lim_{t\to\pm\infty} e^{itH} e^{-i(tH^a + S_{at}(D_a))} E^{pp}(H^a). \tag{1.5}$$

Denote 1.5 by $\Omega_a^\pm$. Then

$$\sum_{a\in\mathcal{A}}^{\oplus} Ran\Omega_a^\pm = L^2(X). \tag{1.6}$$

Note that $\Omega_{a_{max}}^\pm = E^{pp}(H)$. Hence 1.6 can be rewritten in a more familiar way:

$$\sum_{a\neq a_{max}}^{\oplus} Ran\Omega_a^\pm = RanE^c(H).$$

Let us give the most important physical example of a Hamiltonian of the form 1.4 (see also [RS vol III]). Suppose that we want to study a system of N nonrelativistic particles of masses $m_1, \dots, m_N$ and charges $z_1, \dots, z_N$. Such a system is usually described by a Hamiltonian on $L^2(\mathbf{R}^{3N})$ of the form

$$H = -\sum_{i=1}^N \frac{m_i}{2}\Delta_i + \sum_{1\leq i<j\leq N} z_i z_j |x_i - x_j|^{-1}. \tag{1.7}$$

It is easy to see that 1.7 falls into the class 1.4. First of all note that by a linear change of coordinates in the N-body configuration space $\mathbf{R}^{3N}$ we can make the kinetic energy equal to $-\frac{1}{2}\Delta$.

Next define $\mathcal{A}_0$ to be the set of ordered pairs $\{(ij) : 1 \leq i < j \leq N\}$. For any $(i,j) \in \mathcal{A}_0$ we set

$$X_{ij} := \{x \in X : x_i - x_j = 0\}. \tag{1.8}$$

In general the family $\{X_a : a \in \mathcal{A}_0\}$ is not closed with respect to intersection and does not contain $X_{a_{min}} := X$. Therefore we enlarge it by adjoining all the intersections and X. In this way we obtain a larger family of subspaces $\{X_a : a \in \mathcal{A}\}$.

We set

$$V^a := z_i z_j |x_i - x_j|^{-1}$$

if $a = (i,j)$ and

$$V^a := 0$$

if $a \in \mathcal{A} \backslash \mathcal{A}_0$.

It is possible to give a simple description of the family $\{X_a : a \in \mathcal{A}\}$ from this example. It is easy to see that any partition of the set $\{1, \ldots, N\}$ into disjoint subsets $c_1, \ldots, c_n$ determines a unique $a \in \mathcal{A}$. Namely, we will write that $x \in X_a$ where

$$a = \{c_1, \ldots, c_n\} \tag{1.9}$$

iff for any $k = 1, \ldots, n$ and any $i, j \in c_k$ we have $x_i = x_j$. X_a is usually called the collision plane corresponding to the cluster decomposition a and the subsets $c_1, \ldots, c_n$ are called clusters. X^a is the configuration space of inner degrees of freedom of a.

Note that if 1.9 is true then $-\frac{1}{2}\Delta_a$ is the kinetic energy of the relative motion of centers of mass of clusters $c_1, \ldots, c_n$. $-\frac{1}{2}\Delta^a$ is the sum of the kinetic energies of the internal motion of these clusters. V^a is the sum of potentials between all pairs of particles that are contained in the same cluster. Note also that $\#a = n$ (= the number of clusters in a given cluster decomposition)

2 Why $3^{\frac{1}{2}} - 1$ seems to be the natural borderline.

It is well known that in the two-body case the asymptotic completeness holds if potentials decay like $x^{-\mu}$ for an arbitrary $\mu > 0$. In this section we would like to present main ideas of the proof of theorem 1.1 including a heuristic argument indicating that it would be much more difficult, if not impossible, to show the asymptotic completeness for N-body systems with potentials that decay slower than $|x^a|^{-\sqrt{3}+1}$. Of course, it should be possible to show the asymptotic completeness for potentials with slower decay if one imposes some additional assumptions. For instance, the decay of the repulsive part of the potentials can be slower than $|x^a|^{-\sqrt{3}+1}$ and still our proof of the asymptotic completeness will work.

As a preparatory step towards this proof let us mention the following result which was essentially first shown in [De2] and was directly inspired by [Graf].

Theorem 2.1 *Let J be a bounded continuous function on X. Then there exists*

$$s - \lim_{t \to \pm\infty} e^{itH} J\left(\frac{x}{t}\right) e^{-itH}. \tag{2.1}$$

Moreover, there exists a unique self-adjoint operator $P^{\pm}$ such that 2.1 equals $J(P^{\pm})$. $P^{\pm}$ commutes with H and

$$E^{pp}(H) = E_{\{0\}}(P^{\pm}). \tag{2.2}$$

Note that $P^{\pm}$ is a vector of commuting self-adjoint operators; therefore it defines a spectral measure on X. Hence for any measurable $\Theta \subset X$ one can define the spectral projection $E_{\Theta}(P^{\pm})$.

It is easy to see that modified wave operators $\Omega_a^{\pm}$ (if they exist) have the following property:

$$Ran\Omega_a^{\pm} \subset Ran E_{Z_a}(P^{\pm}). \tag{2.3}$$

Moreover,

$$\sum_{a \in \mathcal{A}} E_{Z_a}(P^{\pm}) = 1.$$

Therefore 2.2 will be true if we will show the existence of

$$s - \lim_{t \to \pm\infty} e^{i(tH^a + S_{at}(D_a))} e^{-itH} E_{Z_a}(P^{\pm}). \tag{2.4}$$

Define a new auxiliary Hamiltomian

$$h_a := -\frac{1}{2}\Delta_a + I_a(x_a).$$

To prove the existence of 2.4 it is enough to show that there exists

$$s - \lim_{t \to \pm\infty} e^{it(H^a + h_a)} e^{-itH} E_{Z_a}(P^{\pm}) \tag{2.5}$$

and on the range of 2.5 there exists

$$s - \lim_{t\to\pm\infty} e^{iS_{at}(D_a)}e^{-ith_a}. \tag{2.6}$$

The existence of 2.6 follows by standard two-body arguments (see eg [Hö vol IV,IKi,Sig,De2] and references therein). The proof of the existence of 2.5 is more difficult and is the crucial part of our proof of the asymptotic completeness in the long range case.

It is convenient to make yet another reduction of the problem. Let Θ be a compact subset of Z_a. Consider a function $J \in C_0^\infty(X)$ equal to one on Θ and supported in a sufficiently small neighborhood of Θ. We define a time dependent effective potential

$$W(t,x) := J\left(\frac{x}{t}\right) I_a^l(x).$$

Note that if the support of J is sufficiently close to Θ then

$$\nabla_x W(t,x) = O(t^{-1-\mu}).$$

Define $U(t)$ and $U_0(t)$ to be the evolutions generated by $H_a + W(t,x)$ and $H_a + W(t,x_a)$ respectively. Let

$$\tilde{P}^\pm := \lim_{t\to\pm\infty} U^*(t)\frac{x}{t}U(t);$$

(the limit has to be understood in the sense described in theorem 2.1). A relatively easy argument shows that the existence of the limit 2.5 on $RanE_\Theta(P^\pm)$ follows from the existence of

$$s - \lim_{t\to\pm\infty} U_0^*(t)U(t)E_{X_a}(\tilde{P}^\pm). \tag{2.7}$$

Let us make a side remark about classical N-body systems. This remark will suggest when we can expect 2.7 to exist.

Suppose that our classical Hamiltonian is a function on $X \times X'$ of the form

$$H(x,\xi) = \frac{1}{2}\xi^2 + \sum_{a\in\mathcal{A}} V^a(x^a)$$

where the potentials satisfy the following conditions:

$$|V^a(x^a)| \leq c\langle x^a\rangle^{-\mu}$$

and

$$|\nabla V^a(x^a)| \leq c\langle x^a\rangle^{-1-\mu}.$$

Mimicking arguments used in the proof of theorem 2.1 it is possible to show that for every trajectory $x(t)$ generated by $H(x,\xi)$ there exists

$$\lim_{t\to\pm\infty} \frac{x(t)}{t}. \tag{2.8}$$

If 2.8 belongs to Z_a then a priori we know that

$$x^a(t) = o(t). \tag{2.9}$$

It turns out that one can improve the estimate 2.9, namely one can show that

$$x^a(t) = O(t^\delta) \tag{2.10}$$

where $\delta := 2(2+\mu)^{-1}$ [De4].

In general, one cannot improve 2.10. To see this consider a two-body Hamiltonian

$$h(x,\xi) := \frac{1}{2}\xi^2 - |x|^{-\mu}. \tag{2.11}$$

An explicit calculation shows that for some $c \neq 0$

$$x(t) = ct^\delta$$

solves the equation of motion generated by 2.11.

It turns out that the estimate 2.10 has a quantum analog. Namely, the following proposition is true.

Proposition 2.2 *Let $\phi \in RanE_{X_a}(\tilde{P}^\pm)$. Then for any $c > 0$ and $\delta_1 > \delta$*

$$s - \lim_{t\to\pm\infty} E_{[ct^{\delta_1},\infty)}(|x^a|)U(t)\phi = 0. \tag{2.12}$$

This proposition is a corrollary of our proof of the asymptotic completeness. It can be used to show the main idea of this proof. Note however, that the real proof is much more complicated and uses more precise estimates than 2.12

Let us try to use the Cook method to prove the existence of 2.7. Let $\phi \in RanE_{Z_a}(\tilde{P}^\pm)$. We need to estimate the derivative

$$\frac{d}{dt}U_0(t)U(t)\phi = iU_0(t)(W(t,x) - W(t,x_a))U(t)\phi$$

$$\sim U_0(t)\nabla W(t,x)x^a E_{[0,ct_1^\delta]}(|x^a|)U(t)\phi = O(t^{-1-\mu})O(t^{\delta_1}). \tag{2.13}$$

This is integrable if $\delta_1 < \mu$. Thus we have to solve the inequality

$$2(2+\mu)^{-1} < \mu. \tag{2.14}$$

Now, 2.14 is satisfied if $\mu > \sqrt{3} - 1$. We think that this argument indicates that the number $\sqrt{3} - 1$ is likely to be the natural borderline for the asymptotic completeness of N-body systems.

3 Special observables

The following proposition is proved [De3].

Proposition 3.1 *There exists a function $r \in C^\infty(X)$ which satisfies the following properties:*
a) r is convex;
b) for any $a \in \mathcal{A}$ there exists $\epsilon > 0$ such that r depends just on x_a on $\{x \in X : dist(x, X_a) < \epsilon\}$;
c) for some $c_1, c_2 > 0$

$$c_1\langle x\rangle \leq r(x) \leq c_2\langle x\rangle;$$

d) for any multiindex α

$$|\partial^\alpha(r(x) - \langle x\rangle)| \leq c_\alpha\langle x\rangle^{-1}, \tag{3.1}$$

$$|\partial^\alpha(x\nabla r(x) - \langle x\rangle)| \leq c_\alpha\langle x\rangle^{-1}, \tag{3.2}$$

$$|\partial^\alpha(x\nabla^2 r(x)x)| \leq c_\alpha\langle x\rangle^{-1}. \tag{3.3}$$

To show the main idea of the construction of r let us note that if we set $\rho_j := \epsilon M^{-j}$ with a sufficiently big M then the function

$$\check{r} := max\{\sqrt{\rho_{\#a} + x_a^2} : \; a \in \mathcal{A}\}$$

has the properties a), b) and c). Unfortunately, it is non-differentiable. In [De3] we show how to modify this construction to get a smooth r.

Note also that we do not need the full strength of estimates 3.1, 3.2 and 3.3. We quote them to stress the idea that r is a very close approximation of $|x|$.

Many of the ideas underlying the construction of the function r are due to [Graf]; in fact, the so-called Graf vector field is essentially equal to $\frac{1}{2}\nabla r^2$. The idea of using a function similar to r first appeared in [Ya].

It will be convenient to scale the observable $r(x)$ in a time dependent fashion. Namely, we choose δ such that $0 < \delta < 1$ and for $t > 0$ we set:

$$r_t(x) := t^\delta r\left(\frac{x}{t^\delta}\right).$$

We also define

$$\begin{aligned} b_t &:= \left(\frac{d}{dt} + i[-\frac{1}{2}\Delta, \cdot]\right) r_t \\ &= \frac{1}{2}\left(D\nabla r\left(\frac{x}{t^\delta}\right) + \nabla r\left(\frac{x}{t^\delta}\right) D\right) \\ &+ t^{-1+\delta}\delta\left(r\left(\frac{x}{t^\delta}\right) - \frac{x}{t^\delta}\nabla r\left(\frac{x}{t^\delta}\right)\right). \end{aligned} \tag{3.4}$$

Now let

$$H := -\frac{1}{2}\Delta + V(t,x).$$

We set

$$\mathbf{D} := \frac{d}{dt} + i[H,\cdot].$$

The following identity will play an important role in our proof.

$$\mathbf{D}^2 r_t = \mathbf{D} b_t$$

$$= t^{-\delta} c_t + t^{-2\delta}\frac{1}{4}\left(D\nabla\Delta r\left(\frac{x}{t^\delta}\right) + \nabla\Delta r\left(\frac{x}{t^\delta}\right) D\right)$$

$$+ t^{-2+\delta}\delta(\delta-1)\left(r\left(\frac{x}{t^\delta}\right) - \frac{x}{t^\delta}\nabla r\left(\frac{x}{t^\delta}\right)\right) - \nabla V(t,x)\nabla r\left(\frac{x}{t^\delta}\right) \tag{3.5}$$

where

$$c_t := \left(D - \delta\frac{x}{t}\right)\nabla^2 r\left(\frac{x}{t^\delta}\right)\left(D - \delta\frac{x}{t}\right) \geq 0. \tag{3.6}$$

In other words, the second Heisenberg derivative of r_t is positive up to some error terms which decay with time. Let us describe the magnitude of these terms. The third term on the right of 3.5 is clearly $O(t^{-2+\delta})$. If $V(t,x) = V^l(x) + W(t,x)$ then the fourth term is $O(t^{-(1+\mu)\delta})$. The second term is unbounded because of the presence of the momentum operator D. But if we control it with an energy cut-off then it becomes $O(t^{-\delta})$.

4 Propagation of "exceptional states".

In this section we would like to sketch in more detail the proof of the existence of 2.7. (In a somewhat loose language this is sometimes described as "the existence of the inverse wave operator on exceptional states"). As a byproduct we will obtain a proof of proposition 2.2.

To simplify our notation we will assume that $X_a = \{0\}$. (In fact it is enough to consider just this case, see [SigSof4]). Throughout this section we will also use some additional conventions about notation, which we would now like to describe.

H will be as usual an N-body Schrödinger operator. $W(t,x)$ will be a time dependent potential satisfying

$$|\nabla_x W(t,x)| \leq c\langle t\rangle^{-1-\mu}.$$

$U(t)$ and $U_0(t)$ are evolutions generated by $H + W(t,x)$ and $H + W(t,0)$ respectively. We define $H^+ := \lim_{t\to\infty} U^*(t)HU(t)$ and $P^+ := \lim_{t\to\infty} U^*(t)\frac{x}{t}U(t)$. We will write $\mathbf{D}$ for

$$\frac{d}{dt} + i[H + W(t), \cdot]$$

and ϕ_t for $U(t)\phi$.

We fix an energy interval $[\lambda_1, \lambda_2] \subset \mathbf{R}$ and $\tilde{h} \in C_0^\infty(\mathbf{R})$ such that $\tilde{h} = 1$ on $[\lambda_1, \lambda_2]$. We set

$$\tilde{b}_t := \tilde{h}(H)b_t\tilde{h}(H)$$

and

$$\tilde{c}_t := \tilde{h}(H)c_t\tilde{h}(H).$$

where b_t and c_t were defined in the previous section. Note that $\tilde{b}_t$ and $\tilde{c}_t$ are bounded uniformly in t and $\tilde{c}_t$ is positive. Equation 3.5 yields

$$\mathbf{D}\tilde{b}_t = t^{-\delta}\tilde{c}_t + O(t^{-2+\delta}) + O(t^{-\delta(1+\mu)}). \tag{4.1}$$

We optimize the error term on the right of 4.1 by demanding that

$$2 - \delta = \delta(1 + \mu). \tag{4.2}$$

Henceforth we will assume that δ is the solution of 4.2, that is $\delta := 2(2+\mu)^{-1}$. We also define $\rho := \mu(2+\mu)^{-1}$. The equation 4.1 becomes now

$$\mathbf{D}\tilde{b}_t = t^{-\delta}\tilde{c}_t + O(t^{-1-\rho}). \tag{4.3}$$

Note also the following estimate:

$$ad^n_{\tilde{b}_t}\left(\mathbf{D}\tilde{b}_t\right) = O(t^{-(1+n)\delta}). \tag{4.4}$$

We begin with the following proposition which describes the asymptotic observable obtained from $\tilde{b}_t$.

Proposition 4.1 *There exists*

$$s - \lim_{t\to\infty} U^*(t)\tilde{b}_t U(t). \tag{4.5}$$

4.5 is a self adjoint operator, which we will call $\tilde{b}^+$. *It commutes with* H^+ *and*

$$\tilde{b}^+ = |P^+|\tilde{h}^2(H^+). \tag{4.6}$$

The existence of 4.5 follows easily from the approximate positivity of the Heisenberg derivative of $\tilde{b}_t$ (see 4.3). To see that equality 4.6 is plausible it is enough to replace D with $\frac{x}{t}$ in the definition of $\tilde{b}_t$ (see 3.4). To make this argument rigorous one can use one of the basic estimates due to Graf [Graf,De3].

Next we are going to look more closely at the states that belong to the kernel of $\tilde{b}^+$. The following proposition gives quite precise information on the behavior of the observable $\tilde{b}_t$ on such states.

Proposition 4.2 *Let* $\phi \in E_{\{0\}}(\tilde{b}^+)$ *and* $\rho_1 < \rho$. *Then*

$$\lim_{t\to\infty} E_{[t^{-\rho_1},\infty)}(\tilde{b}_t)\phi_t = 0.$$

Proof. Choose $f \in C_0^\infty(0,\infty)$ with $f > 0$ and $\int f^2(s)ds = 1$. Set

$$F(s) := \int_{-\infty}^{s} f^2(s_1)ds_1.$$

For any $c > 0$ consider

$$k_c(t) := (\phi_t|F(c\tilde{b}_t)\phi_t).$$

Using the formula

$$F(c\tilde{b}_t) = \frac{1}{\pi}\int \hat{F}(s)e^{isc\tilde{b}_t}ds$$

we obtain

$$\mathbf{D}F(c\tilde{b}_t) = cf(c\tilde{b}_t)\mathbf{D}\tilde{b}_t f(c\tilde{b}_t) + c^3R(t) \tag{4.7}$$

where

$$\||R(t)\| \leq \|[\tilde{b}_t,[\tilde{b}_t,\mathbf{D}\tilde{b}_t]]\|.$$

Hence by 4.3

$$\frac{d}{dt}k_c(t) \geq cO(t^{-1-\rho}) + c^3O(t^{-3\delta}). \tag{4.8}$$

Clearly,

$$\lim_{t\to\infty} k_c(t) = 0. \tag{4.9}$$

Now 4.8 and 4.9 yield

$$k_c(t) = -\int_t^\infty \frac{d}{ds}k_c(s)ds \leq cO(t^{-\rho}) + c^3O(t^{1-3\delta}). \tag{4.10}$$

If we set $c := t^{\rho_1}$ then the rhs of 4.10 goes to zero. Therefore

$$\lim_{t\to\infty} k_{t^{\rho_1}}(t) = 0,$$

which ends the proof of the proposition. QED

Our next goal is to make a careful study of the propagation of the observable $\tilde{r}_t$.

Proposition 4.3 *Let $\epsilon > 0$, $h \in C_0^\infty(\mathbf{R})$, $F_\pm \in C^\infty(\mathbf{R})$, $F'_\pm \in C_0^\infty(0,\infty)$, $F_\pm, \pm F'_\pm \geq 0$, $suppF_- \subset (-\infty,\epsilon)$, $suppF_+ \subset (0,\infty)$ and $supph \subset [\lambda_1,\lambda_2]$. Let $0 \leq \rho_2 < \rho_1 < \rho$. Then*

$$\lim_{t\to\infty} \left(\phi_t | h(H)F_+(t^{-1+\rho_2}r_t)F_-^2(t^{\rho_1}\tilde{b}_t)F_+(t^{-1+\rho_2}r_t)h(H)\phi_t\right) = 0. \tag{4.11}$$

Proof. We define $f_\pm = \sqrt{\pm F_\pm^2)'(s)}$. For any $c > 0$ we set

$$k_c(t) := \left(\phi_t | h(H)F_+(cr_t)F_-^2(t^{\rho_1}\tilde{b}_t)F_+(cr_t)h(H)\phi_t\right).$$

Now

$$\mathbf{D}F_+(cr_t) = cb_tF'_+(cr_t) + c^2F''_+(cr_t)|\nabla r_t|^2. \tag{4.12}$$

Note that the second term on the rhs of 4.12 is $c^2O(t^0)$. The first one multiplied by $h(H)$ equals

$$ch(H)F'_+(cr_t)\tilde{b}_t + cO(t^{-\delta}) + c^2O(t^0).$$

We further calculate:

$$\begin{aligned} &ch(H)F'_+(cr_t)\tilde{b}_tF_-^2(t^{\rho_1}\tilde{b}_t)F_+(cr_t)h(H) \\ &= -2ct^{-\rho_1}h(H)f_+(cr_t)(\epsilon - t^{\rho_1}\tilde{b}_t)F_-^2(t^{\rho_1}\tilde{b}_t)f_+(cr_t)h(H) \\ &\quad +2c\epsilon t^{-\rho_1}h(H)f_+(cr_t)F_-^2(t^{\rho_1}\tilde{b}_t)f_+(cr_t)h(H) \\ &\quad + c^2O(t^0) + c^3O(t^{\rho_1}). \end{aligned} \tag{4.13}$$

The second term on the rhs of 4.13 is $cO(t^{-\rho_1})$. We used an estimate similar to 4.7 to commute functions of cr_t through $(\epsilon - t^{\rho_1}\tilde{b}_t)F_-^2(t^{\rho_1}\tilde{b}_t)$.

Now we calculate the Heisenberg derivative of our observable.

$$\begin{aligned} &\mathbf{D}h(H)F_+(cr_t)F_-^2(t^{\rho_1}\tilde{b}_t)F_+(cr_t)h(H) \\ &= -2ct^{-\rho_1}h(H)f_+(cr_t)(\epsilon - t^{\rho_1}\tilde{b}_t)F_-^2(t^{\rho_1}\tilde{b}_t)f_+(cr_t)h(H) \\ &\quad +cO(t^{-\rho_1}) + cO(t^{-\delta}) + c^2O(t^0) + c^3O(t^{\rho_1}) \\ &\quad -\rho_1t^{-1}h(H)F_+(cr_t)f_-^2(t^{\rho_1}\tilde{b}_t)t^{\rho_1}\tilde{b}_tF_+(cr_t)h(H) \\ &\quad -t^{-1-\rho+\rho_1}h(H)F_+(cr_t)f_-(t^{\rho_1}\tilde{b}_t)\tilde{c}_tf(t^{\rho_1}\tilde{b}_t)F_+(cr_t)h(H) \\ &\quad + O(t^{-1-\rho+\rho_1}) + O(t^{-3\delta+3\rho_1}) \end{aligned} \tag{4.14}$$

$$\leq cO(t^{-\rho_1}) + cO(t^{-\delta}) + c^2O(t^0) + c^3O(t^{\rho_1})$$
$$+ O(t^{-1-\rho+\rho_1}) + O(t^{-3\delta+3\rho_1}). \tag{4.15}$$

Hence

$$k_c(t) = k_c(1) + \int_1^t \frac{d}{ds} k_c(s) ds$$
$$= k_c(1) + cO(t^{1-\rho_1}) + cO(t^{1-\delta}) + c^2O(t) + c^3O(t^{1+\rho_1})$$
$$+ O(t^{-\rho+\rho_1}) + O(t^{1-3\delta+3\rho_1}). \tag{4.16}$$

Note that

$$1 - 3\rho_1 - 3\delta < 1 - 3(\mu - 2)(2 + \mu)^{-1} \leq 0$$

and

$$\lim_{c \to 0} k_c(1) = 0.$$

If we set $c = t^{-1+\rho_2}$ then we see that the rhs of 4.16 converges to zero. Hence

$$\lim_{t \to \infty} k_{t^{-1+\rho_2}}(t) = 0$$

which ends the proof of our proposition. QED

Note that propositions 4.2 and 4.3 imply proposition 2.2 (at least in the case $X_a = \{0\}$).

We will also need certain estimates on integrals of observables along the evolution. The first proposition containing such estimates is parallel to proposition 4.2.

Proposition 4.4 *Let $f \in C_0^\infty(0, \infty)$, $h \in C^\infty(\mathbf{R})$ and $h' \in C_0^\infty(\mathbf{R})$. If $0 < \rho_1 < \rho$ then*

$$\int_1^\infty t^{-1} \left\| f(t^{\rho_1} \tilde{b}_t) h(H) \phi_t \right\|^2 \leq c \|\phi\|^2 \tag{4.17}$$

and if $0 \leq \rho_1 < \rho$ then

$$\int_1^\infty t^{-\delta+\rho_1} \left\| \sqrt{\tilde{c}_t} f(t^{\rho_1} \tilde{b}_t) h(H) \phi_t \right\|^2 dt \leq c \|\phi\|^2. \tag{4.18}$$

Proof. For simplicity we set $h = 1$. Otherwise we get additional terms due to $\mathbf{D}h(H)$ which are $O(t^{-1-\mu})$.

We set

$$F(s) := \int_{-\infty}^s f^2(s_1) ds_1. \tag{4.19}$$

Now

$$\mathbf{D}F(t^\delta \tilde{b}_t) = \rho_1 t^{-1} f^2(t^{\rho_1} \tilde{b}_t) t^{\rho_1} \tilde{b}_t$$
$$+ t^{-\delta+\rho_1} f(t^{\rho_1} \tilde{b}_t) \tilde{c}_t f(t^{\rho_1} \tilde{b}_t) + O(t^{-1-\rho+\rho_1}) + O(t^{3\rho_1 - 3\delta}). \tag{4.20}$$

We integrate 4.20 from t_1 to t_2 and we get

$$\begin{aligned}
c\|\phi\|^2 &\geq (\phi_{t_2}|F(t_2^\delta \tilde{b}_{t_2})\phi_{t_2}) - (\phi_{t_1}|F(t_1^\delta \tilde{b}_{t_1})\phi_{t_1}) \\
&= \int_{t_1}^{t_2} (\phi_t|\mathbf{D}F(t^\delta \tilde{b}_t)\phi_t)dt \\
&\geq \int_{t_1}^{t_2} \rho_1 t^{-1}\|\sqrt{t^\delta \tilde{b}_t} f(t^\delta \tilde{b}_t)\phi_t\|^2 dt \\
&+ \int_{t_1}^{t_2} t^{-\delta+\rho_1}\|\sqrt{\tilde{c}_t} f(t^\delta \tilde{b}_t)\phi_t\| dt \\
&+ \int_{t_1}^{t_2} \|\phi_t\|(O(t^{-1-\rho+\rho_1}) + O(t^{3\rho_1-3\delta}))dt. \qquad (4.21)
\end{aligned}$$

Note that the last integral of the rhs of 4.21 is uniformly bounded. The first two integrals yield the estimates 4.17 and 4.18 with $h = 1$. QED

The following proposition is parallel to proposition 4.2. We skip its proof.

Proposition 4.5 *Let $\epsilon > 0$, $h \in C_0^\infty(\mathbf{R})$, $f \in C_0^\infty(0, \infty)$. Let h and $F_\pm$ be as in proposition 4.4. Let $\rho_2 < \rho_1 < \rho$. Then*

$$\int_1^\infty t^{-1} \left\| F_-(t^{\rho_1}\tilde{b}_t) f(t^{-1+\rho_2} r_t) h(H)\phi_t \right\|^2 dt \leq c\|\phi\|^2; \qquad (4.22)$$

$$\int_1^\infty t^{-1+\rho_2-\rho_1} \left\| \sqrt{\epsilon - t^{\rho_1}\tilde{b}_t} F_-(t^{\rho_1}\tilde{b}_t) f(t^{-1+\rho_2} r_t) h(H)\phi_t \right\|^2 dt \leq c\|\phi\|^2; \qquad (4.23)$$

$$\int_1^\infty t^{-1} \left\| f(t^{\rho_1}\tilde{b}_t) F_+(t^{-1+\rho_2} r_t) h(H)\phi_t \right\|^2 dt \leq c\|\phi\|^2; \qquad (4.24)$$

$$\int_1^\infty t^{-\delta+\rho_1} \left\| \sqrt{\tilde{c}_t} f(t^{\rho_1}\tilde{b}_t) F_+(t^{-1+\rho_2} r_t) h(H)\phi_t \right\|^2 dt \leq c\|\phi\|^2. \qquad (4.25)$$

Now we are ready for the proof of the existence of inverse wave operators on exceptional states.

Proposition 4.6 *Suppose that $\mu > \sqrt{3} - 1$. Let $\phi \in RanE_{\{0\}}(P^+)$. Then there exists*

$$\lim_{t\to\infty} U_0^*(t)U(t)\phi. \qquad (4.26)$$

Proof. Let $F_\pm$ be such as in proposition 4.4. We assume in addition that $F_+(s) = 1$ for large positive s and $F_-(s) = 1$ for negative s. Let $0 < \rho_2 < \rho_1 < \rho$.

Suppose that $\phi \in RanE_{\{0\}}(P^+)$. It is enough to assume that there exists $h \in C_0^\infty([\lambda_1, \lambda_2])$ such that

$$\phi_t \sim h^2(H)\phi_t.$$

By propositions 6.1, 6.2 and 6.3

$$h^2(H)\phi_t \sim h(H)F_-(t^{\rho_1}\tilde{b}_t)h(H)\phi_t$$
$$-h(H)F_+(t^{-1+\rho_2}r_t(x))F_-(t^{\rho_1}\tilde{b}_t)F_+(t^{-1+\rho_2}r_t(x))h(H)\phi_t.$$

Hence 4.26 equals

$$\lim_{t\to\infty} U_0^*(t)h(H)\left(F_-(t^{\rho_1}\tilde{b}_t) - F_+(t^{-1+\rho_2}r_t(x))F_-(t^{\rho_1}\tilde{b}_t)F_+(t^{-1+\rho_2}r_t(x))\right)h(H)\phi_t. \quad (4.27)$$

Now note that

$$\left(F_-(t^{\rho_1}\tilde{b}_t) - F_+(t^{-1+\rho_2}r_t(x))F_-(t^{\rho_1}\tilde{b}_t)F_+(t^{-1+\rho_2}r_t(x))\right)(W(t,x) - W(t,0))$$
$$= F_-(t^{\rho_1}\tilde{b}_t)\left(1 - F_+^2(t^{-1+\rho_2}r_t(x))\right)x\int_0^1 \nabla_x W(t,\tau x)d\tau$$
$$+\left[F_+(t^{-1+\rho_2}r_t(x)), F_-(t^{\rho_1}\tilde{b}_t)\right]F_+(t^{-1+\rho_2}r_t(x))(W(t,x) - W(t,0))$$
$$= O(t^{2+\rho_2-\mu}) + O(t^{-1-\mu+\rho_2+\rho_1}). \quad (4.28)$$

We take an arbitrary $\psi \in L^2(X)$. We will write $\psi_t := U_0(t)\psi$. Now by 4.28 we obtain

$$\frac{d}{dt}\left(\psi_t|h(H)(F_-(t^{\rho_1}\tilde{b}_t) - F_+(t^{-1+\rho_2}r_t(x))F_-(t^{\rho_1}\tilde{b}_t)F_+(t^{-1+\rho_2}r_t(x)))h(H)\phi_t\right)$$
$$= \left(\psi_t|\mathbf{D}\left(h(H)(F_-(t^{\rho_1}\tilde{b}_t) - F_+(t^{-1+\rho_2}r_t(x))F_-(t^{\rho_1}\tilde{b}_t)F_+(t^{-1+\rho_2}r_t(x)))h(H)\right)\phi_t\right)$$
$$+ O(t^{-\rho_2-\mu}) + O(t^{-1-\mu+\rho_2+\rho_1}). \quad (4.29)$$

To see that the first term of 4.29 is integrable we just repeat the arguments of the proofs of propositions 4.4 and 4.5. If we choose ρ_1 and ρ_2 close enough to $\rho = \mu(2+\mu)^{-1}$ then we can make the third term integrable for any $\mu > 0$. If in addition $\mu > \sqrt{3} - 1$ then we can make the second term integrable too. Thus the existence of 4.26 follows by a standard argument due to Kato (see [RS vol IV]). QED

5 References

[A] S.Agmon, Lectures on the Exponential Decay of Solutions of Second Order Elliptic Equations, Princeton University Press, 1982.
[De1] J.Dereziński, "A new proof of the propagation theorem for N-body quantum systems", Comm. Math. Phys. 122 (1989) 203-231.
[De2] J.Dereziński, "Algebraic approach to the N-body long range scattering", Rev. Math. Phys. 3 (1991) 1-62.
[De3] J.Dereziński, "Asymptotic completeness of N-body long range quantum scattering", preprint, Ecole Polytechnique.
[De4] J.Dereziński, "Large time behavior of classical N-body systems", preprint, Ecole Polytechnique.
[E1] V.Enss, "Quantum scattering theory of two- and three-body systems with potentials of short and long range", in Schrödinger Operators, S.Graffi (ed.), Lecture Notes in Mathematics, Vol. 1159, Springer, Berlin-Heidelberg-New York, 1985.
[E2] V.Enss, "Two- and three-body quantum scattering: completeness revisited", Proceedings of the "Conference on Partial Differential Equations", Teubner-Texte zur Mathematik, Leipzig and B.-W. Schulze (eds), 1989.
[Fa] L.Faddeev, Mathematical Aspects of the Three Body Problem in Quantum Scattering Theory, Steklov Institute, 1963.
[Graf] G.M.Graf, "Asymptotic completeness for N-body short range systems: a new proof", Comm. Math. Phys. 132 (1990) 73-101.
[Hö] L.Hörmander, The Analysis of Linear Partial Differential Operators, Vol 1,2 (1983) and Vol. 3,4 (1985), Springer, Berlin-Heidelberg-New York-Tokyo.
[IKi] H.Isozaki and H.Kitada, "Modified wave operators with time independent modifiers", J. Fac. Sci. Univ. Tokyo, Sec 1A 32 (1985) 77-104.
[LoSig] M.Loss and I.M.Sigal, "The three-body problem with threshold singularities", ETH preprint.
[RS] M.Reed and B.Simon, Methods of Modern Mathematical Physics, vol. III: Scattering Theory, 1979 and vol. IV: Analysis of Operators, 1978, Academic Press, London.
[Sig] I.M.Sigal, "On the long range scattering" Duke Math. Journ. 60 (1990) 473-492.
[SigSof1] I.M.Sigal and A.Soffer, "The N-particle scattering problem: Asymptotic completeness for short range systems", Anal. Math. 125 (1987) 35-108.
[SigSof2] I.M.Sigal and A.Soffer, "Local decay and velocity bounds", preprint, Princeton University 1988.
[SigSof3] I.M.Sigal and A.Soffer, "Long range many body scattering. Asymptotic clustering for Coulomb type potentials", Invent. Math. 99 (1990) 115-143.
[SigSof4] I.M.Sigal and A.Soffer, "Asymptotic completeness for four-body Coulomb systems", preprint 1991.
[Ya] D.R.Yafaev, "Radiation conditions and scattering theory for three-particle Hamiltonians", preprint, Univ. de Nantes 1991.

A remark on Asymptotic Clustering for N−particle Quantum Systems

J.Derezinski
Department of Mathematical Methods in Physics
Warsaw University Hoza 74, 00-682 Warszawa, Poland

C.Gérard
Centre de Mathématiques
Ecole Polytechnique 91128 Palaiseau Cedex France

July 1991

Abstract

We prove that N−particle Hamiltonians with long range pair potentials decaying at infinity like $< x >^{-\mu}$ for $\mu > 1/2$ are asymptotically clustering at all non-threshold energies.

1 Introduction

We consider in this note N−particle Hamiltonians of the form :

$$H = \sum_{1}^{N} \frac{-1}{2m_i}\Delta_i + \sum_{i<j} V_{ij}(x_i - x_j) \text{ on } L^2(\mathbb{R}^{N\nu})$$

For short range pair potentials, *i.e* decaying at infinity like $|x_i - x_j|^{-\mu}$ for some $\mu > 1$, it is known that H is *asymptotically complete*. This means that the evolution $e^{-itH}u$ of any state u which is not a bound state of H is asymptotic to a superposition of simpler evolutions describing freely moving *stable* clusters of particles. The fact that the clusters are stable means that their internal motion is described by a bound state of the internal Hamiltonian. This has been established for 3−particle systems by Enss (also for long range potentials) (see [E]), and for general N−particle systems by Sigal and Soffer (see [S.S1]). Another very elegant proof has been given by Graf (see [Gr]).

For long range potentials, one can formulate a weaker result on the asymptotic evolution $e^{-itH}u$, called *asymptotic clustering*. This property has been introduced by Sigal and Soffer in [S.S2] and proved there for Coulomb-type potentials. Asymptotic clustering says roughly that for $t \to \pm\infty$, the evolution $e^{-itH}u$ is asymptotic to a superposition of independently moving clusters of particles, such that the relative motion of the center of masses of the clusters is influenced by the long range tail of intercluster potential. This property is weaker than asymptotic completeness in that it does not say that the clusters are in a bound state of their internal Hamiltonian. An important property of these asymptotic evolutions is that they conserve the intercluster momentum. Nevertheless the property of asymptotic clustering has the consequence that the spectrum of H is asymptotically absolutely continuous (see [D]), and is used as an essential step in the proof of asymptotic completeness for four-body Coulomb systems in [S.S3].

Let us now formulate more precisely our result and hypotheses. We will consider a generalization of N−body Hamiltonians introduced by Agmon [Ag]. One considers a finite dimensional real vector space X with a positive definite quadratic form g, and a finite familly $\{X_a\}$, $a \in A$ of linear vector subspaces of X which is closed under intersection and obeying $\cap_{a\in\mathcal{A}} X_a = \{0\}$ and $X \in \{X_a\}$. One denotes by X^a the space $X_a^{\perp}$, by π^a, π_a the orthogonal projections on X^a and X_a.

On $\mathcal{A}$ one puts a partial ordering by saying that $b \subset a$ if $X^b \subset X^a$. With this ordering one gets that $X_{a_{\min}} = X$ and $X_{a_{\max}} = \{0\}$. For $a \in \mathcal{A}$, one denotes by $\sharp a$ the maximal number k such that $a_1 = a \subset_{\neq} a_2 \cdots \subset_{\neq} a_k = a_{\max}$.

If $N = \sharp a_{min}$, one defines an N−body Schrödinger operator by :

$$H = -\frac{1}{2}\Delta_g + V(x)$$

where : $V(x) = \sum_{a\in\mathcal{A}} V_a(\pi^a x)$ and Δ_g is the Laplace -Beltrami operator associated with g. For simplicity of notations, we will simply denote Δ_g by Δ. For $a \in \mathcal{A}$, we denote by H_a the Hamiltonian $H - I_a(x)$, where $I_a(x) =$

$\sum_{b\not\subset a} V_b(x^b)$. One has also $H_a = -\frac{1}{2}\Delta_{x_a} + H^a$, where H^a is the Hamiltonian acting on $L^2(X^a)$ defined by $H^a = -\frac{1}{2}\Delta_{x^a} + V^a(x^a)$ for $V^a(x^a) = \sum_{b\subset a} V_b(x^b)$. One defines then the *threshold set* $\mathcal{T}$ as $\mathcal{T} = \cup_{a\neq a_{\max}} \sigma_{\rm pp}(H^a)$. If S is a subset of $\mathbf{R}$, then $E_S^c(H)$ denotes the spectral projection of H onto the continuous spectrum in S.

We assume that the potentials $V_a(y^a)$ can be written as sums of short range and long range parts :

$$V_a(y^a) = V_{a,s}(y^a) + V_{a,l}(y^a)$$

where the short range part $V_{a,s}$ satisfies for example the conditions in [Gr], namely :

$$(1.1)\qquad \begin{aligned} &\|F(|y^a| > R)V_{a,s}(p^{a2}+1)^{-1}\| \le \text{const } R^{-\mu_1} \\ &\|F(|y^a| > R)\nabla V_{a,s}(p^{a2}+1)^{-1}\| \le \text{const } R^{-1-\mu_2} \end{aligned}$$

for $R > R_0$ and $\mu = \min(\mu_1, \mu_2) > 0$, and $\mu_1 > 1$.

(1.2) $V_{a,s}(p^{a2}+1)^{-1}$, $(p^{a2}+1)^{-1} y^a.\nabla V_{a,s}(y^a)(p^{a2}+1)^{-1}$ are compact.

On the long range part $V_{a,l}$, we make the following assumptions :

$$(1.3)\qquad \begin{aligned} &V_{a,l} \in C^2(\mathbf{R}^{n^a}),\ |< y^a >^{\mu} V_{a,l}(y^a)| \le C \\ &|< y^a >^{1+\mu} \nabla V_{a,l}(y^a)| + |< y^a >^{\mu_1} \Delta_{y^a} V_{a,l}(y^a)| \le C \end{aligned}$$

for some $\mu > \frac{1}{2}$, $\mu_1 > 2$. Let us now state more precisely the property of asymptotic clustering (see [S.S2]). For $a \neq a_{\max}$, we fix a cutoff function $\tilde{q}_a$, supported in the region :

$$\{x \in X \mid |x^a| \le \epsilon_0, |x^b| \ge \epsilon_1 \forall b \not\subset a\}$$

and equal to 1 in :

$$\{x \in X \mid |x^a| \le \epsilon_0/2, |x^b| \ge 2\epsilon_1 \forall b \not\subset a\}$$

Then as in [Ki.Ya], we set :

$$\tilde{I}_a(t,x) = \sum_{b\not\subset a} \tilde{q}_a\Big(\frac{x \log t}{\langle t\rangle}\Big) V_{b,l}$$

Note that $\tilde{I}_a$ satisfies the estimates :

$$(1.4) \qquad \begin{array}{l} |(\langle x\rangle + \langle t\rangle)^{\mu-\epsilon}\tilde{I}_a(t,x)| + |(\langle x\rangle + \langle t\rangle)^{1+\mu-\epsilon}\nabla \tilde{I}_a(x)| \le C_\epsilon \\ |(\langle x\rangle + \langle t\rangle)^{\mu_1-\epsilon}\Delta_x \tilde{I}_a(x)| \le C_\epsilon \end{array}$$

for all $\epsilon > 0$. We denote then by $W_a(t)$ the propagator generated by the time dependent Hamiltonian $H_{a,l}(t) = H_a + \tilde{I}_a(t, x^a, tD_{x_a})$. One says then that the $N-$particle Hamiltonian H is *asymptotically clustering* for the energies in $S \in \mathbf{R}$, if the following holds :

$$\begin{array}{l} \forall \psi \text{ with } E_S^c(H)\psi = \psi,\ \exists\, \psi_a^\pm \in L^2(X) \text{ for } a \neq a_{\max} \text{ such that :} \\ e^{-itH}\chi(H)\psi \ - \ \sum_{a\neq a_{\max}} W_a(t)\psi_a^\pm \to 0 \text{ when } t \to \pm\infty \end{array}$$

Then we have the following result :

Theorem 1 *Under the hypotheses (1.1), (1.2), (1.3) for* $\mu > \frac{1}{2}$, *the Hamiltonian* H *is asymptotically clustering for the energies in* $\mathbf{R}\backslash\mathcal{T}$.

2 proof of the Theorem

In this Section we prove Theorem 1. The proof is divided in three very elementary steps. Let us consider a state $\psi \in L^2(X)$. We assume that ψ is spectrally supported away from the thresholds of H. By density, we can also assume that $\psi = \chi(H)\psi$, for $\chi \in C_0^\infty(\mathbf{R})$ supported away from $\mathcal{T} \cup \sigma_{pp}(H)$. We need to introduce a partition of unity in configuration space. Its construction belongs to the folklore of the $N-$body problem and can be found for example in [Gr]. There exist cutoff functions $q_a(x)$ for $a \in \mathcal{A}$ supported in $\{x \in X \mid |x^a| \le \epsilon_0|x|,\ |x^b| \ge \epsilon_1|x|\ \forall b \not\subset a\}$, for some ϵ_0 and $\epsilon_1 > 0$ and such that :

$$\sum_{a\in\mathcal{A}} q_a(x) = 1$$

Using then [Gr, Theorem 6.1], or [D] we get that :

$$e^{-itH}\psi = \sum_{a\in\mathcal{A}, a\neq a_{\max}} q_a(\frac{x}{ct})e^{-itH}\psi + o(1)$$

for some c small enough. Let us check now that :

$$(2.5) \qquad U_a^\star(t)q_a(\frac{x}{ct})e^{-itH}\psi \text{ has a limit when } t \to +\infty$$

Here $U_a(t)$ is the propagator generated by the time dependent Hamiltonian $H_a + \tilde{I}_a(t,x)$. This can be proven exactly as in [Gr, Theorem 5.3], using the fact that for t large enough, $I_a(x)$ and $\tilde{I}_a(t,x)$ differ only by a short range term on the support of $q_a(\frac{x}{t})$.
In the second step we prove the following propagation estimate :

$$\|(x_a - tD_{x_a})U_a(t) < D_x >^{-1} < x >^{-1} \| = O(< t >^{1-\mu}) \tag{2.6}$$

To prove (2.6), we take a vector $u \in D(< x >) \cap D(< D_x >)$, and put $f(t) = \|(x_a - tD_{x_a})U_a(t)u\|$. The fact that $f(t)$ is finite follows from the well-known fact that $D(< x >) \cap D(< D_x >)$ is invariant under $U_a(t)$. We have :

$$\frac{df}{dt}(t) \leq \|D_t(x_a - tD_{x_a})U_a(t)u\|$$

where $D_t = \frac{d}{dt} + i[H_{a,l}(t), \cdot]$ is the Heisenberg derivative. From (1.3), we get that $\|D_t(x_a - tD_{x_a})\| = O(< t >^{-\mu})$, from which we get (2.6) by integration.
In the third step we prove the Theorem for $\mu > \frac{1}{2}$. Using (2.5), we see that it sufffices to prove that the following wave operators exist :

$$\Omega_a^{\pm} = s.lim_{t \to \pm\infty} W_a^{\star}(t) U_a(t)$$

To prove their existence, we pick a vector $u \in D(< x >) \cap D(< D_x >)$ and we compute the derivative of $W_a^{\star}(t)U_a(t)u$, which is equal to :

$$W_a^{\star}(t)(\tilde{I}_a(t, x^a, tD_{x_a}) - \tilde{I}_a(t,x))U_a(t)u \tag{2.7}$$

By the Baker-Campbell-Hausdorff formula, we get :

$$\begin{aligned} &\tilde{I}_a(t, x^a, tD_{x_a}) - \tilde{I}_a(t,x) = \\ &\int_0^1 \nabla_{x_a} \tilde{I}_a(t, x^a, stD_{x_a} + (1-s)x_a).(tD_{x_a} - x_a)ds + \\ &\frac{t}{2} \int_0^1 \Delta_{x_a} \tilde{I}_a(t, x^a, stD_{x_a} + (1-s)x_a)ds \end{aligned} \tag{2.8}$$

From (1.4) we obtain that :

$$\nabla_{x_a} \tilde{I}_a = O(< t >^{-1-\mu-\epsilon}),\ \Delta_{x_a} \tilde{I}_a = O(< t >^{-\mu'-\epsilon})$$

for $\epsilon > 0$. Using then (2.8) and (2.6), we get that :

$$\|(\tilde{I}_a(t, x^a, tD_{x_a}) - \tilde{I}_a(t,x))U_a(t)u\| \leq C_\epsilon < t >^{-2\mu+2\epsilon} + C_\epsilon < t >^{1-\mu'+\epsilon}$$

For $\mu > \frac{1}{2}$, we obtain then the existence of the wave operators $\Omega_a^{\pm}$ by the Cook argument, which completes the proof of the Theorem. □

References

[Ag] S.Agmon: *Lectures on exponential decay of solutions of second order elliptic equations*, Princeton University Press, Princeton 1982.

[D] J.Derezinski: Algebraic approach to the $N-$body Quantum Long Range Scattering, Preprint 1990, to appear in Rev. Math. Phys.

[E] V.Enss : Quantum scattering theory of two- and three-body systems with potentials of short and long range, Lectures notes in Math. Vol 1159, 1985.

[Gr] G.M.Graf: Asymptotic Completeness for $N-$body Short Range Quantum Systems : A new proof, Comm. in Math. Phys. 132 (1990) p 73-101.

[Ki.Ya] H.Kitada-K.Yajima: A scattering theory for time dependent long range potentials, Duke Math. J. 49 (1982) 341-376.

[S.S1] I.M Sigal-A.Soffer: The $N-$particle scattering problem: asymptotic completeness for short range quantum systems, Ann. of Math. 125 (1987) 35-108.

[S.S2] I.M Sigal-A.Soffer: Long range many-body scattering: Asymptotic clustering for Coulomb-type potentials, Invent.Math. 99, (1990) 115-143.

[S.S3] I.M.Sigal-A.Soffer: Asymptotic completeness for four-body Coulomb systems, preprint.

The Energy Asymptotics of Large Coulomb Systems

C. L. Fefferman [1], V. Ja. Ivrii [1], L. A. Seco [3],

I. M. Sigal [4]

[1]Dept. Mathematics, Princeton University, Princeton NJ 08544 USA.
[2]Centre de Mathématiques, Ecole Polythechnique, F-91128 Palaiseau, France.
[3]Dept. Mathematics, California Institute of Technology, Pasadena CA 91125 USA.
[4]Dept. Mathematics, University of Toronto, Toronto ONTARIO M5S 1A1, Canada.

Introduction.

Consider a molecule consisting of N quantized electrons at positions x_i, and M nuclei of charges $Z = (Z_1, \ldots, Z_M)$ fixed at positions $y = (y_1, \ldots, y_M)$. The Schrödinger Hamiltonian of such a system is given by

$$H_{Z,N} = \sum_{i=1}^{N} (-\Delta_{x_i} + V_{\text{Coulomb}}(x_i)) + \tfrac{1}{2} \sum_{i \neq j} \frac{1}{|x_i - x_j|}$$

$$V_{\text{Coulomb}}(x) = -\sum_{j=1}^{M} \frac{Z_j}{|x - y_j|}$$

acting on $\mathcal{H} = \wedge_{i=1}^{N} L^2 (\mathbf{R}^3 \times \mathbf{Z}_q)$; in this exposition, in order to simplify notation, we neglect spin by putting $q = 1$. Define the ground state of such a system by

$$E(Z, y) = \inf_N E(Z, y; N) \qquad E(Z, y; N) = \inf_{\substack{\|\psi\|=1 \\ \psi \in \mathcal{H}}} \langle H_{Z,N}\psi, \psi\rangle$$

When $M = 1$, this system is an atom. In this case we can assume $y = 0$ and we denote its energy simply by $E_{\text{atom}}(Z)$.

It is a remarkable fact that $E(Z)$ behaves in a very simple way when $Z \to \infty$, as follows:

Theorem 1

$$E(Z, y; |Z|) = -c_0 Z^{7/3} + \tfrac{1}{8} Z^2 + O\left(Z^{2-a}\right) \qquad a > 0$$

$$E_{\text{atom}}(Z) = -c_0 Z^{7/3} + \tfrac{1}{8} Z^2 - c_1 Z^{5/3} + O\left(Z^{\frac{5}{3} - \frac{1}{2835}}\right)$$

In the molecular case, the formula holds for all y such that $|y_j - y_k| > Z^{-\frac{2}{3}+\epsilon}$, and for $Z \to \infty$ along a given direction.

Note that in the molecular case, the asymptotics are asserted for the neutral molecule, whereas in the atomic case, we use the lowest possible energy. The problem of whether both energies are the same is a very subtle one, and although much research has been done about it, not much is known about it. In our setting, however, given the relatively large degree of error we allow ourselves in the energy asymptotics as compared with the delicate phenomenom of neutrality, it is just a technical point. We refer the reader to (4, 5, 20, 23, 27, 28) and references thereof for information on the neutrality problem. We also refer the reader to the contribution of Ruskai and Solovej to appear elsewhere in this same proceedings.

The first term above was conjectured by Thomas (29) and Fermi (14), and proved rigorously by Lieb-Simon (19). Their proof holds for atoms as well as molecules. The second term was conjectured by Scott (22) and proved rigorously by Hughes-Siedentop-Weikard (15 , 24, 25 and 26) in the atomic case; its proof for molecules appears in (16). The third term was conjectured by Schwinger (21), based on the fundamental work of Dirac (3); its proof for the atomic case was announced in (6), and the complete details of it appear in (7, 8, 9, 10, 11, 12 and 13). The proof of the third term for molecules remains open.

The purpose of this paper is to give a brief joint expository presentation of the proof of Theorem 1. We refer the reader to (7, 8, 9, 10, 11, 12, 13 and 16) for accurate statements and detailed proofs of what follows.

Before we preceed to the proof, we recall a few basic facts about Thomas–Fermi theory. We refer the reader to (17) for a comprehensive review.

We denote the Thomas–Fermi density by $\rho^Z_{\text{TF}}(x)$. It satisfies the perfect scalling property

$$\rho^Z_{\text{TF}}(x) = Z^2 \rho^1_{\text{TF}}(Z^{1/3} x)$$

The Thomas–Fermi potential

$$V^Z_{\text{TF}}(x) = \sum_{j=1}^{M} \int \frac{\rho_{\text{TF}}(y_j)}{|x - y_j|}\, dy$$

is smooth away from the origin, and vanishing at infinity. In the atomic case, V_{TF}^{Z} is radially symmetric: this will be of crucial importance for our current proof of the atomic energy asymptotics.

The screened Thomas–Fermi potential

$$W_{\mathrm{TF}}^{Z}(x) = \sum_{j=1}^{M} \frac{Z_j}{|x - y_j|} - V_{\mathrm{TF}}^{Z}(x) \geq 0$$

satisfies the Thomas–Fermi equations

$$6\pi^2 \rho_{\mathrm{TF}}^{Z}(x) = \left|W_{\mathrm{TF}}^{Z}(x)\right|^{3/2} \qquad \Delta W_{\mathrm{TF}}^{Z}(x) = 4\pi\rho_{\mathrm{TF}}^{Z}(x).$$

In the atomic case W_{TF}^{1} can be written as $u(\gamma r)/r$, with an appropiate constant γ, and u satisfying the Thomas–Fermi Ordinary Differential Equation (ODE)

$$\left.\begin{aligned} u''(r) &= \frac{u^{3/2}(r)}{r^{1/2}} \\ u(0) &= 1 \\ u(\infty) &= 0 \end{aligned}\right\}$$

Acknowledgements. The third and fourth authors are grateful to E. Balslev and A. Jensen for a stimulating atmosphere during the conference in Århus and for financial support. The first author is partially supported by an NSF grant. The fourth author is partially supported by NSERC under grant # NA 7901.

1 Mean Field Theory.

The first step in understanding $E(Z)$ is to simplify $H_{Z,N}$. We start by describing Lieb's inequality (18), which was used in (15 , 16 and 25) to prove the Scott term.

Start with the elementary identity

$$\frac{1}{|x-y|} = \frac{1}{\pi} \iint \chi_{B(z,R)}(x)\chi_{B(z,R)}(y)\, \frac{dz\, dr}{R^5}$$

which, except for the constant $\frac{1}{\pi}$, is forced on us by dilation and translation invariance. Summing over pairs of electrons, we arrive at

$$\tfrac{1}{2}\sum_{i\neq j} \frac{1}{|x_i - x_j|} = \frac{1}{2\pi} \iint \mathcal{N}\left(\mathcal{N}-1\right) \frac{dz\, dR}{R^5}$$

where

$$\mathcal{N} = \mathcal{N}(x_1, \ldots, x_N; z, R) = \sum_{i=1}^{N} \chi_{B(z,R)}(x_i)$$

represents the number of x_i in the ball of center z and radius R.

Let us pick now

$$\bar{\mathcal{N}}(z, R) = \int_{B(z,R)} \rho_{\mathrm{TF}}^{Z}(x)\, dx$$

to write

$$\frac{1}{2} \sum_{i \neq j} \frac{1}{|x_i - x_j|} = \frac{1}{2\pi} \iint \left((\mathcal{N} - \bar{\mathcal{N}})^2 + \mathcal{N}\left(2\bar{\mathcal{N}} - 1\right) - \bar{\mathcal{N}}^2 \right) \frac{dz\, dR}{R^5}$$

Since $\bar{\mathcal{N}}$ is independent of the x_i, of the three terms in the integrand above, the first is a two–body potential, the second is one–body and the last is a constant. The two–body part is hard to understand, but its contribution is non–negative. Therefore, dropping the two–body part, we arrive at

$$\frac{1}{2} \sum_{i \neq j} \frac{1}{|x_i - x_j|} \geq \sum_{i=1}^{N} V(x_i) - C$$

where V is essentially V_{TF}^{Z}. This suggests that we should consider the following Schrödinger operator

$$H^{\mathrm{ind}} = \sum_{i=1}^{N} \left(-\Delta_{x_i} - \frac{Z}{|x_i|} + V_{\mathrm{TF}}^{Z}(x_i) \right)$$

This operator describes N independent particles in the effective external potential $-Z/|x| + V_{\mathrm{TF}}^{Z}(x)$, i.e., the nuclear potential screened by the Thomas–Fermi charge density.

To understand the error that such an inequality sacrifices, take a ball of radius $R \sim C\, Z^{-1/3}$, C large, which is expected to contain a fixed, large number of electrons. The term dropped, $(\mathcal{N} - \bar{\mathcal{N}})^2$ is approximately the variance of $\mathcal{N}$, which, if electrons were independent, would be like $\bar{\mathcal{N}} \sim$ large constant. For this ball, $\frac{dz\, dR}{R^5}$ contributes with approximately R^{-1}, this giving an error of about $Z^{1/3}$. Since there are about Z of these balls, the total error is like $Z^{5/3}$.

2 Correlations.

The error obtained in the previous section is acceptable for the asymptotics we announced for molecules, but for the refined atomic asymptotics, we have to do better. The improvement will come from the analysis of electron–electron correlations.

Theorem 2 *Let ψ be the gound state of $H_{Z,N}$. Then*

$$\left\langle \psi, \tfrac{1}{2} \sum_{i \neq j} \frac{1}{|x_i - x_j|} \psi \right\rangle \geq \left\langle \sum_{i=1}^{N} V_{\mathrm{TF}}^{Z}(x_i)\psi, \psi \right\rangle$$
$$- \tfrac{1}{2} \iint \frac{\rho_{\mathrm{TF}}(x)\rho_{\mathrm{TF}}(y)}{|x-y|}\, dx\, dy - c_D \int \rho_{\mathrm{TF}}^{4/3}(x)\, dx + O\left(Z^{\frac{5}{3} - \frac{1}{2835}}\right)$$

The complete proof of this theorem for the atomic case appears in (7). The proof for molecules follows after trivial modifications.

In preparation for the proof of Theorem 2, we split the electron–electron repulsion into a short and long range part, as follows

$$\tfrac{1}{2} \sum_{i \neq j} \frac{1}{|x_i - x_j|} = V_L + V_S = \sum_{i<j} K_L(x_i, x_j) + \sum_{i<j} K_S(x_i, x_j)$$

where

$$K_L(x,y) = \frac{1}{|x-y|} \chi_{|x-y|>d}$$
$$K_S(x,y) = \frac{1}{|x-y|} \chi_{|x-y|\leq d}$$

where d is a variable length–scale, approximately equal to $Z^{-\frac{2}{3}+a_0}$, $a_0 > 0$, throughout most of the atom. Precisely,

$$d = \min(Z^{-\frac{2}{3}+a_0}, c|x|) \qquad \text{c small}$$

Now, in view of the above calculation, if we apply Lieb's inequality to V_L, the error we expect to obtain will behave like

$$\frac{Z}{\text{distance between electrons}} \leq \frac{Z}{Z^{-2/3+a_0}} = Z^{5/3 - a_0}$$

which is good enough for our purposes. Thus, we only need to understand the short range part V_S of the repulsion. This amounts to understanding our wave functions ψ only at short distances. Motivated by this, we put ourselves in the following set–up:

Let $\psi(x_1,\ldots,x_K)$ be an antisymmetric wave function *supported* on $x_i \in Q$, for $i=1,\ldots,K$, Q being a cube of side $L=Z^{-\frac{2}{3}+a_1}$. Consider also $\Lambda(L)=\frac{2\pi}{L}\mathbf{Z}^3$, the lattice dual to Q. For simplicity, we impose on ψ periodic boundary conditions at ∂Q. This condition on ψ will have to be de-localized, in order to allow ψ's of arbitrary support, but this will be entirely omitted in the present exposition.

Furthermore, we assume that ψ satisfies the crucial estimate

$$\|\nabla\psi\|^2 \le c_{\mathrm{TF}} L^{-2} K^{5/3}\left(1+Z^{-a_2}\right) \tag{1}$$

where $c_{\mathrm{TF}}=(\frac{4\pi}{3})^{-5/3}\cdot\frac{4\pi}{5}\cdot(2\pi)^2$ is the semiclassically correct constant, and a_2 is a positive constant to be carefully picked during the proof. Note that if one takes

$$\psi=e^{ix_1\cdot\xi_1}\wedge\cdots\wedge e^{ix_p\cdot\xi_p}$$

for $\xi_j\in B(0,R)\cap\Lambda(L)$ distinct, and as close as possible to the origin to minimize the kinetic energy, then

$$\|\nabla\psi\|^2 \sim \tfrac{4\pi}{5}(\tfrac{L}{2\pi})^3 R^5$$

$$p=\#B(0,R)\cap\Lambda(L)\sim\tfrac{4\pi}{3}R^3(\tfrac{L}{2\pi})^3$$

for p large, which shows

$$R\sim R_{\mathrm{F}}=2\pi(4\pi/3)^{-1/3}p^{1/3}L^{-1}$$

We call R_{F} the Fermi radius, and $B(0,R_{\mathrm{F}})$ the Fermi ball for p particles.

It follows, as a consequence of Lieb–Simon's theorem, that if ψ is a ground state, then assumption (1) (or some version of it) must be satisfied for *most* cubes Q, or else, the first term in the energy asymptotics would be wrong. Therefore, we are justified in assuming (1). Our immediate goal is to conclude that if ψ satisfies (1), then it is very nearly equal to an antisymmetric product of "plane waves", in the following sense:

Consider Q_0, a subcube of Q, of side much smaller than L, but still larger than $Z^{-\frac{2}{3}+a_3}$, $a_3>0$; next, construct an orthonormal set $\{\varphi_\alpha\}_{\alpha=1}^{N_{\mathrm{TF}}}$ of functions supported on Q_0, which are essentially of the form $e^{i\xi_\alpha\cdot x}$, with $\xi_\alpha\in\Lambda(L)$ away from ∂Q_0. Here, N_{TF} is the number of ξ_α inside the Fermi ball for K particles $B(0,2\pi(4\pi/3)^{-1/3}K^{1/3}L^{-1})$. We extend this orthonormal set to a basis for $L^2(Q)$, thus extending α to all integers. Consider also the translates φ_α^z of φ_α to $z\in Q$. Then, in this situation, the following holds:

Lemma 3 *Let ψ satisfy (1). Then, for most $z\in Q$, we can write*

$$\psi=\phi^z_{\mathrm{basic}}\wedge\phi^z_{\mathrm{extra}}+\phi^z_{\mathrm{error}} \tag{2}$$

with $\phi^z_{\mathrm{basic}}=\wedge_{\alpha=1}^{N_{\mathrm{TF}}}\varphi^z_\alpha$, ϕ^z_{extra} strongly orthogonal to ϕ^z_{basic}, i.e.,

$$\int \phi^z_{\text{extra}}(y_1,\dots,y_{K-N_{\text{TF}}})\phi^z_{\text{basic}}(x_1,\dots,x_{N_{\text{TF}}})\delta_{y_j=x_i}\,dx_i\,dy_j = 0$$

for all i, j, and all y_l, x_n with $l \neq j$ and $n \neq i$; ϕ^z_{error} is small in appropriate norms.

Sketch of Proof: We introduce the positive numbers

$$\mu(z,\alpha) = 1 - \sum_{j=1}^{K} \int |\langle\psi,\varphi^z_\alpha\rangle|^2\, dx_1 \cdots \widehat{dx_j} \cdots dx_K$$

where the scalar product is in the x_j variable of ψ, and in the integration we remove the dx_j factor.

These numbers are important mainly for two reasons: first, (1) will imply they are very small. Second, since μ measures how much ψ differs from being an antisymmetric product of the ψ_α, errors of interest in what follows will be bounded by $\mu(z,\alpha)$. We make this *a little* more precise:

If we Fourier-expand ψ, with Fourier coefficients $\xi_i \in \Lambda(L)$, then

$$\mathrm{Av}_{z\in Q}\,\mu(z,\alpha) = \sum_{\xi_1,\dots,\xi_K} \left|\hat\psi(\xi_1,\dots,\xi_K)\right|^2 \left(1 - \sum_{j=1}^{K} |\hat\varphi^z_\alpha(\xi_j)|^2\right)$$

Now, (1) implies that $\hat\psi(\xi_1,\dots,\xi_K)$ is negligible unless most of the ξ_j are inside the Fermi ball; on the other hand, since the φ_α are almost equal to $e^{i\xi_j\cdot x}$, stationary phase implies the term in brackets above is close to zero for all ξ_j in the Fermi ball, when $1 \le \alpha \le N_{\text{TF}}$. Therefore,

$$\mathrm{Av}_{z\in Q}\,\mu(z,\alpha) = O(Z^{-p}) \qquad p > 0$$

This implies that for most z, $\mu(z,\alpha)$ is very small. Then, we can obtain (2) by – essentially – setting

$$\phi^z_{\text{error}} = \sum_{\{1,\dots,N_{\text{TF}}\}\not\subset\{1,\dots,K\}} \hat\psi(\xi_1,\dots,\xi_K)\,\varphi^z{}_1 \wedge\cdots\wedge \varphi^z_K$$

Thus,

$$\begin{aligned}\psi - \phi^z_{\text{error}} &= \sum_{\{1,\dots,N_{\text{TF}}\}\subset\{1,\dots,K\}} \hat\psi(\xi_1,\dots,\xi_K)\varphi^z_1 \wedge\cdots\wedge \varphi^z_K \\ &= \phi^z_{\text{basic}} \wedge \phi^z_{\text{extra}}\end{aligned}$$

as stated. The size of ψ^z_{error} is bounded by $\sum_\alpha \mu(z,\alpha)$, and is therefore small. □

This, with some more work that we will omit here, allows us to simply ignore ϕ^z_{error} from now on. However, ϕ^z_{extra} cannot be ignored in such a way. Fortunately, since $K_S \geq 0$, creating electrons should only increase the energy, which is fine with Theorem 2. Indeed,

$$\langle V_S(\phi^z_{\text{basic}} \wedge \phi^z_{\text{extra}}), (\phi^z_{\text{basic}} \wedge \phi^z_{\text{extra}})\rangle \geq \langle V_S \phi^z_{\text{basic}}, \phi^z_{\text{basic}}\rangle$$

and this is trivial to prove using the strong orthogonality between ϕ^z_{basic} and ϕ^z_{extra}.

In orther words, in order to prove Theorem 2, we can assume that ψ correlates at short distances like the Hartree–Fock wave function ϕ^z_{basic}. Thus, imposing $a_3 < a_0$ we are justified in writing

$$\langle V_S\psi, \psi\rangle \geq \tfrac{1}{2} \sum_{i \neq j} \langle K_S(x_i, x_j)\phi^z_{\text{basic}}, \phi^z_{\text{basic}}\rangle - \text{junk}$$

and the right hand side can be calculated, since we know explicitly the density and correlation function of ϕ_{basic}. This yields

$$\langle V_S\psi, \psi\rangle \geq \tfrac{1}{2} \iint K_S(x,y) \left(\rho^z(x)^2 - |S^z(x,y)|^2\right) dx\, dy - \text{junk}$$

for

$$\rho^z(x) = \sum_{\alpha=1}^{N_{\text{TF}}} |\varphi^z_\alpha(x)|^2 \qquad S^z(x,y) = \sum_{\alpha=1}^{N_{\text{TF}}} \varphi^z_\alpha(x)\overline{\varphi^z_\alpha(y)}$$

Recalling the construction of the φ^z_α, note that

$$\begin{aligned} S^z(x,y) &= \sum_{\alpha=1}^{N_{\text{TF}}} e^{i\xi_\alpha(x-y)} (1 + \text{tiny errors}) \\ &= \int_{\text{Fermi Ball}} e^{iz(x-y)} \frac{dz}{(2\pi)^3} (1 + \text{tiny errors}) \end{aligned}$$

and similarly, ρ^z is very nearly equal to N_{TF}. Therefore,

$$\langle V_S\psi, \psi\rangle \geq \tfrac{1}{2} \iint K(x,y) \left(N_{\text{TF}}^2 - |\widehat{\chi_{\text{Fermi Ball}}}(x-y)|\right) dx\, dy - \text{junk}$$

This last inequality becomes Theorem 2 in passing from ψ supported on cubes Q to arbitrary ψ.

This finishes the exposition of the proof of Theorem 2.

Corollary *Denote by $E^{\text{ind}}(Z,N)$ the lowest eigenvalue of*

$$H^{\text{ind}} = \sum_{i=1}^{N} \left(-\Delta_{x_i} - \frac{Z}{|x_i|} + V^Z_{\text{TF}}(x_i)\right)$$

Then

$$E_{Z,N} \geq E^{\text{ind}}(Z) - \tfrac{1}{2} \iint \frac{\rho_{\text{TF}}(x)\rho_{\text{TF}}(y)}{|x-y|}\,dx\,dy - c_D \int \rho_{\text{TF}}^{4/3}(x)\,dx + O\left(Z^{\frac{5}{3}-\frac{1}{2835}}\right) \tag{3}$$

Note that if E_k are the N lowest negative eigenvalues of

$$-\Delta_x - \frac{Z}{|x|} + V_{\text{TF}}^{Z}(x)$$

then

$$E^{\text{ind}}(Z,N) = \sum_k E_k.$$

3 Upper Bound.

In order to obtain an upper bound for $E(Z,y;N)$ we use the ground state of H^{ind} as a test function. Precisely, if we denote by ψ_k the normalized eigenfunctions of $-\Delta - W_{\text{TF}}^{Z}(x)$, for W_{TF}^{Z} the screened Thomas–Fermi potential, and we put $\psi = \Lambda\psi_k$, then the variational principle yields

$$E(Z) \leq \langle H_{Z,N}\psi, \psi\rangle \tag{4}$$

A simple computation shows that

$$\langle H_{Z,N}\psi,\psi\rangle = \sum_k E_k - \tfrac{1}{2} \iint \frac{\rho_{\text{TF}}(x)\rho_{\text{TF}}(y)}{|x-y|}\,dx\,dy - \iint \frac{|S(x,y)|^2}{|x-y|}\,dx\,dy + \iint \frac{(\rho_{\text{TF}}(x)-\rho(x))(\rho_{\text{TF}}(y)-\rho(y))}{|x-y|}\,dx\,dy$$

where

$$\rho(x) = \sum_k |\psi_k(x)|^2 \qquad S(x,y) = \sum_k \psi_k(x)\overline{\psi_k(y)}$$

Now, it is a very different question whether these bounds (3) and (4) will be any good. It is a remarkable property of the Thomas–Fermi potential that, at least in the atomic case, these upper and lower bounds agree up to an error $O\left(Z^{\frac{5}{3}-a}\right)$ with $a > 0$. Indeed, the following lemma holds:

Lemma 4 *In the atomic case,*

$$\iint \frac{|S(x,y)|^2}{|x-y|}\,dx\,dy \geq c_D \int \rho_{\text{TF}}^{4/3}(x)\,dx + O\left(Z^{\frac{5}{3}-a}\right) \tag{5a}$$

$$\iint \frac{(\rho_{\mathrm{TF}}(x) - \rho(x))\,(\rho_{\mathrm{TF}}(y) - \rho(y))}{|x-y|}\,dx\,dy = O\left(Z^{\frac{5}{3}-a}\right) \tag{5b}$$

Inequality (5a) is a consequence of rather rough asymptotics, and can be obtained by a modification of a result in (16). We remark that in the case of a smooth potential, the formula for the exchange (5a) was previously known (2). Equation (5b) is much more subtle, and requires control of the periodic orbits at zero energy for the classical Hamiltonian.

Combining this lemma with (3), (4) and (5), we note that Theorem 1 is reduced to obtaining the asymptotic formula for the sum of the eigenvalues E_k that implies the expression for $E_{\mathrm{atom}}(Z)$. Note that our result also implies that the Hartree–Fock energy is accurate through the $Z^{5/3}$ term.

The proof of the equivalent of the corollary in the previous section, and of Lemma 4 for molecules remains open. However, there is a recent independent proof of the accuracy of the Hartree–Fock model through the $Z^{5/3}$ term due to V. Bach (1). His proof also holds for molecules, is very concise and, in the atomic case, leads to better error terms than ours. He also obtains some of the correlation estimates of the previous section.

4 The One–Electron Problem.

The goal of this Section is to understand the spectrum of a certain class of operators in $L^2(\mathbf{R}^3)$ given by

$$H = -\Delta - W$$

Again, we omit a detailed description of the conditions on W, and we simply say that W is a large, slowly varying potential, which is allowed Coulomb singularities, and that our screened Thomas–Fermi potential belongs to this class.

In the present exposition we will take care of the eigenvalue sum only. At exactly this point, the proofs for atoms and molecules start to be very different: for molecules, one does not have radial symmetry. For atoms, one has radial symmetry, but one wants to go beyond Z^2 and obtain the $Z^{5/3}$ term. A common feature, however, remains to be the presence of singularities in the potential.

4.1 Quasiclassical Asymptotics: General Case.

We consider in this section the negative eigenvalue sum for a Schrödinger operator with symbol

$$h(x,\xi) = |\xi|^2 - W_{\mathrm{TF}}^Z(x)$$

It follows by the perfect scaling of the Thomas–Fermi potential, that this problem can be viewed as a quasiclassical limit for the Operator

$$P = -\hbar^2\Delta - \phi(x) \qquad \phi(x) = W_{\mathrm{TF}}^1(x)$$

for $\hbar = Z^{-1/3}$. We denote the symbol of P by $p(x,\xi) = |\xi|^2 - \phi(x)$.

If ϕ were smooth, then one can use standard results based on the pseudo–differential calculus to obtain sharp spectral asymptotic for this problem. This smooth case is related to the work of Duistermaat, Guillemin, Helffer, Hörmander, Ivrii, Melrose, Petkov, Robert, among many others. Our problem is that ϕ is has singularities.

Theorem 5 *Let $Z \to \infty$ along a given direction and let the mutual distances between the y_j's $\geq$ const$|Z|^{-\frac{2}{3}+\epsilon}$. Then*

$$\sum_{i=1}^{N} E_i \;=\; \mathrm{Weyl} + \mathrm{Scott} + o(|Z|^2)\;,$$

where

$$\mathrm{Weyl} \;=\; \iint_{p\leq 0} h(x,\xi)\,dx\,d\xi$$

and

$$\mathrm{Scott} \;=\; \frac{1}{8}\sum Z_j^2\;.$$

Proof. There are two ingredients in our proof.

Microlocal Spectral Analysis. We estimate global quantities through local ones. For instance, we study

$$\mathrm{tr}\big(\psi(x)g(P)\big) \tag{6}$$

where $g(\lambda) = \lambda$ for $\lambda \leq 0$ and 0 for $\lambda \geq 0$. If $\psi \equiv 1$, then the trace above is just the sum of negative eigenvalues of P. We take for ψ smooth functions localized outside of the singularities of the potentials. Then it is not difficult to obtain asymptotic expansion in the quasiclassical parameter $\hbar$ of the trace (6).

The traditional approach to obtaining spectral information, is first, to study a particular P.D.E. $f(P,t)$, and then use a tauberian theorem to pass information to the eigenvalues, Different P.D.E.'s and the corresponding tauberian techniques that one can use are summarized in the following table:

P.D.E.	Tauberian Technique
Heat	Laplace Transform
Wave Eqn.	sin transform
Schrödinger Eqn.	Fourier Transform
Stationary Schrödinger Eqn.	Hilbert Transform
Powers H^z	Mellin Transform

The tauberian analysis is generally easy and sharp, while the P.D.E. analysis is generally hard and subject to improvement. This justifies the choice in our approach of the best tauberian technique (Fourier Transform) and the corresponding –hardest– P.D.E., the Schrödinger Equation. In general, this equation is very hard to study for second order operators, which is our case, due to the infinite speed of the propagation of singularities. However, microlocally, we have

$$|\xi| \sim \sqrt{E + \phi(x)}$$

and since we are interested only in the negative eigenvalues, we obtain the bound –always microlocally–

$$|\xi| \leq \sup |\phi_+(x)|^{1/2}$$

where the sup ranges over the support of ψ. This gives us a bound for speed of propagation of singularities. And as a result we are left with a domain in phase space given by

$$\{(x,\xi) \mid |\xi| \leq |\phi_+(x)|^{1/2} \quad x \in \operatorname{supp} \psi\}$$

outside of which we get a negligible contribution to the negative part of the spectrum. We denote by $\varphi(x,\xi)$ a smooth cut–off function adapted to a small neighborhood of such domain.

Pseudodifferential calculus provides convenient tools for carrying out our program. Adapting a standard technique, one represents $g(P)$ as

$$g(P) = \frac{1}{2\pi\hbar} \int \hat{g}(t) e^{-iPt/\hbar}\, dt$$

where $\hat{g}(t)$ is the $\hbar$–Fourier transform of g,

$$\hat{g}(t) = \int e^{-it\lambda/\hbar}\, g(\lambda)\, d\lambda \tag{7}$$

The evolution operator $e^{-iPt/\hbar}$ is then approximated locally to any power in $\hbar$ by Fourier integral operators in the spirit of the geometrical optics

$$e^{-iPt/\hbar} \approx F_t$$

on the appropiate part of phase space, for

$$F_t u = \hbar^{-d} \iint e^{i\left(S(t,x,\xi)-z\cdot\xi\right)/\hbar} a(t,x,\xi,\hbar)\, u(z)\; dz\, d\xi \tag{8}$$

where S satisfies the Hamilton–Jacobi equation

$$\left.\begin{aligned} \partial_t S + p(x, \nabla_x S) = 0 \\ S(0,x,\xi) = x\cdot\xi \end{aligned}\right\}$$

and

$$a = \sum_{j=0}^{N} \hbar^j a_j(t,x,\xi)$$

where the a_j satisfy the transport equations

$$\partial_t a_j + \nabla_x S\cdot\nabla_x a_j + \tfrac{1}{2}\Delta_x S\, a_j = -\tfrac{i}{2}\Delta_x a_{j-1} \qquad a_{-1} \equiv 0$$

$$a_j\big|_{t=0} = \delta_{j,0}\varphi(x,\xi)$$

The initial conditions on the a_j imply that $F_0 = \varphi(x, -i\partial_x)$ and since $\varphi \equiv 1$ on the support of ψ, we have that $\psi F_t \approx F_t$ for small enough time. For sufficiently small time, the classical trajectories starting in suppφ do not reach the singularities of $\phi(x)$. Thus, the singularities of $\phi(x)$ are decoupled, and all the objects we are dealing with are smooth.

Expressions like the one in (8) allow us to analyze the trace in (6) by the method of stationary phase. Note that after the localization introduced with φ, all integrals are absolutely convergent, so that the analysis involved is rather easy. Since in this part of the analysis we stay away from the singularities, this analysis is quasiclassical in nature and in our asymptotics we will obtain only the contribution to the Weyl term, with a good error bound. The problem here is that the remainder estimates here depend on supp ψ and on estimates of $\partial^\nu \phi$ on supp ψ. To remedy this we introduce the second ingredient of our analysis.

Multiscale Analysis. There are three scales in the problem: momentum scale determined by the quasiclassical parameter $\hbar = O(|Z|^{-\frac{1}{3}})$, space scale, $\ell(x)$, determined by the behavior of the potential under differentiation, and the energy scale, $f(x)^2$, determined by the size of the potential. The first scale is constant while the other two depend on x. Specifically, assume there are positive functions $f(x)$ and $\ell(x)$ such that

1. $|\nabla\ell(x)| \le M$.
2. $cf(y) \le f(x) \le Cf(y)$ for $x \in B\big(y, \ell(y)\big)$.
3. $|\partial_x^\alpha \phi(x)| \le C_\alpha f(x)^2 \ell(x)^{-|\alpha|}$.

In dealing with the Coulomb singularities of $\phi(x)$ we set

$$\ell(x) \;=\; \text{dist of } x \text{ to the singularities}$$

and $f(x) = \ell(x)^{-1/2}$. At each point outside of the singularities we rescale the problem using the scales at this point in such a way that the problem is mapped into a model one, i.e. the one with a potential U obeying $|\partial^\alpha U(x)| \le C_\alpha$ on a unit ball, with the effective quasiclassical parameter

$$\hbar_{\text{eff}}(x) \;=\; \frac{\hbar}{\ell(x)f(x)}$$

which depends on all the scales.

What is important here is that quantities of interest (eigenvalues, traces, density) commute with rescalings and that the new –rescaled– problem admits a quasiclassical expansion discussed above with a remainder bound independent of the singular structure of $\phi(x)$. This implies an expansion for the original problem outside of small balls around the singularities. The term "singularity" now has to be understood in a broad sense, since the singularities in the flow caused by stationary points also affect our error bounds. We are thus confronted with two kinds of singularities: those coming from the points where $\nabla_x\phi$ vanishes, and the Coulomb singularity. The choice of $\ell(x)$ and $f(x)$ is different in these cases.

Away from the Coulomb singularities, we have the following result:

Theorem 6 *Consider the Schrödinger operator P in $\mathbf{R}^d$, $d \ge 2$, and for $0 \le s \le 1$ let*

$$g_s(\lambda) = \begin{cases} (-\lambda)^s & \text{if } \lambda < 0 \\ 0 & \text{otherwise} \end{cases}$$

Then

$$\left|\operatorname{tr}(\psi g_s(P)) - (2\pi\hbar)^{-d}\iint \psi(x)\, g_s(p(x,\xi))\, dx\, d\xi\right| \le C\hbar^{-d+1+s}\int_\Omega f(x)^{d+s-1}\ell(x)^{-s-1}\, dx$$

for any ψ smooth supported in $\{x \mid \ell(x)f(x) \ge \hbar\}$ and such that

$$\partial_x^\alpha\psi(x) = O\left(\ell(x)^{-|\alpha|}\right)$$

and for any Ω which is a union of balls $B(y,\ell(y))$ covering $\operatorname{supp}\psi$.

Using the scales we can see that the critical points of the Hamiltonian have essentially no effect on the eigenvalue asymptotics when $d \ge 2$.

Inside those balls around the Coulomb singularities, we analyze the problem differently. Namely, we replace the potential by its leading term near the singularity, which is pure Coulomb, and solve the quasiclassical problem for this truncated operator more accurately. Since the Schrödinger for the Coulomb potential admits an exact solution, we can analyze the crucial part of this problem explicitly. Leading terms near and outside the singularities combine into a single Weyl term over $\mathbf{R}^3$ which gives the Thomas-Fermi energy. Precise expansion of low-lying eigenvalues of the truncated problem near the singularities yields the Scott correction.

4.2 Quasiclassical Asymptotics: Radial Case.

As before, consider the Schrödinger Operator given by

$$H = -\Delta - W(x)$$

with negative eigenvalues E_k. Our next task is to obtain the next term for the sum of the negative eigenvalues. It is known that this problem is related to the presence of periodic orbits for the hamiltonian flow, in contrast with the previous asymptotic analysis where no dynamical properties of H where required. In addition to this, as before, we also have to deal with the singularities of the potential.

The main theorem we try to proof now is the following:

Theorem 7 *Let W be radial. Denote by E_k the negative eigenvalues of H. Then,*

$$\sum_k E_k = -\tfrac{1}{15\pi^2}\int (W(x))_+^{5/2}\,dx + \tfrac{1}{8}Z^2 + \tfrac{1}{48\pi^2}\int (W(x))_+^{1/2}\,\Delta W(x)\,dx$$
$$+\text{lower order terms}$$

assuming that zero-energy classical periodic orbits of the Hamiltonian given by $p^2 - W(x)$ have measure 0.

A crucial property of H in what follows is that it is radially symmetric; thus, separation of variables into spherical harmonics yields that, if we denote by $E_{k,l}$ the k'th eigenvalue of

$$H^{\mathrm{ODE}} = -\frac{d^2}{dr^2} - W(r) + \frac{l(l+1)}{r^2}$$

then the eigenvalues of H come in a strightforward way from the $E_{k,l}$, each taken with multiplicity $2l+1$. Here, l is a non-negative integer.

The analysis of H is thus reduced to the analysis of ODE's

$$\frac{d^2}{dr^2} + p_E(r,l) = 0 \tag{9}$$

$$p_E(r,l) = W(r) - \frac{l(l+1)}{r^2} + E \qquad E<0 \quad l=0,1,\ldots$$

in the sense that if we can find a solution of (9) which vanishes at 0, at infinity, and at $k-1$ other points, then $E = E_{k,l}$. The main ingredient in the proof of Theorem 7 is the following eigenvalue formula.

Theorem 8 *If $l \geq Z^{a_4}$, a_4 small,*

$$\phi(E_{k,l}) + \tfrac{1}{48}\psi(E_{k,l}) = \pi(k+\tfrac{1}{2}) + O\left(\|p\|^{-1+\epsilon_2}\right) \tag{10}$$

where

$$\phi(E) = \int (p_E(x;l))_+^{1/2}\, dx$$

$$\psi(E) = \lim_{\delta\to 0}\left(\int (p_E(x;l))_+^{-3/2} p''(x;l)\, dx - G(E,l)\delta^{-1/2}\right)$$

where $G(E,l)$ is uniquely determined by requiring the finiteness of the limit.

When $l \leq Z^{a_4}$, then

$$\phi(E_{k,l}) = \pi(k+\tfrac{1}{2}) + O(1)$$

The proof of this theorem is the main content of (8). We also refer the reader to (8) for the exact meaning of $\|p\|$ above, and for the exact formulation of the theorem. The strategy of the proof is as follows:

Note first the following change of variables formula for ODE's. Consider two ODE's

$$F''(x) + p(x)F(x) = 0 \tag{11a}$$

$$A''(y) + V(y)A(y) = 0 \tag{11b}$$

Then, the change of variables $y = y(x)$ given by

$$(y')^2\, V(y) + \{y,x\} = p(x) \qquad \{y,x\} = -\tfrac{3}{4}\left(\frac{y''(x)}{y'(x)}\right)^2 + \tfrac{1}{2}\frac{y'''(x)}{y'(x)} \tag{12}$$

makes (11a) and (2) equivalent via the transformation law

$$F(x) = (y'(x))^{-1/2} A(y(x))$$

What we will do then is, first, take V a simple potential for which (2) admits an explicit solution; then, a virtue of (12) is that we can understand the change of variables as an explicit formal power series expansion in $\|p\|^{-1}$; then, we combine both $A(y)$ and $y(x)$ to understand the solution of (11a). The choices of V will be as follows:

a. Near a zero of p, $V(x) = \lambda^2 x$ for λ large. Then, (2) becomes the well known Airey equation.

b. When $p > 0$, we take $V = \lambda^2$. When $V < 0$, we take $V = -\lambda^2$, λ large.

The previous two choices apply for generic values of l and E. For special values, we need the two additional choices:

c. When l is small, say 1, our two previous choices will not work for x near the Coulomb singularity. The correct choice will be, of course, $V(x) = \frac{\lambda^2}{x}$; then (2) becomes the well known Hydrogen equation. This is very important, since it gives rise to the non–semiclassical term Z^2 in the energy asymptotics in Theorems 1 and 7. Smooth potentials don't have it.

d. When E is very near $\min p_0$, then p_E and its first derivative are almost zero when p_E is negative; as a consequence, the previous choices will not work nicely, and we compare in this case with $V(x) = \lambda^2 x^2$, the equation of the Harmonic Oscillator.

When performing the WKB approximation, one needs to estimate error terms carefully, and understand their dependence on E and l, so that we can see that in all cases discussed above, they are small. An effective way of doing this is, as in the previous section, to introduce functions $f(x)^2$ and $\ell(x)$, which, following the notation in (8) we denote by $S(x)$ and $B(x)$, and assume the estimates

$$\left|\left(\frac{d}{dx}\right)^n p(x)\right| \leq C_n S(x) B(x)^{-n}$$

on a suitable interval I where we intend to perform the WKB approximation.

For instance, if $p(x) = \lambda^2 p_0(x)$, then we can take

$$S(x) = \lambda^2 \qquad B(x) = 1$$

If $p(x) = \frac{1}{x} - \frac{l(l+1)}{x^2}$, then we can take

$$S(x) = \frac{1}{x} \qquad B(x) = x \qquad \text{on } \left(\frac{l(l+1)}{10}, \infty\right)$$

Then our error bounds for the eigenvalue equation will essentially be given in terms of the effective semiclassical parameter

$$\hbar = \int_I \frac{dx}{S^{1/2}(x)\, B^2(x)}$$

In the case that $p(x) = \lambda^2 p_0(x)$, note that $\hbar \sim \lambda^{-1}$, and this makes the WKB approximation optimal. However, if $p(x) = \frac{1}{x} - \frac{l(l+1)}{x^2}$, then $\hbar \sim l^{-1}$.

This expresses the fact that for large l, one expects the first eigenvalue formula in the statement of Theorem 8 to be accurate, whereas one expects to obtain only the second crude eigenvalue formula for l a small integer (the case $l = 0$ is a hybrid). This is no surprise, since the exact eigenvalues for the Hydrogen Hamiltonian satisfy the relation

$$\phi(E_k, l) = \pi(k + l - \sqrt{l(l+1)})$$

in clear violation of formula (10) for l small.

Note that this method gives control over the eigenfunctions, too.

The next step is to lift Theorem 8 in order to say something about *radial* potentials in $\mathbf{R}^3$. We do this by summing Theorem 8 in k and l, recalling the multiplicity of each eigenvalue, obtaining

$$\sum_k E_k = -\tfrac{1}{15\pi^2} \int W(x)_+^{5/2}\, dx + \tfrac{1}{8} Z^2 + \tfrac{1}{48\pi^2} \int W(x)_+^{1/2} \Delta W(x)\, dx$$
$$+ G(W) + \text{junk}$$

where junk is negligible, and $G(W)$ is a term that comes from a careful replacement of sums by integrals: it has the form

$$\sum_{l=1}^{l_{\max}} \frac{2l+1}{n_l} \beta(\phi_l)$$

for $\beta(x) = \{x\}^2 - \frac{1}{12}$, $\{\cdot\}$ meaning the distance to the closest integer, and

$$\phi_l = \frac{1}{\pi} \int \left(W(r) - \frac{l(l+1)}{r^2} \right)_+^{1/2} dr$$

$$n_l = \frac{1}{\pi} \int \left(W(r) - \frac{l(l+1)}{r^2} \right)_+^{-1/2} dr$$

However, G is far from being just a technical term: it captures the main contribution of zero–energy classical closed orbits to semiclassical asymptotics.

Define

$$F(t) = \int \left(W(r) - \frac{t^2}{r^2} \right)_+^{1/2} dr$$

It is rather easy to see ([Ar]) that a potential W such that $F(t)$ is a linear function of t, with F' a rational multiple of π, then all classical orbits at zero energy are periodic, our term $G(W)$ is oscillating, and has the same size as the third term in the formula in Theorem 7, which would then be wrong in this case; for the atom, this term G would then be of size $O\left(Z^{5/3}\right)$, and therefore Theorem 1 would also be wrong if the screened Thomas–Fermi potential were of this type. Potentials of this type are, for example, those of the Hydrogen atom and the Harmonic Oscillator.

The opposite case is a potential W for which $|F''(t)| \geq c > 0$. In this case, classical periodic at zero energy orbits have measure 0, and $G(W)$ can be seen to be of size $O\left(Z^{\frac{5}{3}-b}\right)$, $b > 0$. An intermediate case of potentials are those for which F'' vanishes to finite order. In this case, closed orbits also have measure 0, and the size of $G(W)$ continues to be $O\left(Z^{5/3-b}\right)$, but the value of b depends on the order of the zeros of F''.

It is therefore very important to determine to which of these classes of potentials the screened Thomas–Fermi potentials belongs to. It is rather easy to see that F'' vanishes only to finite order; therefore, Theorem 1 holds, possibly with a bad error term if F'' vanishes very much. However, more is true, and it can be seen that $|F''| \geq c > 0$; this is a very delicate fact since arbitrary small perturbations of V_{TF} in a natural topology allow F'' to vanish as many times as we want. The analysis of F'' is taken care of in (13).

We end by pointing out the similarity between problems in Analytic Number Theory like the Lattice Point Problem, and the effect of the zeros of F'' on the semiclassics for $E(Z)$. In retrospect, this is not so surprising, since the Lattice Point Problem also comes from the semiclassics of a Laplacian on the Thorus.

References

1. Bach, V. "*Error Bound for the Hartree–Fock Energy of Atoms and Molecules*" Comm. Math. Phys. *To appear.*

2. Conlon, J. "*Semiclassical Limit Theorems for Hartree–Fock Theory*" Comm. Math. Phys. 88 no 1 (1983)

3. Dirac, P. (1930) "*Note on Exchange Fenomena in the Thomas–Fermi Atom*" Proc. Cambridge Philos. Soc. **26**, 376 – 385.

4. Fefferman, C., Seco, L. (1989) "*An Upper Bound for the Number of Electrons in a Large Ion*" Proceedings of the Nat. Acad. Sci., USA Vol. **86**, no. 10, May 1989, pp 3464–3465.

5. Fefferman, C., Seco, L. (1990) "*Asymptotic Neutrality of Large Ions* " Comm. Math. Phys. Vol. 128, pp 109 – 130.

6. Fefferman C., Seco, L. "*The Ground State of a Large Atom*" Bull. AMS, (1990).

7. Fefferman C., Seco, L. "*On the Dirac and Schwinger Corrections to the Ground–State Energy of an Atom*" Advances in Math (To appear.)

8. Fefferman C., Seco L. "*Eigenvalues and Eigenfunctions of Ordinary Differential Operators*" Advances in Math. (To appear.)

9. Fefferman C., Seco L. "*The Density in a One–Dimensional Potential*" Advances in Math. (To appear.)

10. Fefferman C., Seco L. "*The Eigenvalue Sum for a One–Dimensional Potential*" Advances in Math. (To appear.)

11. Fefferman C., Seco L. "*The Density in a Three–Dimensional Radial Potential*" Advances in Math. (To appear.)

12. Fefferman C., Seco L. "*The Eigenvalue Sum in a Three–Dimensional Potential*" Advances in Math. (To appear.)

13. Fefferman C., Seco L. "*Aperiodicity of the Hamiltonian Flow in the Thomas–Fermi Potential*" Advances in Math. (To appear.)

14. Fermi, E. (1927) "*Un Metodo Statistico per la Determinazione di alcune Priorieta dell'Atome*" Rend. Accad. Naz. Lincei **6**, 602 – 607.

15 . Hughes, W. "*An Atomic Energy Lower Bound that Agrees with Scott's Correction*" Advances in Mathematics 79, 213 – 270 (1990).

16. Ivrii V., Sigal I.M. To appear in Annals of Mathematics.

17. Lieb, E. "*Thomas–Fermi and Related Theories of Atoms and Molecules*" Reviews of Modern Physics Vol 53 no. 4. (1981)

18. Lieb, E. H. (1979) "*A Lower Bound for Coulomb Energies* " Phys. Lett. **70A** 444 – 446.

19. Lieb E., and Simon B. (1977) "*Thomas–Fermi Theory of Atoms, Molecules and Solids*" Adv. Math. 23, pp 22 – 116.

20. Ruskai, M.B. (1982) "*Absence of Discrete Spectrum in Highly Negative Ions* **I & II**" Comm. Math. Phys. 82, 457 – 469 and 85, 325 – 327.

21. Schwinger, J. (1981) "*Thomas–Fermi Model: The Second Correction*" Physical Review A24, **5**, 2353 – 2361.

22. Scott, J. M. C. (1952) "*The Binding Energy of the Thomas–Fermi Atom*" Phil. Mag. **43** 859 – 867.

Quantum Stability

James S. Howland[1]

Abstract: Stability of the motion of a discrete quantum system under the influence of a time-dependent perturbation is discussed. Various results are reviewed, and a few new ones obtained, notably the absence of absolutely continuous spectrum for the pulsed rotor.

When we are young, energetic, and inexperienced, then every time we are asked to speak, we think up a new title and prepare a new talk for the occasion. But with advancing age, more economical measures are called for, and many of us adopt one of two strategies. We always give either the same talk with different titles, or different talks with the same title.

I believe that I have now found my title, because most of the work that I have ever done--and I suspect that this applies to my audience as well--can be spoken of under the topic of "Quantum Stability."

In the actual event, however, it is necessary to narrow oneself to a subject that can be discussed in an hour, and this I shall now do.

The particular problem that I have in mind is the following. Let H_0 be a *discrete* Hamiltonian on $\mathcal{H}$, with eigenvalues $\lambda_0 \le \lambda_1 \le \lambda_2 \le \cdots$. For example, consider one of the following simple quantum mechanical systems:

(a) The Harmonic Oscillator,

$$H_0 = \frac{1}{2} p^2 + \frac{\omega^2}{2} x^2;$$

(b) The Anharmonic Oscillator,

$$\frac{1}{2} p^2 + \frac{\omega^2}{2} x^2 + gx^4;$$

(c) The Rotor,

---

[1]Supported by NSF Contract DMS-9002357.

23 Seco, L., Sigal, I. M., Solovej, J. P., "*Bound on the Ionization Energy of Large Atoms* " To appear in *Comm. Math. Phys.*

24. Siedentop, H., Weikard, R. (1987) "*On the Leading Energy Correction for the Statistical Model of the Atom: Interacting Case*" Communications in Mathematical Physics **112** 471-490

25. Siedentop, H., Weikard, R. (1989) "*On the Leading Correction of the Thomas–Fermi Model: Lower Bound*" Inv. Math., Vol., 97, pp 159 – 193.

26. Siedentop, H., Weikard, R. (1990) "*A Lower Bound of Scott Type by a New Microlocalization Technique* " To appear.

27. Sigal, I. M. (1982) "*Geometric Methods in the Quantum Many–Body Problem. Nonexistence of Very Negative Ions* " Comm. Math. Phys. 85, 309–324.

28. Solovej, J.P. (1991) "*Proof of the Ionization Conjecture in a Reduced Hartree–Fock Model*" Inv. Math. 104, 291–311.

29. Thomas, L. H. (1927) "*The Calculation of Atomic Fields*" Proc. Cambridge Philos. Soc. **23** 542 – 548.

$$H_0 = -\alpha \frac{d^2}{d\theta^2} = \alpha p^2$$

on $L_2(\mathcal{T}^1)$, where $\mathcal{T}^1$ is the circle, and $\alpha > 0$;

(d) <u>The Particle in a Box</u>,

$$H_0 = - \frac{d^2}{dx^2}$$

on $L_2[0,a]$ with Dirichlet boundary conditions;

(e) <u>The Particle in a Box in dimension $d > 1$</u>.

Suppose that the system H_0 is driven by a time-dependent perturbing energy $V(t)$, for simplicity assumed to be a *bounded* operator on $\mathcal{H}$.

<u>Question</u>: *Is the motion under*

$$H(t) = H_0+V(t)$$

a 'stable' motion?

The answer clearly depends on the nature of $V(t)$. For example, if the perturbation is of *finite duration*, then its only effect is to induce transitions between the states of H_0, which would not be described as instability. So we want a perturbation which persists substantially throughout time.

The simplest case is when $V(t)$ is *constant in time*:

$$V(t) \equiv V.$$

The motion is then governed by

$$H = H_0+V$$

and the question of stability is just whether H is another discrete operator--a question which has been studied extensively, to say the least.

The simplest case where something *new* happens is when $V(t)$ is *periodic in time*:

$$V(t+T) = V(t).$$

As is well known [1,2], the proper object to consider is *quasienergy*

$$K = D + H_0 + V(t)$$

on $L_2[0,T]\otimes\mathcal{H}$, where

$$D = i\,\frac{d}{dt}$$

with periodic boundary condition $u(0) = u(T)$. One then has the unitary equivalence

$$e^{iKT} \cong I\otimes U(T,0)$$

where $U(t,s)$ is the propagator of $H_0+V(t)$. A point eigenvalue λ of K therefore signifies the existence of a vector ϕ in $\mathcal{H}$ such that

$$U(T,0)\phi = e^{i\lambda T}\phi.$$

This means that the state represented by ϕ returns to itself after motion through a period T of the perturbation.

One version of the stability question is therefore to ask:

Does K have pure point spectrum?

This would, for example, signify an almost periodic motion of the state vector.

We say only 'pure point' and not 'discrete' for the following reason. If $V(t)$ is identically zero, the operator

$$K_0 = D+H_0$$

has pure point spectrum with eigenvalues

$$\lambda_{nk} = \omega n + \lambda_k$$

$(k = 0,1,\cdots;\ n = 0,\pm 1,\cdots)$, where $\omega = 2\pi T^{-1}$. Unless there is some rational relationship between ω and the λ_k's, this is a *dense set*. In fact, even if this should happen for some ω, it will fail if ω is changed slightly. We are therefore faced with a simple example of *perturbation of a dense point spectrum*.

One can, of course, consider more general time behavior of $V(t)$; for example, (a) almost periodic or quasi-periodic, or (b) random ergodic. From the point of view of Kotani [4], periodicity and almost periodicity are, in fact, special cases of ergodicity. However, Combescure [21] has considered a perturbation regulated by the Thue-Morse sequence. Most of the work to date, though, has been done on the periodic case, to which we shall now turn.

§2. Explicitly Solvable Models

A few systems of this sort are explicitly solvable, almost all relying on the fact that for Hamiltonians which are quadratic in q and p the quantum evolution is determined by the classical. The most detailed treatment is [6], with applications to special time-periodic problems in [7,8,9,11,12]. Combescure, in [10], also uses the special properties of the $1/x^2$ potential in one dimension. Bunimovich, et al., use this method to study *randomly* driven oscillators [13], while Duclos and Karner [14] consider some problems on intervals with periodicity moving endpoints.

I will quote just two characteristic results as illustrations of what may happen. The *forced harmonic oscillator*

$$\frac{1}{2}p^2 + \frac{\omega^2}{2}x^2 + Fx \sin \omega_0 t$$

is pure point unless $\omega = \omega_0$, in which case it is absolutely continuous.

The harmonic oscillator with periodically varying period

$$\frac{1}{2}p^2 + \frac{\omega^2(t)}{2}x^2$$

is either absolutely continuous or pure point according as the solutions of Hill's equation

$$u'' + \omega^2(t)u = 0$$

are bounded or not.

There is also the Maryland model of Fishman, Graempel and Prange--a lattice Schrödinger model related to our problem. See [19] for references and details.

§3. Kicked Systems

The case

$$H(t) = H_0 + V\Delta(t),$$

where V is bounded on $\mathcal{H}$, and

$$\Delta(t) = \sum_{n=-\infty}^{+\infty} \delta(t-nT)$$

describes a system H_0 subject to a periodic kick by V. The system is solvable in a sense, in that

$$U(T,0) = e^{iH_0T} e^{iV}.$$

One is thus led to the *spectral theory of a product of unitary operators*, which is by no means a simple question.

Several such systems have been studied. Bellissard [19] gives a relatively complete analysis of

$$H_0 = \alpha p = -\alpha \ i \frac{d}{d\Theta}$$

on $L_2(T^1)$ and period $T = 2\pi$, where V is multiplication by a holomorphic function $v(\Theta)$. Very roughly, he finds that

(1) for α rational, U is absolutely continuous for most $v(\Theta)$;

(2) for a.e. α, U is purely singular; and

(3) for 'most' irrational α's, U is simple pure point, although there are cases in which U is purely singular continuous.

Combescure [20] considers a general pure point Hamiltonian H_0, kicked by a perturbation

$$V = \lambda \ \langle \cdot, \phi \rangle \phi$$

of rank one. This is essentially a rank one perturbation problem for a unitary operator, which is treated by methods like those of Simon-Wolff [32] discussed in Section 7 below.

The case

$$H_0 = \alpha p^2 = -\alpha \frac{d^2}{d\Theta^2}$$

--the *kicked rotor*--has been the subject of a great many studies, mostly numerical (see [19] for references). It is thought to behave more or less as the αp case, but rigorous analytical results are not plentiful. It is known that U is absolutely continuous for α rational ($T = 2\pi$) [15,16], and that for certain α's well approximated by rationals, there must be *some* continuous spectrum [17]. To give a complete rigorous analysis of the kicked rotor is a major problem of the theory.

Systems have also been kicked randomly by Guarneri [18], and according to the Thue-Morse sequence by Combescure [21].

§4. Operator KAM Theory

Let us now consider the *pulsed rotor*:

$$-\alpha \frac{d^2}{d\theta^2} + \beta\, V(t,\theta)$$

on the circle. One method for dealing with this system, which indicates the problems involved, is an operator-theoretic version of KAM theory, due originally to Dinaberg-Sinai [22], and subsequently used by W. Craig [23], among others.

A *very crude* sketch of the method is the following. Let $H_0(\alpha)$ be a pure point Hamiltonian depending on a parameter α, V a perturbation, and consider

$$H(\alpha,\beta) = H_0(\alpha)+\beta V.$$

For the pulsed rotor, $H_0(\alpha)$ will be the quasi-energy

$$D + \alpha p^2$$

and V multiplication by a *holomorphic* function $v(t,\theta)$.

Suppressing α for a moment, transform $H(\beta)$ by the unitary

$$U_1(\beta) = e^{i\beta A_1}$$

to obtain

$$U_1(\beta)(H_0+\beta V)U_1^*(\beta) = H_0+\beta V+i\beta[A_1,H_0]+O(\beta^2).$$

If we choose A_1 to be the solution of

$$i[H_0,A_1] = V,$$

then $U_1(\beta)H(\beta)U(\beta)^*$ is diagonal in the same basis as H_0 to order β^2.

Continuing, one obtains

$$U_N(\beta)\cdots U_1(\beta)H(\beta)U_1(\beta)^*\cdots U_N(\beta)^* = H_N(\beta)+O(\beta^{2N}),$$

where $H_N(\beta)$ is diagonal.

It would be nice to think that the limit

$$U(\beta) = \lim U_N(\beta)\cdots U_1(\beta)$$

exists, since then $H(\beta)$ would be equivalent to the diagonal matrix

$$\lim H_N(\beta).$$

For that, we remember α again, and note that, for $N = 1$,

$$A_{k\ell}^{(1)} = -i \frac{V_{k\ell}}{\lambda_k(\alpha)-\lambda_\ell(\alpha)} .$$

The problem in estimating $A^{(N)}$ is that if the $\lambda_k(\alpha)$'s are *dense*, as here, the denominator may be very small indeed. This is a typical problem of *resonance* or *'small divisors'*. One tries to show that if α is excluded from a small set of resonant values, then the method converges for small β.

This is done in [19] by Bellisard for the pulsed rotor, and one find that this system is pure point for small β, if α is excluded from a set of small measure.

A similar theory for perturbations of the harmonic oscillator was given by Combescure [24].

One difference that should be noted between classical and quantum KAM is that classical KAM gives the existence for small coupling of a canonical transformation on a large, but porous, subset of phase space. Thus, there are stable solutions for most initial conditions. In the quantum case there exists a unitary transformation--the quantum analogue of a canonical map--for a large, but porous set of *parameters* in the initial Hamiltonian $H_0(\alpha)$. The classical theory provides information for most *initial conditions*, and the quantum for most *Hamiltonians*. So for a good α, the quantum theory provides information for the entire system, and for bad α, no information at all.

This is perhaps less surprising if one reflects that *quantization selects certain orbits*, which will all be stable if they lie in the preferred region of phase space.

§5. Operator Gauge Transformations

Bellissard's KAM result leaves open the following question: *Is there a transition from pure point spectrum for β small to absolutely continuous spectrum for β large?*

The answer is negative, because, regardless of α and β, $K(\alpha,\beta)$ has no absolutely continuous spectrum.

5.1. Theorem. *Let H_0 be positive and discrete, with simple eigenvalues $\lambda_0 < \lambda_1 < \lambda_2 < \cdots$. Let $V(t)$ be bounded and strongly C^∞, and assume that the gap between eigenvalues satisfies*

$$(5.1) \qquad \Delta\lambda_n = \lambda_{n+1}-\lambda_n \geq c\, n^{\gamma}$$

for some $\gamma > 0$. Then

$$(5.2) \qquad K = D + H_0 + V(t)$$

has no absolutely continuous spectrum.

The method of proof may be called the *method of operator gauge transformations*.

Suppose first that $V(t)$ is of trace class for each t. By a result of E.G.P.G. Schmidt [2,26], it follows from scattering theory that the absolutely continuous parts of K and $D+H_0$ are unitarily equivalent. But $D+H_0$ is pure point! So, in fact, K has no absolutely continuous part. This method of *Vacuous Scattering Theory* was also used by Simon-Spencer [27].

Next, we claim that the general K is unitarily equivalent to an opeator of the form (5.2) with $V(t)$ of trace class. To show this, we proceed as follows. Regard $V(t)$ and H_0 as matrices in the basis in which H_0 is diagonal.

Step 1. Eliminate the diagonal terms of $V(t)$ by incorporating them into H_0. On the diagonal, we have

$$D + \lambda_n + V_{nn}(t)$$

which by a gauge transformation is equivalent to

$$D + \lambda_n + \mu_n,$$

where

$$\mu_n = \frac{1}{2\pi}\int_0^{2\pi} V_{nn}(t)dt.$$

Step 2. Transform K by $e^{iG(t)}$ to obtain

$$e^{iG(t)}Ke^{-iG(t)} = (I+iG(t)+\cdots)(D+H_0+V(t))(I-iG(t)+\cdots)$$

$$= D+H_0+\dot{G}(t)+V(t)+i[G(t),H_0]+\cdots.$$

If G(t) is obtained by solving

$$i[G(t),H_0] = -V(t),$$

then K is equivalent to

$$D+H_0+\dot{G}(t)+\cdots.$$

Now

$$G_{k\ell}(t) = \frac{iV_{k\ell}(t)}{\lambda_\ell-\lambda_k} \qquad (k \neq \ell).$$

The gap condition implies a certain growth in the denominator, and so the elements of $G_{k\ell}(t)$ decay more rapidly than those of $V_{k\ell}(t)$. In fact, if $\beta > \alpha/2$, then

$$k^\beta G_{k\ell}(t)\ell^\beta$$

are the matrix elements of a bounded operator. So V(t) has been replaced by a compact operator $\dot{G}(t)$, at the expense of one time derivative. Iterating this procedure improves compactness of V(t) each time, roughly by $\alpha/2$. We eventually obtain a trace class operator, and the theorem. In fact, for $\gamma = 1$, two iterations suffice.

This does not quite justify the statement that the pulsed rotor has singular spectrum, since--except for the ground state--the rotor eigenvalues have multiplicity *two*. Because of the interest in this problem, we complete the proof of the following theorem in the Appendix.

__5.2. Theorem__. *Let* $v(t,\Theta)$ *be in* $C^2(T^2)$. *Then the pulsed rotor*

$$K = D+\alpha p^2+v(t,\Theta)$$

has no absolutely continuous part.

§6. A Simple System

That the gap condition is necessary in Theorem 5.1 may be seen by considering a simple example, discussed, by a somewhat different method by Bellissard [28].

Consider

$$K(\alpha) = i\frac{d}{dt} + \alpha p + v(t,\Theta)$$

on $L_2(T^2)$, where $v(t,\Theta)$ is periodic in t, and Θ:

$$v(t,\Theta) = \sum_{nm} v_{nm}e^{int}e^{im\Theta}.$$

The eigenvalues of $H_0 = \alpha p$ are

$$\lambda_n = \alpha n, \tag{6.1}$$

so that $\Delta\lambda_n = \alpha$, and the gap exponent $\gamma = 0$. However, one has:

__Theorem 6.1__. *For* α *rational,* $K(\alpha)$ *is absolutely continuous for 'generic'* $v(t,\Theta)$.

In this case, 'generic' that (6.3) below is not constant. This is clearly 'generic' by most standards.

__Proof__. Consider first the case

$$K_1 = i\frac{\partial}{\partial\Theta_1} + f(\Theta_1,\Theta_2) \tag{6.2}$$

on the torus. By gauge transformation, this is unitarily equivalent to

$$K_2 = i\frac{\partial}{\partial\Theta_1} + \overline{f}(\Theta_2)$$

where

$$\overline{f}(\Theta_2) = \int_0^{2\pi} f(\Theta_1,\Theta_2)\frac{d\Theta_1}{2\pi}.$$

But K_2 is uniformly equivalent to the direct sum

$$\sum_n^{\oplus}[n+\bar{f}(\theta_2)]$$

which is absolutely continuous, provided $\bar{f}(\theta_2)$ is not constant on a set of positive measure.

In general, if α is rational, we can clear fractions by multiplication and obtain

$$K_3 = i\, n_1 \frac{\partial}{\partial t} + i\, n_2 \frac{\partial}{\partial \theta} + f(t,\theta).$$

Let

$$t = n_1\theta_1+\theta_2 ,$$

$$\theta = n_1\theta_1 .$$

The map $g\colon (\theta_1,\theta_2) \to (t,\theta)$ maps the torus to itself and is an n_2-fold covering map. The induced map

$$W\, u(\theta) = u(g(\theta))$$

maps $L_2(T^2)$ *into* $L_2(T^2)$ isometrically, and carries K_3 into

$$K_4 = i \frac{\partial}{\partial \theta_1} + f(n_1\theta_1+\theta_2, n_2\theta_1).$$

To be more precise, since W is only an into map, K_3 is equivalent under W to the restriction of K_4 to the range of W, which is the space of all functions invariant under covering transformations of the (θ_1,θ_2) variables.

Thus, if

$$\bar{f}(\theta_2) = \frac{1}{2\pi}\int_0^{2\pi} f(n_1\theta_1+\theta_2, n_2\theta_1)d\theta_1 \tag{6.3}$$

is not constant on a set of positive measure, K_4 and hence also K_3 is absolutely continuous. ∎

The generic behavior of $K(\alpha)$, however, is *dense pure point* for smooth $v(t,\theta)$.

6.2. Theorem. *Assume that*

(6.4) $$\sum_{nm} |v_{nm}|\ |\log|v_{nm}|| < \infty.$$

Then $K(\alpha)$ *is unitarily equivalent to*

$$D + \alpha p + v_{00}$$

for a.e. α.

Proof. Assume $v_{00} = 0$, and let α be irrational. Define

$$g(t,\Theta) = \sum_{nm} \frac{v_{nm}}{n+\alpha m} e^{int} e^{im\Theta}.$$

6.3. Lemma. *Let* $a_n \geq 0$, *and*

(6.5) $$\sum_{n} a_n |\log a_n| < \infty.$$

Then

$$F(t) = \sum_{n} \frac{a_n}{|t-t_n|}$$

is finite for a.e. t, *for any sequence* $\{t_n\}$.

The proof is the same as for [35, Theorem 3.1].

For a.e. α, we therefore have

$$\sum_{nm} \frac{|v_{nm}|}{|n+\alpha m|} = \sum_{nm} \frac{m^{-1}|v_{nm}|}{|\alpha + \frac{n}{m}|} < \infty$$

by Lemma 6.3, since

$$\sum_{nm} m^{-1}|v_{nm}|\ |\log|v_{nm}m^{-1}||$$

$$\leq \sum_{nm} |v_{nm}|\ |\log|v_{nm}|| + |v_{nm}|m^{-1} \log m < \infty.$$

Hence, $g(t,\Theta)$ is absolutely convergent. Moreover, it is easily checked that $g \in C^1(T^2)$ and

$$(D+\alpha p)g(t,\Theta) = v(t,\Theta),$$

since the series for $v(t,\Theta)$ is also absolutely convergent by (6.). By gauge transformation, we therefore have

$$e^{ig(t,\Theta)}K(\alpha)e^{-ig(t,\Theta)} = D+\alpha p. \quad ■$$

Thus, $K(\alpha)$ is *generic pure point*, but *absolutely continuous in case of resonance*. Indeed,

$$\sum_{nm} \frac{|v_{nm}|}{|n+\alpha m|} < \infty$$

may be interpreted as a non-resonant condition on α.

This example becomes even more interesting when it is remarked that the quasi-periodic Hamiltonian

$$H(t) = v(t,\alpha t)$$

on the *one-dimensional* space $\mathcal{H} = \mathbb{C}$, leads, via the formalism of [3], to the Hamiltonian $K(\alpha)$ on L_2 of the hull T^2 of $v(t,\alpha t)$. Thus, again, Theorem 5.1 cannot hold in the quasi-periodic case.

§7. Resonance and Radomization

Resonance is a phenomenon of cooperation between two or more elements. *Randomness* is based on independence--non-cooperation --between elements. It should not prove surprising, therefore, that resonance in a system can be removed by randomizing parameters or functions in the system.

One example is the theory of Schrödinger operators with random potentials, where the randomness may prevent the cooperation necessary for traveling waves, resulting in the phenomenon of *localization*. I merely allude to this theory here, and refer to the monograph [29] of Cormona and LaCroix for random potentials, and to Cycon, et al. [30], for information on the related almost periodic potentials.

The particular line of thought I wish to pursue briefly here goes back to work of Aronszajn on boundary conditions for Sturm-Liouville operators, and of Donoghue [31] on rank one perturbations. More recently, it stems from Kotani's work on *random* boundary conditions, and the important paper of Simon-Wolff [32] on rank one perturbations. These ideas were generalized considerably in [35].

A rough description of the method is as follows. Let $H(\omega)$ be some random Hamiltonian, which one may think of as arising from randomization of some parameters of the system. Suppose that we are able to prove three things:

(1) $H(\omega)$ has no absolutely continuous spectrum a.s.; and

(2) the singular continuous spectrum of $H(\omega)$ is a.s. supported by a set N of Lebesgue measure zero, which is independent of ω.

Let it suffice to say that these can frequently be proved by *stationary scattering theory* [35].

The third condition will look a bit strange. Define the operator

$$\mathbb{H}u(\omega) = H(\omega)u(\omega)$$

on $L_2(\Omega)\otimes\mathcal{H}$, where Ω is the sample space for the process $H(\omega)$. We are therefore making the family $H(\omega)$ of operators into one big operator $\mathbb{H}$. The third condition is

(3) $\mathbb{H}$ is absolutely continuous.

About the only useful way that I have found to prove such a thing is to produce a bounded operator having a positive commutator with $\mathbb{H}$, and invoke the Kato-Putnam theorem. For this approach, see [35], and for a somewhat simpler construction, [38]. Every time I give this talk, I appeal to the audience for new ideas, but so far, I have drawn only blanks.

But I digress. The point is that we then have

7.1. Theorem. ("Kotani's Trick") $H(\omega)$ *is a.s. pure point.*

Proof. Let $u(\omega) = u$ be a constant function. If $\mathbb{E}$ is the spectral measure of $\mathbb{H}$, then

$$0 = |\mathbb{E}[N]u|^2 = \int_\Omega |E_\omega[N]u|^2 P(d\omega)$$

so that

$$E_\omega[N]u = 0 \quad \text{a.s.}$$

The exceptional set of ω's depends on u, but if we take u in a countable dense set, we obtain

$$E_\omega[N] = 0.$$

But since N supports the continuous part of $H(\omega)$ a.s., we have $H(\omega)$ pure point. ■

This idea has been applied to localization in higher dimensions [32], random Jacobi matrices [33], random Schrödinger equations [34,39], operator theory [37,38].

In [25] it was applied to the stability problem in the following way. Let the operator H be written as a diagonal matrix

$$H = \mathrm{diag}\{\lambda_n\}$$

and define

$$A(\omega) = \mathrm{diag}\{a_n X_n(\omega)\},$$

where $a_n > 0$ and $X_n(U)$ are i.i.d. random variables, uniformly distributed on $[-1,1]$. If

$$\sum_n a_n^{1/2} < \infty,$$

the set N exists, and one finds that

$$K(\omega) = D+H+A(\omega)+V(t)$$

is pure point a.s. This says that a slight jiggling of the eigenvalues of H to remove resonances leads to a stable system.

One obtains in this way, a theorem on 'generic' behavior in the sense of *measure* rather than *topology*.

The trouble with this result, of course, is that--in the cases of interest, $K(\omega)$ is no longer a *differential* operator because $A(\omega)$ is *trace class*. One would like to randomize in the class of Schrödinger operators as well. A result of this kind would be of great interest.

§8. Questions

I will conclude with a short list of interesting questions. I have already mentioned two:

(1) *Analysis of the kicked rotor*; and

(2) *Proof that the particle in a box is pure point with the addition of a generic small background potential.*

To these, let me add three more:

(3) Particle in a box in dimension $d > 1$. Leading experts have assured the author that this is "like one dimension." The problem is that the number of states less than E is of order $E^{d/2}$, so that the gap condition fails if $d > 1$, and in fact; for $d \geq 3$, the states *grow closer together* as E increases.

(4) Almost periodic systems. What is the spectral theory of

$$K = i \frac{d}{dt} + H(t)$$

where $H(t)$ is a 2×2 quasi-periodic matrix?

(5) Transitions of spectral type. For what type of systems

$$H + \beta V(t)$$

is there a transition from pure point spectrum for; say, $0 \leq \beta < \beta_c$ to absolutely continuous spectrum for $\beta > \beta_c$? Or, the same thing with respect to some other parameters of the system? I close with a problem which may--or may not--show this phenomenon.

Trapping problem. Consider the discrete Laplacian

$$(\Delta u)_n = u_{n+1}+u_{n-1}$$

on $\ell_2(Z)$. For fixed $a \in Z$, let F_a be the rank two operator

$$F_a = \langle \cdot, e_a \rangle e_{a+1} + \langle \cdot, e_{a+1} \rangle e_a$$

which decouples the left and right of a. Pick a *bilateral* sequence a_n of points, and let

$$V = -\sum_n F_{a_n} .$$

The operator

$$\Delta + V$$

is the direct sum of finite matrices and has dense point spectrum if the gaps between a_n's grow.

Finally, let $V_\beta(t)$ be 2π-periodic with

$$V_\beta(t) = V, \quad 0 \leq t < \beta$$

$$= 0, \quad \beta \leq t < 2\pi .$$

and

$$H_\beta(t) = \Delta + V_\beta(t).$$

Clearly, K_0 is pure absolutely continuous, while $K_{2\pi}$ is pure point.

The is effectively a kicked system, with

$$U_\beta(2\pi,0) = e^{i(1-\beta)\Delta} e^{i(\Delta+V)\beta}$$

but the physical interpretation is clear. The particle is confined to one of the intervals (a_n, a_{n+1}) for $0 \leq t \leq \beta$, and otherwise move freely. Is the particle ever trapped, or does it always leak out to infinity, and if so is the spectrum ever singular continuous? Does the answer depend on the sequence a_n?

Appendix

Let H_0 be discrete, with eigenvalues $0 < \lambda_0 < \lambda_1 < \cdots$ of *simple* multiplicity. We say that H_0 has *gap exponent* $\gamma \geq 0$ iff for some $c > 0$,

$$\Delta\lambda_n = \lambda_{n+1} - \lambda_n \geq cn^\gamma. \tag{A.1}$$

More generally, let H_0 be discrete with eigenvalues $0 < \mu_1 \leq \mu_2 \leq \cdots$, with repeats according to multiplicity.

Let $1 = n_1 < n_2 < \cdots$, be a sequence of integers, and define the *eigenvalue blocks*

$$B_k = \{\mu_{n_k}, \mu_{n_k+1}, \cdots, \mu_{n_{k+1}-1}\}; \tag{A.2}$$

their *centers*

$$\lambda_n = \frac{1}{(n_{k+1}-n_k)} \cdot \sum_{j=n_k}^{n_{k+1}-1} \mu_j \tag{A.3}$$

and *radii*

$$r_n = \max\{\mu_{n_k} - \mu_{n_k-1}; \mu_{n_{k+1}} - \mu_{n_{k+1}-1}\}. \tag{A.4}$$

We say that H_0 has *block gap exponent* $\gamma \geq 0$ if

$$\Delta\lambda_n = \lambda_{n+1} - \lambda_n \geq c\, n^\gamma \tag{A.5}$$

and

$$\mu_j - \lambda_n = o(r_n) \tag{A.6}$$

for all $\mu_j \in B_n$. The *multiplicity* $|B_k|$ of a block B_k is just the number of eigenvalues in it; i.e., $n_{k+1}-n_k$. We shall consider only the case of *bounded multiplicity*, where

$$\sup_k |B_k| < \infty. \tag{A.7}$$

Let P_k be the projection onto the span of the eigenvectors belonging to eigenvalues of B_k , and define

$$\Lambda = \sum_k \frac{1}{k} P_k. \tag{A.8}$$

This is--in a basis in which H_0 is diagonal--a matrix with 1/k's down the diagonal, repeated according to the multiplicity of the block.

If $V \in \mathcal{L}(\mathcal{H})$, we define the *operator*

$$V_{nk} = P_n V P_k. \tag{A.9}$$

The following definition is from [25]. We say that an operator $A \in \mathcal{L}(\delta)$, $\delta \geq 0$, if $0 < \gamma, \gamma'$ and $\gamma+\gamma' < 2\delta$ imply that

$$B = \Lambda^{-\gamma} A \Lambda^{-\gamma'} \in \mathcal{L}(\mathcal{H}). \tag{A.10}$$

For $\delta = 0$, $\mathcal{L}(0)$, $= \mathcal{L}(\mathcal{H})$. Thus, for increasing δ, the operators $\mathcal{L}(\delta)$ are increasingly compact. If $\delta > 2/p$, $p \geq 1$, then

$$\mathcal{L}(\delta) \subset \mathcal{I}_p , \tag{A.11}$$

the trace ideals (provided (A.7) holds).

Let $V(t)$ be bounded and strongly C^{r+1}, $r \geq 0$, and define

$$K = D+H_0+V(t)$$

on $L_2(T^1)\otimes\mathcal{H}$. We write $V(t) \in C^{r+1}(\mathcal{L}(\delta))$ iff $B(t) = \Lambda^{-\gamma} V(t) \Lambda^{\gamma'}$ is strongly C^{r+1} for $\gamma+\gamma' < 2\delta$.

<u>**Theorem A.1**</u>. Let H_0 be non-negative and discrete, and have block gap exponent $\gamma > 0$, for blocks with bounded multiplicity. Let $V(t)$ be in

$C^{r+1}(\mathcal{L}(\delta))$, $\delta \geq 0$. Then K is unitarily equivalent to an operator of the form

$$D + H_1(t) + V_1(t),$$

where H_1 has block exponent γ for the same blocks as H_0, and is *block diagonal* in these blocks; and $V_1(t) \in C^r(\mathcal{L}(\delta+\eta))$ for any η, $0 < \eta < \alpha/2$.

Proof. The proof of this theorem is, *mutatis mutandis*, the same as that of [25, Theorem 2.3], except that we do not reduce the time-dependent diagonal operator to constant form by a gauge transformation. This process would be complicated by eigenvalue crossings; and is in any case unnecessary for our purposes.

Some extra terms appear. Write

$$V(t) = B(t)+V^{(1)}(t)$$

with B(t) block diagonal and $V^{(1)}$ block off-diagonal, and let

$$i[H_0,V(t)] = V^{(1)}(t).$$

Transforming by $e^{iG(t)}$ gives, to first order

$$D+H_0+B(v)+V^{(1)}(t)+i[H_0,G(t)]+i[B(t),G(t)]+\dot{G}(t)$$

$$= D+H_0+B(t)+i[B(t),G(t)]+\dot{G}(t)+\cdots .$$

The new term is [B(t),G(t)]. Now Λ is block diagonal and a multiple of the identity in each block, so it commutes with B(t). Thus

$$\Lambda^{-\gamma}[B(t),G(t)]\Lambda^{-\gamma'} = [B(t),\Lambda^{-\gamma}G(t)\Lambda^{-\gamma'}]$$

is in $\mathcal{L}(\delta+\eta)$. The higher order terms are estimated as in [25]. There is one rather embarrassing caveat: [25, II, eqn. (3.13)] is in error. The right side should contain some obvious additional terms which can be estimated by Proposition 3.3. ∎

Remark. There is no need here to invoke the adiabatic method of [25,I], as was done in the proof of [25, Theorem 2.3; p. 238]. The operator gauge method suffices from the beginning.

The adiabatic method is useful, however, for dealing with perturbations V(t) which are only *relatively bounded*.

Theorem A.2. If $r \geq [\alpha^{-1}]+1$, *then K is purely singular*.

Proof. After $[\alpha^{-1}]+1$ iterations, $V_1(t)$ is trace class for each t, so K and

$$K_0 = D+H_0+B(t)$$

have unitarily equivalent absolutely continuous parts. But on each block K_0 is just

$$D + F(t),$$

with F(t) a continuous, finite matrix, and such an operator is discrete. ■

Corollary. *The pulsed rotor*

$$-\frac{d^2}{d\theta^2} + v(\theta,t)$$

is purely singular if $v(\theta,t)$ *is* C^2.

References

1. Introduction

[1] Yajima, K. Scattering theory for Schrödinger equations with potential periodic in time. J. Math. Soc. Japan 29 (1977), 729-743.

[2] Howland, J. S. Scattering theory for Hamiltonians periodic in time. Indiana J. Math. 28(1979), 471-494.

[3] Jauslin, H. R. and J. L. Lebowitz. Spectral and stability aspects of quantum chaos. Chaos 1(1991) (to appear).

[4] Kotani, S. Support theorems for random Schrödinger operators. Comm. Math. Phys. 97(1985), 443-452.

[5] Thomas, L. E. and C. E. Wayne. On the stability of dense point spectra for self-adjoint operators. J. Math. Phys. 27(1986), 71-75.

2. Explicitly solvable models

[6] Hagedorn, G., M. Loss, and J. Slawny. Non-stochasticity of time-dependent quadratic Hamiltonians and the spectra of canonical transformation. J. Phys. A 19(1986), 521-531.

[7] Combescure, M. Ann. Inst. H. Poincare, Sec. A 44(1986), 293-314.

[8] Combescure, M. Trapping of quantum particles for a class of time-periodic potentials. A semi-classical approach. Ann. Phys. 173(1987), 210-225.

[9] Combescure, M. The quantum stability problem for some class of time-dependent Hamiltonians. Ann. Phys. 185(1988), 86-110.

[10] Combescure, M. Crystalization of trapped ions--a quantum approach. Preprint.

[11] Huang, M. J. On stability for time-periodic perturbations of harmonic oscillators. Ann. Inst. H. Poincare 17(1989), 229-238.

[12] Mielnik, B. and D. J. Fernandez. Is there an instability in standing wave traps? Lett. Math. Phys. 17(1989), 87-94.

[13] Bunimovich, L., H. R. Jauslin, J. L. Lebowitz, A. Pellegrinotti, and P. Nielaba. Diffusive energy growth in classical and quantum mechanics. J. Stat. Phys. 62(1991), 793-817.

[14] Duclos, P. and G. Karner. The Schrödinger equation on oscillating domains--a solvable example. Preprint, 1991.

3. Kicked Systems

[15] Israelev, F. M. and D. Shepelanski. Quantum resonances for a rotator in a nonlinear periodic field. Theo. Mat. Fiz. 43(1990), 553-560.

[16] Chirikov, B. V., F. M. Israelev, and D. L. Shepelanski. Somet Scientific Review C2(1981),

[17] Casati, G. and I. Guarneri. Non-recurrent behavior in quantum dynamics, Comm. Math. Phys. 95(1984), 121-127.

[18] Guarneri, I. Lett. Nuovo Cimento 40(1984), 171.

[19] Bellissard, J. Stability and instability in quantum mechanics in "Trends and Developments in the Eighties." S. Albevario and Ph. Blanchard, eds., World Scientific, Singapore, 1985.

[20] Combescure, M. Spectral properties of a periodically kicked quantum Hamiltonian. Preprint.

[21] Combescure, M. Recurrent versus diffusive dynamics for a kicked quantum system. J. Stat. Phys. 62(1991), 779-791.

4. Operator KAM Theory

[22] Dinaberg, E. I. and Ya. Sinai. On the one-dimensional Schrödinger equation with a quasi-periodic potential. Funct. Anal. Appl. 9(1985), 279.

[23] Craig, W. Dense pure point spectrum for the almost periodic Hill's equation, Comm. Math. Phys. 88(1983), 113-131.

[24] Combescure, M. The quantum stability problem perturbations of the harmonic oscillator. Ann. Inst. H. Poincare 47(1987), 63-83.

5. Operator Gauge Transformations

[25] Howland, J. S. Floquet operators with singular spectrum. I. Ann. Inst. H. Poincare, Sec. A 50(1989), 309-323; II, 325-334.

[26] Schmidt, E. G. P. G. On scattering by time-dependent perturbations, Indiana J. Math. 24(1975), 925-935.

[27] Simon, B. and T. Spencer. Trace class perturbations and the absence of absolutely continuous spectrum. Comm. Math. Phys. 125(1989), 113-125.

6. A Simple Example

[28] Bellissard, J. Stability and instability in quantum mechanics in "Schrödinger Operators," S. Graffi, ed., Lecture Notes in Math., v. 1159. Springer-Verlag, New York, 1985.

7. Randomization

[29] Carmona, R. and J. LaCroix. "Spectral Theory of Random Schrödinger Operators." Birkhauser, Boston, 1990.

[30] Cycon, H. L., R. G. Froese, W. Kirsh, and B. Simon. "Schrödinger Operators with Applications to Quantum Mechanics and Global Geometry." Springer-Verlag, New York, 1987.

[31] Donoghue, W. F. On the perturbation of spectra. Comm. Pure Appl. Math. 18 (1965), 559-579.

[32] Simon, B. and T. Wolff. Singular continuous spectrum under rank one perturbations and localization for random Hamiltonians. Comm. Pure Appl. Math. 39(1986), 75-90.

[33] Simon, B. and S. Kotani. Localization in general one-dimensional systems, I: Jacobi matrices. Comm. Math. Phys. 102(1985), 327-336.

[34] Simon, B. and S. Kotani. Localization in general one-dimensional systems, II. Comm. Math. Phys. 112(1987), 103-120.

[35] Howland, J. S. Perturbation theory of dense point spectrum. J. Funct. Anal. 74(1987), 52-80.

[36] Howland, J. S. Random perturbation theory and quantum chaos in "Differential Equations and Mathematical Physics" (I. W. Knowles and Y. Saito, eds.), Lec. Notes in Math., v. 1285, Springer-Verlag, New York, 1987, 197-204.

[37] Howland, J. S. On a theorem of Carey and Pincus, J. Math. Anal. Appl. 145(1990), 562-565.

[38] Howland, J. S. Random perturbations of singular spectra. Proc. Amer. Math. Soc. 112(1991), 1009-1011.

[39] Howland, J. S. A localization theorem for one-dimensional Schrödinger operators. Preprint, 1988.

[40] Kato, T. Positive commutators i[f(P),g(Q)], J. Func. Anal. 96(1991), 117-129.

On the S-matrix for three body Schrödinger Operators

Hiroshi ISOZAKI

Department of Mathematics, Osaka University, Toyonaka, 560, JAPAN

1 Introduction and summary

In recent years, there has been a considerable progress in scattering theory for many body Schrödinger operators. Namely, the completeness of wave operators was proved for 3-body systems by Enss [4] and for the general N-body systems by Sigal-Soffer [13]. See also an elegant proof of Graf [7].

Although the asymptotic completeness is one of the principal problem in scattering theory, it seems that many problems, which require detailed analysis of solutions of Schrödinger equations, are left open in this field. In this paper, we shall study some properties of the scattering matrix associated with the three body Schrödinger operators.

1.1 Collision processes in the three-body problem

We consider collision processes of quantum mechanical three particles labelled by 1, 2, 3. Suppose in the initial state the two of them form a bound state, denoted by (1,2), and the third particle collides with this pair. Then there occurs one of the following five phenomena:

$$(1,2)+(3) \Longrightarrow \begin{cases} (a) & (1,2)+(3), \\ (b) & (1,2)^* + (3)_*, \\ (c) & (1,2)' + (3), \\ (d) & (1,3)+(2), \\ (e) & (1)+(2)+(3). \end{cases}$$

(a) is an elastic process. In (b), the energy of the pair changes. In (c), the energy of the pair does not change, but this pair takes a different state (which happens when the eigenvalue is degenerate). (d) is a rearrangement process. Finally in (e), all of the three particles move freely after the collision. The first four cases are treated in essentially the same way as in the 2-body problem. In this paper, we study properties of the S-matrix associated with the case (e).

1.2 S-matrix

In $\mathbf{R}^3$ we consider three particles with mass m_i and position x^i. We choose a pair (i,j) and denote it by α. Let

$$\frac{1}{m_\alpha} = \frac{1}{m_i} + \frac{1}{m_j}, \quad \frac{1}{n_\alpha} = \frac{1}{m_k} + \frac{1}{m_i + m_j}$$

be the reduced masses and

$$x^\alpha = \sqrt{2m_\alpha}(x^i - x^j), \quad x_\alpha = \sqrt{2n_\alpha}(x^k - \frac{m_i x^i + m_j x^j}{m_i + m_j})$$

be the relative coordinates. Let

$$X = \{(x^1, x^2, x^3); \sum_{i=1}^{3} m_i x^i = 0\}.$$

Then in $L^2(X)$ the Schrödinger operator is given by

$$H = H_0 + \sum_\alpha V_\alpha(x^\alpha), \qquad H_0 = -\Delta_{x^\alpha} - \Delta_{x_\alpha}. \tag{1.1}$$

If the pair potentials are short-range, the wave operators are known to exist:

$$W_0^\pm = \text{s} - \lim_{t\to\pm\infty} e^{itH} e^{-itH_0}, \tag{1.2}$$

$$W_\alpha^\pm = \text{s} - \lim_{t\to\pm\infty} e^{itH} e^{-itH_\alpha} J_\alpha, \tag{1.3}$$

where

$$H_\alpha = H_0 + V_\alpha, \qquad (J_\alpha f)(x^\alpha, x_\alpha) = u_\alpha(x^\alpha) f(x_\alpha), \tag{1.4}$$

u_α being a normalized eigenfunction of $h^\alpha = -\Delta_{x^\alpha} + V_\alpha(x^\alpha)$ with eigenvalue $E^\alpha < 0$. The scattering operator $S_{0\alpha}$ is defined by

$$S_{0\alpha} = (W_0^+)^* W_\alpha^-. \tag{1.5}$$

To introduce the S-matrix, we use unitary operators $\mathcal{F}_0 : L^2(\mathbf{R}^6) \to L^2((0,\infty); L^2(S^5))$ and $\mathcal{F}_\alpha : L^2(\mathbf{R}^3) \to L^2((E^\alpha,\infty); L^2(S^2))$ defined by

$$\begin{aligned} (\mathcal{F}_0 f)(\lambda, \theta) &= C_0(\lambda) \int_{\mathbf{R}^6} e^{-i\sqrt{\lambda}\theta\cdot x} f(x) dx, \\ C_0(\lambda) &= (2\pi)^{-3} 2^{-1/2} \lambda, \end{aligned} \tag{1.6}$$

$$(\mathcal{F}_\alpha f)(\lambda,\omega) = C_\alpha(\lambda)\int_{\mathbf{R}^3} e^{-i\sqrt{\lambda-E^\alpha}\omega\cdot x} f(x)dx,$$
$$C_\alpha(\lambda) = (2\pi)^{-3/2}2^{-1/2}(\lambda - E^\alpha)^{1/4}. \tag{1.7}$$

Let

$$\hat{S}_{0\alpha} = \mathcal{F}_0 S_{0\alpha}\mathcal{F}_\alpha^*. \tag{1.8}$$

Then as is well-known, $\hat{S}_{0\alpha}$ is decomposable, namely, for any $\lambda > 0$, there exists a bounded operator $\hat{S}_{0\alpha}(\lambda) \in \mathbf{B}(L^2(S^2); L^2(S^5))$ such that

$$(\hat{S}_{0\alpha}f)(\lambda,\theta) = (\hat{S}_{0\alpha}(\lambda)f(\lambda,\cdot))(\theta)$$

for a.e. $\lambda > 0, \theta \in S^5$ and all $f \in L^2((E^\alpha,\infty); L^2(S^2))$. This $\hat{S}_{0\alpha}(\lambda)$ is called the S-matrix. Note that this definition contains a sort of ambiguity. Two families of operators $\{\hat{S}_{0\alpha}(\lambda)_i\}_{\lambda=0}^\infty$ (i = 1, 2) define the same scattering operator $S_{0\alpha}$, if $\hat{S}_{0\alpha}(\lambda)_1 = \hat{S}_{0\alpha}(\lambda)_2$ for a.e. $\lambda > 0$. The study of this family of operators $\{\hat{S}_{0\alpha}(\lambda)\}_{\lambda=0}^\infty$ is not an easy problem. The general result known so far is that of Amrein-Pearson-Sinha [1] and Enss-Simon [5] asserting that $\hat{S}_{0\alpha}(\lambda)$ is a Hilbert-Schmidt operator for *a.e.*$\lambda > 0$, if the pair potentials decay faster than $|x^\alpha|^{-2}$. In this paper, we shall show that there is a representative $\{\hat{S}_{0\alpha}(\lambda)\}_{\lambda=0}^\infty$ continuous in $\lambda > 0$ and investigate its detailed properties.

1.3 Main results (1)

We assume that V_α is a real C^∞-function such that for a constant $\rho > 0$

$$|\partial_y^m V_\alpha(y)| \le C_m(1+|y|)^{-\rho-m}, \quad m = 0,1,2,..., \tag{1.9}$$

where ∂_y^m denotes an arbitrary derivative of m-th order with respect to y, and C_m is a constant. This assumption is stronger than actually needed. One can also allow certain local singularities for V_α. Let

$$X_\beta = \{x \in X; x^\beta = 0\},$$

and define

$$M = S^5 \setminus \cup_\beta X_\beta, \quad N = S^5 \cap (\cup_\beta X_\beta).$$

Theorem 1.1 *(1) Suppose $\rho > 4+1/2$. Then $\hat{S}_{0\alpha}(\lambda)$ has a continuous kernel outside N :*

$$\hat{S}_{0\alpha}(\lambda;\theta,\omega) \in C((0,\infty)\times M \times S^2).$$

(2) Suppose $\rho > 5 + 1/2$. Let β be any pair and decompose $\theta \in S^5$ as $\theta = (\theta^\beta, \theta_\beta)$ in accordance with the choice of the Jacobi-coordinates. Then as $|\theta^\beta| \to 0$,

$$\hat{S}_{0\alpha}(\lambda;\theta,\omega) \simeq |\theta^\beta|^{-1}A_{\beta,-1}(\lambda;\frac{\theta^\beta}{|\theta^\beta|},\theta_\beta,\omega) + A_{\beta,0}(\lambda;\frac{\theta^\beta}{|\theta^\beta|},\theta_\beta,\omega),$$

where

$$
\begin{aligned}
&A_{\beta,-1}(\lambda; \frac{\theta^\beta}{|\theta^\beta|}, \theta_\beta, \omega) \\
&= \sum_j^{finite} C_{\beta 1}^{(j)}(\lambda; \theta_\beta, \omega) \times \int_{\mathbf{R}^3} \frac{\theta^\beta}{|\theta^\beta|} \cdot x^\beta V_\beta(x^\beta) u_\beta^{(j)}(x^\beta) dx^\beta \\
&+ C_{\beta 2}(\lambda; \theta_\beta, \omega) \times \int_{\mathbf{R}^3} V_\beta(x^\beta) \varphi_\beta(x^\beta) dx^\beta,
\end{aligned}
$$

$u_\beta^{(j)}$ being the eigenfunction with zero eigenvalue for h^β, and φ_β the zero-resonance. $A_{\beta,0}$ is continuous with respect to all of its arguments. $A_{\beta,-1} = 0$, if 0 is neither an eigenvalue nor the resonance for h^β. In this case, $\hat{S}_{0\alpha}(\lambda; \theta, \omega)$ is continuous at $\theta^\beta = 0$.

In the course of the proof, we shall see that $\hat{S}_{0\alpha}(\lambda)$ is a $\mathbf{B}(L^2(S^2); L^2(S^5))$ -valued continuous function of $\lambda > 0$ if $\rho > 5 + 1/2$. The zero-resonance φ_β is the solution of the equation $h^\beta \varphi_\beta = 0$ which behaves like $\varphi_\beta \simeq C/|x^\beta|$ as $|x^\beta| \to \infty, C \neq 0$. From our proof given below, one can see that if the pair potentials decay sufficiently rapidly, $\hat{S}_{0\alpha}(\lambda; \theta, \omega)$ is smooth on $M \times S^2$, but the zero eigenvalue and the zero-resonance are known to exist even if the potential is compactly supported. (See e.g. [3]). For the S-matrix from 2 to 3 cluster scattering, it is therefore the zero-eigenfunctions and the zero-resonances of subsystems that determines its singularities.

As for the coefficients $C_{\beta 1}^{(j)}(\lambda; \theta_\beta, \omega)$ and $C_{\beta 2}(\lambda; \theta_\beta, \omega)$, we have

Theorem 1.2 *Up to a multiplicative constant depending only on λ and E^α, $C_{\beta 1}^{(j)}(\lambda; \theta_\beta, \omega)$ and $C_{\beta 2}(\lambda; \theta_\beta, \omega)$ coincide with the scattering amplitudes for 2 cluster scattering.*

More precisely, $C_{\beta 1}^{(j)}(\lambda; \theta_\beta, \omega)$ and $C_{\beta 2}(\lambda; \theta_\beta, \omega)$ are the scattering amplitudes for 2-cluster scattering in which, after the collision, the pair β becomes the bound state with zero energy or the zero-resonance, respectively. One should note that the notion of 2-cluster scattering associated with the zero-resonance is somewhat ambiguous since, as far as the author knows, this notion has not yet been introduced in mathematical literatures. So, we shall begin by explaining the meaning of Theorem 1.2.

1.4 Generalized eigenfunctions

We first recall the 2-body problem. The generalized eigenfunction $\varphi(x,\xi)$ of $-\Delta+V$ in $\mathbf{R}^n$ is written as

$$\varphi(x,\xi)=e^{ix\xi}-v,$$
$$v=v(x,\xi)=(-\Delta+V-|\xi|^2-i0)^{-1}(V(x)e^{ix\xi}).$$

The first term, $e^{ix\xi}$, represents the incident wave, and the second term, v, the scattered wave. The scattering amplitude $A(\lambda;\theta,\omega)=\hat{S}(\lambda;\theta,\omega)-\delta(\theta-\omega)$ is derived from v in the following way :

$$v(x,\sqrt{\lambda}\omega)\sim C(\lambda)r^{-(n-1)/2}e^{i\sqrt{\lambda}r}A(\lambda;\theta,\omega),$$
$$\theta=x/|x|,\quad r=|x|\to\infty.$$

In the case of the three body problem, our generalized eigenfunction is given by

$$\varphi(x,\lambda,\omega)=e^{i\sqrt{\lambda-E^\alpha}\omega\cdot x_\alpha}u_\alpha(x^\alpha)-v, \tag{1.10}$$

$$v=R(\lambda+i0)f,\quad R(z)=(H-z)^{-1}, \tag{1.11}$$

$$f=f(x,\lambda,\omega)=\sum_{\gamma\neq\alpha}V_\gamma(x^\gamma)u_\alpha(x^\alpha)e^{i\sqrt{\lambda-E^\alpha}\omega\cdot x_\alpha}. \tag{1.12}$$

We relate the coefficients $C^{(j)}_{\beta 1}$, $C_{\beta 2}$ appearing in Theorem 1.1 to the asymptotic properties of v.

For small positive constants ϵ and ϵ_2, we take smooth functions $\chi_\beta(x)$ and $\psi_\beta(t)$ defined by

$$\chi_\beta(x)=\begin{cases}1 & \text{if}\quad \dfrac{|x^\beta|}{|x|}<2\epsilon,|x|>1,\\[2ex] 0 & \text{if}\quad \dfrac{|x^\beta|}{|x|}>3\epsilon,\end{cases} \tag{1.13}$$

$$\psi_\beta(t)=\begin{cases}1 & \text{if}\quad |t-\lambda|<\epsilon_2,\\ 0 & \text{if}\quad |t-\lambda|>2\epsilon_2.\end{cases} \tag{1.14}$$

We put

$$w(x_\beta)=\int_{\mathbf{R}^3}u^{(j)}_\beta(x^\beta)\psi_\beta(D_{x_\beta})\chi_\beta(x)v(x)dx^\beta, \tag{1.15}$$

where $\psi_\beta(D_{x_\beta})$ is a pseudo-differential operator (Ps.D.Op. in short) with symbol $\psi_\beta(|\xi_\beta|^2)$, and $u^{(j)}_\beta(x^\beta)$ is an eigenfunction of h^β with zero eigenvalue. Note that the integral is convergent since it is actually performed on the set $\{|x^\beta|\leq 2\epsilon(1-4\epsilon^2)^{-1}|x_\beta|\}$. Then we have

Lemma 1.3 *The following asymptotic expansion holds :*

$$w(r\theta_\beta) \sim C(\lambda) r^{-1} e^{i\sqrt{\lambda} r} C^{(j)}_{\beta 1}(\lambda; \theta_\beta, \omega),$$

as $r = |x_\beta| \to \infty$ *in* $L^2(S^2)$. *Moreover, up to a constant depending only on* λ, $C^{(j)}_{\beta 1}(\lambda; \theta_\beta, \omega)$ *coincides with the scattering amplitude introduced by the time-dependent method.*

One should note that it is not an obvious problem to construct the integral kernel of the 2-cluster scattering amplitude when the final state has the zero energy. The above lemma means that the kernel does exist and is equal to $C^{(j)}_{\beta 1}$.

Now, one can think of the collision process, in which, after the collision, the pair β takes the zero-resonance state. It is not easy to define the associated scattering amplitude by the time-dependent method, since the zero-resonance, φ_β, does not belong to $L^2(\mathbf{R}^3)$. However, the stationary method explained above works equally well for this case. We define $\tilde{w}$ by

$$\tilde{w}(x_\beta) = \int_{\mathbf{R}^3} \varphi_\beta(x^\beta) \psi_\beta(D_{x_\beta}) \chi_\beta(x) v(x) dx^\beta. \tag{1.16}$$

By the same reasoning as above, this makes sense. We then have

Lemma 1.4

$$\tilde{w}(r\theta_\beta) \sim C(\lambda) r^{-1} e^{i\sqrt{\lambda} r} C_{\beta 2}(\lambda; \theta_\beta, \omega),$$

as $r = |x_\beta| \to \infty$ *in* $L^2(S^2)$.

Comparing the above two lemmas, it seems to be natural to call $C_{\beta 2}(\lambda; \theta_\beta, \omega)$ the scattering amplitude associated with the zero-resonance, up to a constant factor. Actually, the above two lemmas are the content of Theorem 1.2.

1.5 Main results (2)

We continue the above point of view of studying the spatial asymptotic properties of v.

Theorem 1.5 *If* $\rho > 4 + 1/2$, *for any* $\lambda > 0$,

$$\mathrm{s} - \lim_{r \to \infty} r^{5/2} e^{-i\sqrt{\lambda} r} v(r\cdot) = C(\lambda) \hat{S}_{0\alpha}(\lambda; \cdot, \omega),$$

$$C(\lambda) = e^{\pi i/4} 2\pi \lambda^{-1/4} (\lambda - E^\alpha)^{-1/4},$$

in $L^2_{loc}(M)$.

So, the scattering matrix $\hat{S}_{0\alpha}(\lambda)$ is obtained from the asymptotic behavior of v in the same way as in the two body problem. However, it is not easy to replace M by S^5 in the above theorem, since the behavior of v in a neighborhood of N is rather complicated. What we can expect is the limit in an averaged sense.

Let $\chi_\beta(x)$ be as in (1.13). We take $\rho_+(t) \in C^\infty(\mathbf{R}^1)$ such that $\rho_+(t) = 1$ if $t > 1 - \epsilon$, $\rho_+(t) = 0$ if $t < 1 - 2\epsilon$. We also take $\rho(t) \in C_0^\infty((0,\infty))$ such that $\rho(t) = 1$ near $t = 1$ and $\int_0^\infty \rho(t)dt = 1$. In order to facilitate the proof, the pair potentials are assumed to be rapidly decreasing, but the following two theorems can of course be proved for more slowly decreasing potentials.

Theorem 1.6 *Suppose that V_α's are rapidly decreasing functions. Then*

$$\begin{aligned} &\mathrm{s}-\lim_{R\to\infty} \frac{1}{R}\int_{\mathbf{R}^6} e^{-i\sqrt{\lambda}\theta\cdot x}\theta_\beta \cdot \hat{x}_\beta \rho_+\Big(\frac{\theta_\beta}{|\theta_\beta|}\cdot\frac{x_\beta}{|x_\beta|}\Big)\chi_\beta(x)\rho\Big(\frac{|x_\beta|}{R}\Big)v(x)dx \\ &= C(\lambda)\hat{S}_{0\alpha}(\lambda;\theta,\omega), \\ &C(\lambda) = -(4\pi)^{-1}\lambda^{-1/2}C_0(\lambda)^{-1}C_\alpha(\lambda)^{-1}, \end{aligned}$$

in $L^2(\tilde{N}_\beta)$, where $\tilde{N}_\beta$ is a small neighborhood of $N \cap X_\beta$ and $\hat{x}_\beta = x_\beta/|x_\beta|$.

In the neighborhood of the X_β-plane, there are two sorts of scattering, the 3-cluster scattering and the 2-cluster scattering. We can distinguish between them by changing the way of taking the limit at infinity of v.

Let $u_{\beta l}(x^\beta)$ be a normalized eigenfunction of h^β with eigenvalue $E_l^\beta \le 0$. Let $A_{\beta l}(\lambda;\theta_\beta,\omega)$ be the 2-cluster scattering amplitude associated with the process in which, after the colloision, the pair β takes the bound state $u_{\beta l}$.

Theorem 1.7 *Suppose that V_α's are rapidly decreasing functions. For a small $\epsilon > 0$, let $\psi_\beta(t) \in C^\infty(\mathbf{R}^1)$ be such that $\psi_\beta(t) = 1$ if $t > \lambda - \epsilon$, $\psi_\beta(t) = 0$ if $t < \lambda - 2\epsilon$. Let $\psi_\beta(D_{x_\beta})$ be the Ps.D.Op. with symbol $\psi_\beta(|\xi_\beta|^2)$. Fix $a > 0$ arbitrarily. Then as $r = |x_\beta| \to \infty$, we have the following asymptotic expansion :*

$$\begin{aligned} &\psi_\beta(D_{x_\beta})v \simeq \sum_l C_{\beta l}(\lambda)u_{\beta l}(x^\beta)r^{-1}e^{i\sqrt{\lambda - E_l^\beta}\,r}A_{\beta l}(\lambda;\theta_\beta,\omega), \quad \theta_\beta = x_\beta/r, \\ &C_{\beta l}(\lambda) = 2\pi i(\lambda - E^\alpha)^{-1/4}(\lambda - E_l^\beta)^{-1/4}, \end{aligned}$$

uniformly for $|x^\beta| < a, \theta_\beta \in S^2$.

1.6 Some remarks

Amrein, Pearson and Sinha [1] showed that, for the N-body problem, the total cross-section with 2-cluster initial state is finite for almost all energy and derived its asymptotic properties in an averaged sense under the assumption that the potentials decay faster than $|x^\alpha|^{-2}$. See also Enss-Simon [5]. Amrein and Sinha [2] also showed that, for the three body problem, the total cross-section is finite for all $\lambda > 0$ under the assumption that each 2-body subsystem has neither the zero eigenvalue nor the zero-resonance. Ito and Tamura [11] studied the semi-classical asymptotics for the total cross-section in distributional sense. All of these works treats the case of the initial state of 2-clusters, while Newton [18] and Yafaev [20] studied the structure of the S-matrix coressponding the collision process from 3-clusters to 3-clusters. Asymptotic properties of generalized eigenfunctions of three-body Schrödinger operators have been so far studied mainly by physicists. In the work of Newton [17], Theorems 1.5 and 1.7 were derived by intuitive arguments. If we consider the collision process of initial state of 3-clusters, we are led to consider the generalized eigenfunction formally defined by

$$\psi(x,\lambda,\theta) = e^{i\sqrt{\lambda}\theta\cdot x} - R(\lambda + i0)\sum_{\alpha} V_\alpha e^{i\sqrt{\lambda}\theta\cdot x}.$$

The rigorous study of this generalized eigenfunction seems to be much harder. Nuttal [19] and Newton [17] gave precise explanations. One should also note the work of Mercuriev [16] of the three-body scattering theory for the Coulomb potential based on the stationary theory.

To prove the above theorems, following Isozaki-Kitada [10], we first localize the S-matrix in the phase space. An important role is played by the estimate of the resolvent of H multiplied by Ps.D.Op's. For that purpose we utilize the estimate of Skibsted [15] on propagation properties in the phase space of e^{-itH}, whose method is based on the work of Sigal-Soffer [14]. These estimates have been further refined by Gérard [6]. The singularities of $\hat{S}_{0\alpha}(\lambda)$ arise from the low-energy asymptotics of 2-body subsystems studied by Jensen and Kato [12]. Another important tool is the spectral representation theory developed for 2-body Schrödinger operators, the key idea of which is to relate the generalized eigenfunction to the spatial symptotics of the resolvent of Schrödinger operators .

The notation used in this paper is almost standard. In particular, $L^{2,s}$ denotes the usual weighted L^2-space :

$$f \in L^{2,s} \Longleftrightarrow \|f\|_s^2 = \int_{\mathbf{R}^n} <x>^{2s} |f(x)|^2 dx < \infty,$$

$<x> = (1+|x|^2)^{1/2}$. For two Banach spaces X_1 and X_2, $\mathbf{B}(X_1; X_2)$ denotes the totality of bounded operators from X_1 and X_2.

In this paper, we explain only the outline of the proof. The details are discussed in [8] and [9].

2 Localization of the S-matrix

The first step of the proof of Theorem 1.1 is to derive a formula of localization of $\hat{S}_{0\alpha}$ in the phase space. To localize the direction of propagation of particles, we take a real-valued function $\psi(\theta) \in C^{\infty}(S^5)$. To localize the energy, we take a compact interval $I \subset (0,\infty)$ and a real-valued function $\psi_0(t) \in C_0^{\infty}((0,\infty))$ such that $\psi_0(t) = 1$ on I. Let $\chi(x,\xi)$ be a smooth function satisfying

$$|\partial_x^m \partial_\xi^n \chi(x,\xi)| \leq C_{mn} < x >^{-m}, \quad \text{for any} \quad m,n \geq 0$$

and $\chi(x,\xi) = 1$, if $|\xi|^2 \in supp\psi_0$ and $|\hat{x}\cdot\hat{\xi}| > 1-\epsilon$ for some $0<\epsilon<1$ and sufficiently large $|x|$. Letting

$$p(x,\xi) = \chi(x,\xi)\psi_0(|\xi|^2)\psi(\xi/|\xi|), \tag{2.1}$$

we consider a Ps.D.Op. P with symbol $p(x,\xi) : P = p(x,D_x)$, and define

$$G = HP - PH_0,$$

$$Q_\alpha = (H - H_\alpha)J_\alpha = \sum_{\gamma\neq\alpha} V_\gamma J_\alpha.$$

We define the trace of the Fourier transform by $(\mathcal{F}_0(\lambda)f)(\theta) = (\mathcal{F}_0 f)(\lambda,\theta)$ and $(\mathcal{F}_\alpha(\lambda)f)(\omega) = (\mathcal{F}_\alpha f)(\lambda,\omega)$ for $f \in L^{2,s}, s > 1/2$. Let $<,>$ be the inner product of $L^2(S^5)$ and $\varphi(\xi) = \psi_0(|\xi|^2)\psi(\xi/|\xi|)$.

Lemma 2.1 *Let $\hat{f} \in C_0^{\infty}((E^\alpha,\infty);C^{\infty}(S^2))$ and $\hat{g} \in C_0^{\infty}((0,\infty);C^{\infty}(S^5))$. Then*

$$\begin{aligned}
&(\mathcal{F}_0\varphi(D_x)S_{0\alpha}\mathcal{F}_\alpha^*\hat{f},\hat{g}) \\
&= -2\pi i\int_0^\infty < \mathcal{F}_0(\lambda)P^*Q_\alpha\mathcal{F}_\alpha^*(\lambda)\hat{f}(\lambda),\hat{g}(\lambda) > d\lambda \\
&+ \lim_{\epsilon\downarrow 0} 2\pi i\int_0^\infty < \mathcal{F}_0(\lambda)G^*R(\lambda+i\epsilon)Q_\alpha\mathcal{F}_\alpha^*(\lambda)\hat{f}(\lambda),\hat{g}(\lambda) > d\lambda.
\end{aligned}$$

The proof of this lemma is essentially the same as in Theorem 3.3 of [10]. If one can give a definite meaning to the operator $G^*R(\lambda+i0)Q_\alpha$ and exchange the order of integration in λ and $\lim_{\epsilon\downarrow 0}$, one obtains

$$\begin{aligned}
\psi(\theta)\hat{S}_{0\alpha}(\lambda) = &-2\pi i\mathcal{F}_0(\lambda)P^*Q_\alpha\mathcal{F}_\alpha^*(\lambda) \\
&+ 2\pi i\mathcal{F}_0(\lambda)G^*R(\lambda+i0)Q_\alpha\mathcal{F}_\alpha^*(\lambda),
\end{aligned}$$

if $\lambda \in I$.

3 Proof of Theorem 1.2

We first split $\hat{S}_{0\alpha}(\lambda)$ into two parts. For sufficiently small $\epsilon > 0$, we set

$$\begin{aligned} X_\beta^\epsilon &= \{x \in X; |x^\beta|/|x| < \epsilon\}, \\ M^\epsilon &= S^5 \setminus \cup_\beta X_\beta^\epsilon, \\ N^\epsilon &= S^5 \cap (\cup_\beta X_\beta^\epsilon). \end{aligned}$$

We take $\psi_M(\theta), \psi_N(\theta) \in C^\infty(S^5)$ such that

$$\psi_M(\theta) + \psi_N(\theta) = 1,$$

$$\psi_M(\theta) = \begin{cases} 1 & \theta \in M^{2\epsilon}, \\ 0 & \theta \in N^\epsilon, \end{cases}$$

$$\psi_N(\theta) = \begin{cases} 1 & \theta \in N^\epsilon, \\ 0 & \theta \in M^{2\epsilon}. \end{cases}$$

We put

$$\begin{aligned} \hat{S}_M(\lambda) &= \psi_M(\theta)\hat{S}_{0\alpha}(\lambda), \\ \hat{S}_N(\lambda) &= \psi_N(\theta)\hat{S}_{0\alpha}(\lambda). \end{aligned}$$

We also prepare localizations in x-space. We take $\chi_M(x), \chi_N(x) \in C^\infty(X)$ homogeneous of degree 0 for $|x| > 1$ and

$$\chi_M(x) = \begin{cases} 1 & \text{if} \quad \hat{x} \in M^{\epsilon/2}, |x| > 1, \\ 0 & \text{if} \quad \hat{x} \in N^{\epsilon/4}, \end{cases}$$

$$\chi_N(x) = \begin{cases} 1 & \text{if} \quad \hat{x} \in N^{2\epsilon}, |x| > 1, \\ 0 & \text{if} \quad \hat{x} \in M^{3\epsilon}, \end{cases}$$

where $\hat{x} = x/|x|$. The important properties of these localizations are as follows.

Lemma 3.1 *(1) supp* $\nabla\chi_M$, *supp* $\nabla\chi_N \subset X \setminus \cup_\beta X_\beta$.
(2) supp $\nabla\chi_M \cap$ *supp* $\psi_M = \emptyset$, *supp* $\nabla\chi_N \cap$ *supp* $\psi_N = \emptyset$, *if* $|x| > 1$.

Note that in Lemma 3.1 (2), we extend ψ_M, ψ_N on $X - \{0\}$ as homogeneous functions of degree 0. One should also note that if ϵ is sufficiently small, $\chi_N(x)$ is split into three parts

$$\chi_N(x) = \sum_\beta \chi_\beta(x),$$

$$\chi_\beta(x) = \begin{cases} 1 & \text{if} \quad x \in X_\beta^{2\epsilon}, |x| > 1, \\ 0 & \text{if} \quad x \notin X_\beta^{3\epsilon}. \end{cases} \tag{3.1}$$

We next localize the energy. We fix $\lambda > 0$ arbitrarily and for small $\epsilon_1 > 0$ we take $\psi_1(t) \in C_0^\infty(\mathbf{R}^1)$ such that

$$\psi_1(t) = \begin{cases} 1 & \text{if} \quad |t - \lambda| < \epsilon_1, \\ 0 & \text{if} \quad |t - \lambda| > 2\epsilon_1. \end{cases}$$

3.1 Continuity of the kernel of $\hat{S}_M(\lambda)$

With χ_M, ψ_M, ψ_1 as above, we put

$$p(x,\xi) = \chi_M(x)\psi_M(\xi/|\xi|)\psi_1(|\xi|^2). \tag{3.2}$$

Let P be a Ps.D.Op. with symbol $p(x,\xi)$ and $G = HP - PH_0$. We use the formula in Lemma 2.1. The main problem here is to give a definite meaning to the operator $G^* R(\lambda + i0)Q_\alpha$. For this purpose, we use the following resolvent estimate essentially due to Skibsted [15].

Theorem 3.2 *Let P be a Ps.D.Op. with symbol $p_-(x,\xi)$ satisfying*

$$|\partial_x^m \partial_\xi^n p_-(x,\xi)| \le C_{mn} < x >^{-m}, \quad \textit{for any} \quad m, n \ge 0,$$

there exists a closed cone $\Gamma \subset X \setminus \cup_\alpha X_\alpha$ such that
$supp_x p_-(x,\xi) \subset \Gamma$ for all ξ.

Moreover, we assume that

there exists a constant μ_- such that $-1 < \mu_- < 1$ and
$p_-(x,\xi) = 0$ if $\hat{x} \cdot \hat{\xi} > \mu_-$,

where $\hat{x} = x/|x|$ and $\hat{\xi} = \xi/|\xi|$. Then we have

$$< x >^s P_- R(\lambda + i0) < x >^{-s-t} \in \mathbf{B}(L^2(X); L^2(X)),$$

if $s > -1/2, t > 1, \lambda \in \sigma_e(H) - \Lambda$, where $\sigma_e(H)$ is the essential spectrum of H and Λ is the threshold of H.

Now, since $G = [H_0, P] + \sum_\beta V_\beta P$, by virtue of our localization $\chi_M(x)$ and Lemma 3.1, we have

$$G^* = < x >^{-1} P_- + O(< x >^{-\rho}),$$

where P_- verifies the assumptions in Theorem 3.2. Let f be defined by (1.12). Then by Theorem 3.2,

$$\int_{\mathbf{R}^6} e^{-i\sqrt{\lambda}\theta\cdot x} G^* R(\lambda + i0) f dx$$

is continuous with respect to $\lambda > 0, \theta \in S^5$ and $\omega \in S^2$, which implies that $\mathcal{F}_0(\lambda) G^* R(\lambda + i0) Q_\alpha \mathcal{F}_\alpha^*(\lambda)$ has a continuous kernel.

We note that, since $P^* - G^* R(\lambda + i0) = (H_0 - \lambda) P^* R(\lambda + i0)$, the kernel of $\hat{S}_M(\lambda)$ is given by

$$-2\pi i C_\alpha(\lambda) \mathcal{F}_0(\lambda)(H_0 - \lambda) P^* R(\lambda + i0) f. \tag{3.3}$$

3.2 Singularities of the kernel of $\hat{S}_N(\lambda)$

To calculate the kernel of $\hat{S}_N(\lambda)$, we again make use of Lemma 2.1. However, in this case, we must be careful in choosing $p(x, \xi)$. First we note that on the support of $\psi_1(|\xi|^2)\psi_N(\xi/|\xi|)$, $|\xi^\beta|/|\xi| < 2\epsilon$, $||\xi|^2 - \lambda| < 2\epsilon_1$. We take $\psi_\beta(t) \in C_0^\infty(\mathbf{R}^1)$ such that

$$\psi_\beta(t) = \begin{cases} 1 & |t - \lambda| < \epsilon_2, \\ 0 & |t - \lambda| > 2\epsilon_2. \end{cases}$$

Then by an appropriate choice of ϵ's, we have

$$\psi_1(|\xi|^2)\psi_N(\xi/|\xi|) = \sum_\beta \psi_\beta(|\xi_\beta|^2)\psi_1(|\xi|^2)\psi_N(\xi/|\xi|).$$

With $\chi_\beta(x)$ introduced in (3.1), we define

$$p(x, \xi) = \sum_\beta \chi_\beta(x)\psi_\beta(|\xi_\beta|^2)\psi_1(|\xi|^2)\psi_N(\xi/|\xi|). \tag{3.4}$$

For a suitable choice of ϵ's, $supp\, \nabla\chi_\beta$ and $supp\, \psi_\beta(|\xi_\beta|^2)\psi_1(|\xi|^2)$ are disjoint. Let P be a Ps.D.Op. with symbol $p(x, \xi)$ defined by (3.4) and $G = HP - PH_0$. Then, since $G = [H_0, P] + \sum_\beta V_\beta P$, we have

$$G^* = < x >^{-1} P_- + O(< x >^{-k}) + P^* \sum_\beta V_\beta,$$

where P_- satisfies the assumption of Theorem 3.2, and k can be chosen large enough. We consider the third term. Note that P can be written as

$$P = \sum_\beta A_\beta,$$

where A_β is the Ps.D.Op. with symbol

$$\chi_\beta(x)\psi_\beta(|\xi_\beta|^2)\psi_1(|\xi|^2)\psi_N(\xi/|\xi|).$$

The most difficult part is $\lim_{\kappa\downarrow 0}\mathcal{F}_0(\lambda)A_\beta^* V_\beta R(\lambda+i\kappa)Q_\alpha\mathcal{F}_\alpha^*(\lambda)$. We must be careful in taking this limit, since $A_\beta^* V_\beta$ does not decay in x_β.

The important trick we use here is to replace $(H-z)^{-1}$ by $(H_\beta - z)^{-1}$ near the X_β-plane. Letting $\Psi_N(\xi) = \psi_1(|\xi|^2)\psi_N(\xi/|\xi|)$, we have

$$A_\beta^* = \Psi_N(D_x)\psi_\beta(D_{x_\beta})\chi_\beta(x),$$

where $\psi_\beta(D_{x_\beta})$ denotes the Ps.D.Op. with symbol $\psi_\beta(|\xi_\beta|^2)$. Noting that $\psi_\beta(D_{x_\beta})\chi_\beta(x)$ commutes with V_β, we consider $\psi_\beta(D_{x_\beta})\chi_\beta(x)R(\lambda+i\kappa)Q_\alpha\mathcal{F}_\alpha^*(\lambda)$.

Lemma 3.3 *Let f be defined by (1.12), $z = \lambda + i\kappa$ and $g = (H_\beta - z)\psi_\beta(D_{x_\beta})\chi_\beta(x)R(z)f$. Then $g \in L^{2,s}$ for any $s < \rho - 3/2$ uniformly in $\kappa \geq 0$.*

Proof. Letting $u = R(z)f$, we have

$$g = \psi_\beta(D_{x_\beta})\chi_\beta(x)f - \psi_\beta(D_{x_\beta})\chi_\beta(x)\sum_{\gamma\neq\beta} V_\gamma u + \psi_\beta(D_{x_\beta})[H_0,\chi_\beta]u$$

$$= g_1 + g_2 + g_3.$$

It easily follows from our assumption that $g_1 \in L^{2,s}$, $\forall s < \rho - 3/2$ and $g_2 \in L^{2,s}$, $\forall s < \rho - 1/2$. We take $\varphi \in C_0^\infty(\mathbf{R}^1)$ such that for small $\epsilon_3 > 0$,

$$\varphi(t) = \begin{cases} 1J & |t-\lambda| < \epsilon_3, \\ 0J & |t-\lambda| > 2\epsilon_3. \end{cases}$$

Then as is well-known,

$$(1-\varphi(H))R(\lambda+i\kappa) \in \mathbf{B}(L^{2,s};L^{2,s})$$

for any $s > 1/2$ uniformly in $\kappa \geq 0$. Therefore

$$\psi_\beta(D_{x_\beta})[H_0,\chi_\beta](1-\varphi(H))R(\lambda+i\kappa)f \in L^{2,s}$$

for any $s < \rho - 1/2$ uniformly in $\kappa \geq 0$. We also have

$$[H_0,\chi_\beta]\varphi(H)R(\lambda+i\kappa) = [H_0,\chi_\beta]\varphi(H_0)R(\lambda+i\kappa) + O(< x >^{-\rho-1})R(\lambda+i\kappa).$$

Now, on the support of the symbol of $\psi_\beta(D_{x_\beta})[H_0,\chi_\beta]\varphi(H_0)$, $||\xi_\beta|^2 - \lambda| \leq 2\epsilon_2$, $||\xi|^2 - \lambda| \leq 2\epsilon_1$ and $|x_\beta|/|x| \geq 2\epsilon$. If ϵ_1 and ϵ_2 are chosen small enough, ξ is localized near the X_β-plane, which shows that $< x > \psi_\beta(D_{x_\beta})[H_0,\chi_\beta]\varphi(H_0)$ satisfies the assumptions of Theorem 3.2. Then we have $g_3 \in L^{2,s}$ for any $s < \rho - 3/2$ uniformly in $\kappa \geq 0$. Q.E.D.

Using Lemma 3.3, letting $R_\beta(z) = (H_\beta - z)^{-1}$, we have

$$\psi_\beta(D_{x_\beta})\chi_\beta(x)R(\lambda+i\kappa)f = R_\beta(\lambda+i\kappa)g, \tag{3.5}$$

whence

$$\mathcal{F}_0(\lambda)A_\beta^* V_\beta R(\lambda + i\kappa)f = \mathcal{F}_0(\lambda)\Psi_N(D_x)V_\beta R_\beta(\lambda + i\kappa)g. \tag{3.6}$$

Our next trick is to take the partial Fourier transformation with respect to x_β and to reduce the problem to that of 2-body subsystems. Note that $\Psi_N(\sqrt{\lambda}\theta) = 1$ if $|\theta^\beta|$ is sufficiently small. Recall that $\mathcal{F}_0(\lambda)$ is the Fourier transformation followed by the restriction to the sphere of radius $\sqrt{\lambda}$. Thus integrating $V_\beta R_\beta(\lambda+i\kappa)g$ with respect to x_β first and to x^β later and letting $r_\beta(z) = (-\Delta_{x^\beta} + V_\beta - z)^{-1}$, we see that (3.6) is equal to

$$C_0(\lambda)\int_{\mathbf{R}^3} e^{-i\sqrt{\lambda}\theta^\beta\cdot x^\beta} V_\beta r_\beta(\lambda|\theta^\beta|^2 + i\kappa)\hat{g}\,dx^\beta,$$

$$\hat{g} = \hat{g}(x^\beta;\lambda,\theta_\beta,\omega) = \int_{\mathbf{R}^3} e^{-i\sqrt{\lambda}\theta_\beta\cdot x_\beta} g(x^\beta, x_\beta)dx_\beta.$$

If $\rho > 5 + 1/2$, $\hat{g}(\cdot;\lambda,\theta_\beta,\omega)$ is an $L^{2,s}(\mathbf{R}^3)$-valued continuous function of λ, θ_β and ω for some $s > 5/2$.

Here we recall the result of Jensen-Kato on the low-energy asymptotics of the 2-body Schrödinger operators. Let P_0 be the projection onto the eigenspace of h^β with zero eigenvalue. Let B_{-2} and B_{-1} be dfined by

$$B_{-2} = -P_0,$$

$$B_{-1} = -iP_0V_\beta K V_\beta P_0 + i < \cdot, \varphi_\beta > \varphi_\beta,$$

where K is the integral operator with kernel $|x^\beta - y^\beta|^2/(24\pi)$. Then, by [12], if $\rho > 5$ and $s > 5/2$, we have the following asymptotic expansion :

$$r_\beta(z) = \frac{B_{-2}}{z} + \frac{B_{-1}}{\sqrt{z}} + O(1),$$

as $z \to 0$ in $\mathbf{B}(L^{2,s}; L^{2,-s})$.

We can now find the singularities of the kernel of $\hat{S}_N(\lambda)$. By using Jensen-Kato, we have

$$\begin{aligned}
&\int_{\mathbf{R}^3} e^{-i\sqrt{\lambda}\theta^\beta\cdot x^\beta} V_\beta r_\beta(\lambda|\theta^\beta|^2 + i\kappa)\hat{g}\,dx^\beta \\
&\simeq (\lambda|\theta^\beta|^2)^{-1}\int_{\mathbf{R}^3} e^{-i\sqrt{\lambda}\theta^\beta\cdot x^\beta} V_\beta B_{-2}\hat{g}\,dx^\beta \\
&+ (\lambda|\theta^\beta|^2)^{-1/2}\int_{\mathbf{R}^3} e^{-i\sqrt{\lambda}\theta^\beta\cdot x^\beta} V_\beta B_{-1}\hat{g}\,dx^\beta
\end{aligned} \tag{3.7}$$

as $\kappa \to 0$. Let $u_\beta^{(j)}$ be normalized eigenfunctions of h^β with zero eigenvalue. Then $B_{-2}\hat{g} = \sum_j A_{\beta 1}^{(j)}(\lambda;\theta_\beta,\omega)u_\beta^{(j)}(x^\beta)$ with

$$A_{\beta 1}^{(j)}(\lambda;\theta_\beta,\omega) = -\int_{\mathbf{R}^3} \hat{g}(x^\beta;\lambda,\theta_\beta,\omega)u_\beta^{(j)}(x^\beta)dx^\beta.$$

A simple calculation shows that

$$\int_{\mathbf{R}^3} V_\beta(x^\beta)u_\beta^{(j)}(x^\beta)dx^\beta = 0.$$

Therefore, as $|\theta^\beta| \to 0$, we see that the first term of the right-hand side of (3.7) behaves like

$$-i(\sqrt{\lambda}|\theta^\beta|)^{-1}\sum_j A_{\beta 1}^{(j)}(\lambda;\theta_\beta,\omega)\int_{\mathbf{R}^3}\frac{\theta^\beta}{|\theta^\beta|}\cdot x_\beta V_\beta(x^\beta)u_\beta^{(j)}(x^\beta)dx^\beta + O(1).$$

So we have found the contribution of the zero-eigenvalue of h^β to the singularities. By a similar calculation one can also find the contribution of the zero-resonances. In particular, we see that $C_{\beta 1}^{(j)}(\lambda;\theta_\beta,\omega)$ and $C_{\beta 2}(\lambda;\theta_\beta,\omega)$ are given by

$$C_{\beta 1}^{(j)}(\lambda;\theta_\beta,\omega) = 2\pi\lambda^{-1/2}C_0(\lambda)C_\alpha(\lambda)\int_{\mathbf{R}^3}\hat{g}(x^\beta;\lambda,\theta_\beta,\omega)u_\beta^{(j)}(x^\beta)dx^\beta, \quad (3.8)$$

$$C_{\beta 2}(\lambda;\theta_\beta,\omega) = 2\pi\lambda^{-1/2}C_0(\lambda)C_\alpha(\lambda)\int_{\mathbf{R}^3}\hat{g}(x^\beta;\lambda,\theta_\beta,\omega)\varphi_\beta(x^\beta)dx^\beta. \quad (3.9)$$

We end this section by noting that as in (3.3) the kernel of $\hat{S}_N(\lambda)$ is given by

$$-2\pi i C_\alpha(\lambda)\mathcal{F}_0[(H_0-\lambda)P^*R(\lambda+i0)f](\lambda), \quad (3.10)$$

P being defined by (3.4).

4 Spatial asymptotics of v

We explain the proof of Theorems 1.5, 1.6 and 1.7.

4.1 Proof of Theorem 1.5.

Let v be defined by (1.11). We take $\chi(x) \in C^\infty(X)$ homogeneous of degree 0 for $|x| > 1$ and

$$\chi(x) = \begin{cases} 1 & \text{if} \quad \hat{x} \in M^{4\epsilon}, |x| > 1, \\ 0 & \text{if} \quad \hat{x} \notin M^{3\epsilon}. \end{cases}$$

To prove Theorem 1.5, we consider the limit of $\chi(x)v$ as $r = |x| \to \infty$. Let us introduce the following notation:

$$u_1 \sim u_2 \Longleftrightarrow r^{5/2}(u_1(r\cdot) - u_2(r\cdot)) \to 0 \quad \text{in} \quad L^2(S^5) \quad \text{as} \quad r \to \infty.$$

We introduce the space $H^{1,s}(\mathbf{R}^6)$ by

$$w \in H^{1,s}(\mathbf{R}^6) \Longleftrightarrow w, \nabla w \in L^{2,s}(\mathbf{R}^6).$$

As can be checked easily, $w \sim 0$ if $w \in H^{1,s}(\mathbf{R}^6)$ for some $s > 1/2$.

Recall that $v = R(\lambda + i0)f$ and that

$$|f| \leq C_k < x^\alpha >^{-k} < x_\alpha >^{-\rho}, \quad \forall k \geq 0. \tag{4.1}$$

Let $\psi_1(t)$ be as above. Choose $s > 1/2$ sufficiently close to $1/2$. Then $v - \psi_1(H)v \in H^{1,s}(\mathbf{R}^6)$. Then, $\chi(x)(\psi_1(H) - \psi_1(H_0))v \in H^{1,s}(\mathbf{R}^6)$. So we have

$$\chi(x)v \sim \chi(x)\psi_1(H_0)v.$$

We set $\psi(\xi) = \psi_1(|\xi|^2)\psi_M(\xi/|\xi|)$. Since the support of $\chi(x)$ and that of $(1 - \psi_M(\xi/|\xi|))\psi_1(|\xi|^2)$ are disjoint, one can use Theorem 3.2 to see

$$\chi(x)\psi_1(H_0)v \sim \chi(x)\psi(D_x)v. \tag{4.2}$$

Since $\chi(x) = \chi(x)\chi_M(x)$, we have

$$\chi(x)\psi(D_x)v \sim \chi(x)\psi(D_x)\chi_M(x)v. \tag{4.3}$$

Noting that P is a Ps.D.Op. with symbol $\chi_M(x)\psi(\xi)$, we let $g = (H_0 - \lambda)P^*v$. Then $g \in L^{2,s}$ for some $s > 3/2$.

Here we recall the following well-known result on the relation between the Fourier transformation and the resolvent of the Laplacian.

Lemma 4.1 *Let $R_0(z)$ be the resolvent of $-\Delta$ in $\mathbf{R}^n$ and $f \in L^{2,3/2}$. Let $C(\lambda) = e^{(n-3)\pi i/4}\pi^{-1/2}\lambda^{1/4}$. Then the following strong limit exists in $L^2(S^{n-1})$ for any $\lambda > 0$:*

$$\mathcal{F}_0(\lambda)f = \mathrm{s} - \lim_{r\to\infty} C(\lambda)r^{(n-1)/2}e^{-i\sqrt{\lambda}r}(R_0(\lambda + i0)f)(r\cdot),$$

where $\mathcal{F}_0(\lambda)$ is the Fourier transformation defined in the same way as in (1.6) with $C_0(\lambda) = (2\pi)^{-n/2}2^{-1/2}\lambda^{(n-2)/4}$. The right-hand side converges uniformly on S^{n-1}, if f is rapidly decreasing.

We rewrite P^*v as $P^*v = (H_0 - \lambda - i0)^{-1}g$ and apply the above lemma. Then

$$P^*v \sim C(\lambda)r^{-5/2}e^{i\sqrt{\lambda}r}\mathcal{F}_0(\lambda)(H_0 - \lambda)P^*v,$$
$$C(\lambda) = e^{-3\pi i/4}\pi^{1/2}\lambda^{-1/4}.$$

(3.3) shows that

$$\mathcal{F}_0(\lambda)(H_0 - \lambda)P^*v = i(2\pi)^{1/2}2^{1/2}(\lambda - E^\alpha)^{-1/4}\psi_M(\theta)\hat{S}_{0\alpha}(\lambda; \theta, \omega).$$

We have shown that

$$\chi(x)v \sim C(\lambda)r^{-5/2}e^{i\sqrt{\lambda}r}\psi_M(\theta)\hat{S}_{0\alpha}(\lambda; \theta, \omega),$$
$$C(\lambda) = e^{\pi i/4}2\pi\lambda^{-1/4}(\lambda - E^\alpha)^{-1/4}, \quad \theta = x/|x|.$$

This completes the proof of Theorem 1.5.

4.2 Proof of Theorem 1.6.

We first recall the following well known fact.

Lemma 4.2 *Let $T_0(\lambda)$ be defined by*

$$(T_0(\lambda)f)(\theta) = \int_{\mathbf{R}^6} e^{-i\sqrt{\lambda}\theta\cdot x} f(x)dx, \quad \theta \in S^5.$$

Then for any $\lambda > 0$, we have

$$\|T_0(\lambda)f\|_{L^2(S^5)} \le C_s\|f\|_s, \quad s > 1/2.$$

It is easy to see that if θ is very close to $N \cap X_\beta$, $\hat{S}_{0\alpha}(\lambda;\theta,\omega)$ is represented as

$$-2\pi i C_\alpha(\lambda)\mathcal{F}_0[(H_0 - \lambda)A_\beta^* R(\lambda + i0)f](\lambda) \tag{4.4}$$

We take $\rho(t) \in C_0^\infty((0,\infty))$ such that $\rho(t) = 1$ in a neighborhood of 1 and $\int_0^\infty \rho(t)dt = 1$. Let v be defined by (1.11).

Lemma 4.3 *Let $\tilde{N}_\beta$ be a small neighborhood of $N \cap X_\beta$. Then*

$$\frac{1}{R}\int_{\mathbf{R}^6} e^{-i\sqrt{\lambda}\theta\cdot x}\theta_\beta \cdot \hat{x}_\beta \rho(\frac{|x_\beta|}{R})A_\beta^* v dx \to C(\lambda)\hat{S}_{0\alpha}(\lambda;\theta,\omega),$$

$$C(\lambda) = -(4\pi)^{-1}\lambda^{-1/2}C_0(\lambda)^{-1}C_\alpha(\lambda)^{-1},$$

in $L^2(\tilde{N}_\beta)$, where $\hat{x}_\beta = x_\beta/|x_\beta|$.

We breifly explain the proof. Let $\rho_1(t) = \int_t^\infty \rho(s)ds$ and $w = A_\beta^* v$. Then by integration by parts, we have

$$\begin{aligned} &\frac{2i\sqrt{\lambda}}{R}\int_{\mathbf{R}^6} e^{-i\sqrt{\lambda}\theta\cdot x}\theta_\beta \cdot \hat{x}_\beta \rho(\frac{|x_\beta|}{R})w(x)dx \\ &= \int_{\mathbf{R}^6} e^{-i\sqrt{\lambda}\theta\cdot x}\rho_1(\frac{|x_\beta|}{R})(\Delta + \lambda)w(x)dx \\ &- \int_{\mathbf{R}^6} e^{-i\sqrt{\lambda}\theta\cdot x}(\Delta\rho_1(\frac{|x_\beta|}{R}))w(x)dx. \end{aligned} \tag{4.5}$$

We show that the right-hand side of (4.5) converges in $L^2(\tilde{N}_\beta)$ as $R \to \infty$. It is easy to see that $(\Delta\rho_1(|x_\beta|/R))w(x) \to 0$ in $L^{2,s}$ for some $s > 1/2$ as $R \to \infty$. By Lemma 4.2, the second term of the right-hand side of (4.5) tends to 0 in $L^2(S^5)$ as $R \to \infty$. The first term tends, formally, to

$$\int_{\mathbf{R}^6} e^{-i\sqrt{\lambda}\theta\cdot x}(\Delta + \lambda)w(x)dx,$$

which is equal to $2i\sqrt{\lambda}C(\lambda)\hat{S}_{0\alpha}(\lambda;\theta,\omega)$ by virtue of (4.4). Therefore, the proof of Lemma 4.3 will be completed if we show the convergence of the first term of the right-hand side of (4.5). But this term must be treated carefully, since $(\Delta+\lambda)w(x) \notin L^{2,s}$ for $s > 1/2$.

Let $\Psi_N(\xi) = \psi_1(|\xi|^2)\psi_N(\xi/|\xi|)$ and $\psi_\beta(D_{x_\beta})$ be the Ps.D.Op. with symbol $\psi_\beta(|\xi_\beta|^2)$. We set

$$g_\beta = (H_\beta - \lambda)\psi_\beta(D_{x_\beta})\chi_\beta(x)R(\lambda+i0)f.$$

By Lemma 3.3, g_β is rapidly decreasing. A straightforward calculation shows that

$$\begin{aligned}(\Delta+\lambda)w &= -\sum_\beta \Psi_N(D_x)(H_0-\lambda)\psi_\beta(D_{x_\beta})\chi_\beta(x)R(\lambda+i0)f \\ &= -\sum_\beta \Psi_N(D_x)g_\beta + \sum_\beta \Psi_N(D_x)V_\beta R_\beta(\lambda+i0)g_\beta.\end{aligned}$$

So, we have only to show the existence of the limit

$$\mathrm{s}-\lim_{R\to\infty}\int_{\mathbf{R}^6} e^{-i\sqrt{\lambda}\theta\cdot x}\rho_1(\frac{|x_\beta|}{R})V_\beta R_\beta(\lambda+i0)g_\beta dx, \tag{4.6}$$

in $L^2(\tilde{N}_\beta)$. For this purpose, we again make use of the result of Jensen-Kato to represent $R_\beta(\lambda+i0)$ by the zero eigenfunctions and the zero-resonance of the 2-body subsystem and calculate each term separately. We omit the details, since they are long and complicated.

We have localized v in the region $\{x; a < |x_\beta|/R < b, |x^\beta| \le 3\epsilon r\}$. The next aim is to localize v in the region where θ_β and x_β have almost the same directions.

Theorem 4.4 *Let P_β be the Ps.D.Op. with symbol $p(x_\beta,\xi_\beta)$ having the following properties:*

$$|\partial_{x_\beta}^m\partial_{\xi_\beta}^n p(x_\beta,\xi_\beta)| \le C_{mn} < x_\beta >^{-m} \quad \forall m,n,$$

there exist constants $0 < a < b < \infty$ such that

$$supp_{\xi_\beta} p(x_\beta,\xi_\beta) \subset \{a < |\xi_\beta| < b\},$$

there exists a constant μ_- such that $-1 < \mu_- < 1$ and

$$p(x_\beta,\xi_\beta) = 0 \text{ if } \quad x_\beta\cdot\xi_\beta > \mu_-|x_\beta||\xi_\beta|.$$

Let $\chi_\beta(x)$ be as above. Suppose that V_α satisfies (1.9) with $\rho > 0$. Then we have

$$< x >^s P_\beta\chi_\beta(x)R(\lambda+i0) < x >^{-s-t} \in \mathbf{B}(L^2(X);L^2(X)),$$

for $s > -1/2, t > 1, \lambda \in \sigma_e(H)\setminus\Lambda$.

This theorem is proved by the commutator technique based on the calculus of Ps.D.Op.'s.

Theorem 1.6 then follows from Lemma 4.3 and Theorem 4.4.

4.3 Proof of Theorem 1.7.

For a small $\epsilon > 0$ we take $\psi_\beta(t) \in C^\infty(\mathbf{R}^1)$ such that $\psi_\beta(t) = 1$ if $t > \lambda - \epsilon$, $\psi_\beta(t) = 0$ if $t < \lambda - 2\epsilon$. Let $\psi_\beta(D_{x_\beta})$ be the Ps.D.Op. with symbol $\psi_\beta(|\xi_\beta|^2)$ and put $w = \psi_\beta(D_{x_\beta})\chi_\beta(x)v$, where v is defined by (1.11) and $\chi_\beta(x)$ is given in §3. Note that by our assumption, f defined by (1.12) is rapidly decreasing. A simple calculation shows that for any $N > 0$,

$$\chi_\beta(x)\psi_\beta(D_{x_\beta})v = \psi_\beta(D_{x_\beta})\chi_\beta(x)v + O(|x_\beta|^{-N}), \tag{4.7}$$

as $|x_\beta| \to \infty$ uniformly for $|x^\beta| < a$. So, we have only to consider the asymptotic behavior of w.

Let $g = (H_\beta - \lambda)w$. Then by the same arguments as in the proof of Lemma 3.3, we have $g \in L^{2,s}$ for all $s \geq 0$. Hence

$$w = R_\beta(\lambda + i0)g.$$

Let $P_p(h^\beta)$ denote the projection onto the point spectrum of h^β. Then

$$P_p(h^\beta)w = \sum_l P_l^\beta \otimes (-\Delta_{x_\beta} - (\lambda - E_l^\beta + i0))^{-1}g,$$

where P_l^β denotes the projection onto the eigenspace of h^β with eigenvalue E_l^β. The 2-cluster scattering amplitudes are derived from the asymptotic behavior of $P_p(h^\beta)w$.

Lemma 4.5 *Let $r = |x_\beta|, \theta_\beta = x_\beta/r$. Then we have the following expansion*

$$\begin{aligned} &P_l^\beta \otimes (-\Delta_{x_\beta} - (\lambda - E_l^\beta + i0))^{-1}g \\ &\simeq C_{\beta l}(\lambda)r^{-1}e^{i\sqrt{\lambda - E_l^\beta}\,r}A_{\beta l}(\lambda;\theta_\beta,\omega)u_{\beta l}(x^\beta), \\ &C_{\beta l}(\lambda) = 2\pi i(\lambda - E^\alpha)^{-1/4}(\lambda - E_l^\beta)^{-1/4}, \end{aligned}$$

as $r = |x_\beta| \to \infty$ uniformly for $\theta_\beta \in S^2$.

Let $P_{ac}(h^\beta)$ denote the projection onto the absolutely continuous subspace for h^β. It remains to show that the asymptotic behavior of $P_{ac}(h^\beta)w$ is negligible compared with that of $P_p(h^\beta)w$. Let $h_0^\beta = -\Delta_{x^\beta}$, $h^\beta = h_0^\beta + V_\beta(x^\beta)$ and $r_\beta(z) = (h^\beta - z)^{-1}$. Let T_0 and T_β be defined by

$$(T_0 f)(\xi^\beta) = (2\pi)^{-3/2}\int_{\mathbf{R}^3} e^{-ix^\beta\cdot\xi^\beta} f(x^\beta)dx^\beta,$$

$$(T_\beta f)(\xi^\beta) = (2\pi)^{-3/2} \int_{\mathbf{R}^3} e^{-ix^\beta \cdot \xi^\beta} (f(x^\beta) - V_\beta(x^\beta) r_\beta(|\xi^\beta|^2 + i0) f) dx^\beta.$$

Let $\Omega^\beta = (T_\beta)^* T_0$. We choose $\varphi_0(t), \varphi_1(t) \in C^\infty(\mathbf{R}^1)$ such that $\varphi_0(t) + \varphi_1(t) = 1$, $\varphi_0(t) = 1$ if $t < 1/2$, $\varphi_1(t) = 1$ if $t > 2$. We define the operators $F(h^\beta \le \epsilon)$ and $F(h^\beta \ge \epsilon)$ by $F(h^\beta \le \epsilon) = \varphi_0(h^\beta/\epsilon)$, $F(h^\beta \ge \epsilon) = \varphi_1(h^\beta/\epsilon)$. We split $P_{ac}(h^\beta)w$ into two parts :

$$P_{ac}(h^\beta)w = P_{ac}(h^\beta)F(h^\beta \le \epsilon)w + F(h^\beta \ge \epsilon)w.$$

We first note that $R_\beta(z) = \Omega^\beta R_0(z)(\Omega^\beta)^*$, where $R_0(z)$ is the resolvent of $-\Delta$ in $\mathbf{R}^6$. Then by the intertwining property, we have

$$F(h^\beta \ge \epsilon)w = F(h^\beta \ge \epsilon)\Omega^\beta R_0(\lambda + i0)(\Omega^\beta)^* \tilde{F}(h^\beta \ge \epsilon/4)g,$$

where $\tilde{F}(h^\beta \ge \epsilon/4)$ is an operator similar to $F(h^\beta \ge \epsilon)$ which cuts off the part $h^\beta \le \epsilon/4$. By the Sobolev inequality, we have

$$\begin{aligned} &\|F(h^\beta \ge \epsilon)\Omega^\beta (h_0^\beta + 1)^{-2} < x^\beta >^s \|_{\mathbf{B}(L^2(\mathbf{R}^3); L^\infty(|x^\beta| < a))} \\ &\le C \|F(h^\beta \ge \epsilon) h^\beta \Omega^\beta (h_0^\beta + 1)^{-2} < x^\beta >^s \|_{\mathbf{B}(L^2(\mathbf{R}^3); L^2(|x^\beta| < a+1))}, \end{aligned}$$

the last term being finite, since for $s' > s > 0$

$$< x^\beta >^{-s'} F(h^\beta \ge \epsilon)(h^\beta)^{-1} \Omega^\beta < x^\beta >^s \in \mathbf{B}(L^2(\mathbf{R}^3); L^2(\mathbf{R}^3)),$$

which follows from the differentiability properties of generalized Fourier transforms. Therefore,

$$\begin{aligned} &\|F(h^\beta \ge \epsilon)w\|_{L^\infty(|x^\beta| < a)} \\ &\le C \| < x^\beta >^{-s} R_0(\lambda + i0)(1 + h_0^\beta)^2 (\Omega^\beta)^* \tilde{F}(h^\beta \ge \epsilon/4) g \|_{L^2(\mathbf{R}^3_{x^\beta})}. \end{aligned}$$

Let $\tilde{g} = (1 + h_0^\beta)^2 (\Omega^\beta)^* \tilde{F}(h^\beta \ge \epsilon/4) g$. Since g is rapidly decreasing, so is $\tilde{g}$. We then have

$$|(R_0(\lambda + i0)\tilde{g})(x)| \le C(1 + |x^\beta| + |x_\beta|)^{-5/2},$$

by the well-known property of the Hankel function. Hence

$$\|F(h^\beta \ge \epsilon)w\|_{L^\infty(|x^\beta| < a)} \le C(1 + |x_\beta|)^{-5/2}. \tag{4.8}$$

We next consider $P_{ac}(h^\beta)F(h^\beta \le \epsilon)w$. We define the trace of the generalized Fourier transform by $(T_\beta(\lambda)f)(\omega) = 2^{-1/2}\lambda^{1/4}(T_\beta f)(\sqrt{\lambda}\omega)$. Passing to the generalized Fourier transform, we have

$$\begin{aligned} &P_{ac}(h^\beta)F(h^\beta \le \epsilon)w \\ &= \int_0^{2\epsilon} \varphi_0(k/\epsilon)(-\Delta_{x_\beta} - \lambda + k - i0)^{-1} T_\beta(k)^* T_\beta(k) g dk. \end{aligned}$$

By the Sobolev inequality, we have

$$\begin{aligned}&\|P_{ac}(h^\beta)F(h^\beta \le \epsilon)w\|_{L^\infty(|x^\beta|<a)}\\ &\le C\|P_{ac}(h^\beta)F(h^\beta \le \epsilon)w\|_{L^2(|x^\beta|<a+1)}\\ &\le C\int_0^{2\epsilon}\varphi_0(k/\epsilon)\|(-\Delta_{x_\beta}-\lambda+k-i0)^{-1}T_\beta(k)^*T_\beta(k)g\|_{L^2(|x^\beta|<a+1)}dk,\end{aligned}$$

C being independent of ϵ as $\epsilon \to 0$. We then use the decay property of the Green's function of $-\Delta_{x_\beta}$ and see that the above integral is dominated by

$$(1+|x_\beta|)^{-1}\int_0^{2\epsilon}\| < x^\beta >^{-s} < x_\beta >^s T_\beta(k)^*T_\beta(k)g\|_{L^2(\mathbf{R}^6)}dk,$$

for some $s > 5/2$. Now, by the result of Jensen-Kato, we have

$$\|T_\beta(k)^*T_\beta(k)\|_{\mathbf{B}(L^{2,s};L^{2,-s})} \le C/\sqrt{k},$$

if $0 < k < 1$ and $s > 5/2$. Therefore, the above integral is dominated by

$$(1+|x_\beta|)^{-1}\int_0^{2\epsilon}k^{-1/2}dk = C(1+|x_\beta|)^{-1}\sqrt{\epsilon}. \tag{4.9}$$

(4.8) and (4.9) show that if $|x_\beta| < a$, $P_{ac}(h^\beta)w = o(r^{-1})$ as $r = |x_\beta| \to \infty$, which completes the proof of Theorem 1.7.

References

1 W.O.Amrein, D.B.Pearson and K.B.Sinha, Bounds on the total scattering cross-section for N-body systems, *Nuovo Cimento* 52 A (1979), 115-131.

2 W.O.Amrein and K.B.Sinha, On the three body scattering cross sections, *J. Phy.A : Math. Gen.* 15 (1982), 1567-1586.

3 C.L.Dolph, J.B.Macleod and D.Thoe, The analytic continuation of the resolvent kernel and scattering operator associated with the Schrödinger operator, *J. Math. Anal. Appl.*, 16 (1966), 311-332.

4 V.Enss, Quantum scattering theory of two and three body systems with potentials of short and long range, *Lecture Notes in Math.* 1159, Springer, Berlin-Heidelberg-New York, (1985).

5 V.Enss and B.Simon, Finite total cross sections in non-relativistic quantum mechanics, *Commun. Math. Phys.* 76 (1980), 177-209.

6 C.Gérard, Sharp propagation estimates for N-particle systems, *preprint* (1991).

7 G.M.Graf, Asymptotic completeness for N-body short-range quantum systems : A new proof, *Commun. Math. Phys.* 132 (1990), 73-101.

8 H.Isozaki, Structures of S-matrices for three body Schrödinger operators, (to appear in *Commun. Math. Phys.*).

9 H.Isozaki, Asymptotic properties of generalized eigenfunctions for three body Schrödinger operators, *preprint* (1991).

10 H.Isozaki and H.Kitada, Scattering matrices for two-body Schrödinger operators, *Scientific papers of the college of arts and sciences*, Tokyo Univ. 35 (1985), 81-107.

11 H.T.Ito and H.Tamura, Semi-classical asymptotics for total scattering cross sections of 3-body systems, *preprint* (1990).

12 A.Jensen and T.Kato, Spectral properties of Schrödinger operators and time decay of the wave functions, *Duke Math. J.* 46 (1979), 583-611.

13 I.M. Sigal and A. Soffer, The N-particle scattering problem : Asymptotic completeness for short range quantum systems, *Ann. of Math.* 125 (1987), 35-108.

14 I.M.Sigal and A.Soffer, Local decay and propagation estimates for time-dependent and time-independent Hamiltonians, preprint (1988).

15 E.Skibsted, Propagation estimates of N-body Schrödinger operators, *Commun. Math. Phys.* 142 (1991), 67-98.

16 S.P.Mercuriev, On the three-body Coulomb scattering problem, *Ann. Phys.* 130 (1980), 395-426.

17 R.G.Newton, The asymptotic form of the three-particle wave functions and the cross sections, *Ann. Phys.* 74 (1972), 324-351.

18 R.G.Newton, The three particle S-matrix, *J. Math. Phys.* 15 (1974), 338-343.

19 J.Nuttal, Asymptotic form of the three-particle scattering wave functions for free incident particles, *J. Math. Phys.* 12 (1971), 1896-1899.

20 D.Yafaev, On the multichannel scattering in two spaces, *Theo. Math. Phys.* 37 (1978), 867-874.

EIGENVALUES AND RESONANCES OF POLYATOMIC MOLECULES IN THE BORN-OPPENHEIMER APPROXIMATION

André MARTINEZ

University of Paris 13
Department of Mathematics
Institut Galilée
Avenue Jean-Baptiste Clément,
93430 Villetaneuse - FRANCE

Abstract

We describe some results of Klein-Martinez-Seiler-Wang [12] and Martinez-Messirdi [15], concerning the study of the discrete spectrum and the resonances of molecular systems in the Born-Oppenheimer approximation.

1. – Introduction

Consider a quantum system of $n+p+1$ particles, among which $n+1$ (the nuclei) are assumed to be heavy (with a mass of order $M \gg 1$), and the p other ones (the electrons) are light (i.e. with a mass of order 1). Removing the center of mass motion of this system, and choosing properly the coordinates, one can describe it with an hamiltonian of the type :

$$P = -h^2\Delta_x - \Delta_y + V(x,y) + h^2 p(\partial_y) \tag{1.1}$$

where h is a small parameter proportional to $M^{-1/2}$, V is the sum of all the interactions between the particles, and $p(\partial_y)$ is just a second order operator (the so-called isotropic term) that will have no importance for us in the sequel (and thus will be omitted). Moreover, $x \in \mathbf{R}^{3n}$ denotes the relative positions of the nuclei, and $y \in \mathbf{R}^{3p}$ those of the electrons.

Then, one defines the so-called electronic levels to be the discrete eigenvalues $\lambda_1(x) < \lambda_2(x) \leq \cdots$ of the operator

$$Q(x) = -\Delta_y + V(x,y)$$

on $L^2(\mathbf{R}_y^{3p})$.

The deep intuition of Born and Oppenheimer [3] was to realize that the study of P can be approximately reduced, when h is small, to the one of the family of operators

$$-h^2\Delta_x + \lambda_j(x) \qquad (j = 1, 2, \ldots)$$

on $L^2(\mathbf{R}_x^{3n})$.

In particular (using semiclassical arguments), when for example $\lambda_1(x)$ admits a non degenerate point well at some energy level E_0, the eigenvalues of P near E_0 should admit a complete asymptotic expansion in (half)-powers of h.

Until now, such a result was proved only for smooth interactions (Hagedorn [8], Martinez [13]), or only up to some finite order in the asymptotics (Combes-Duclos-Seiler [4]).

As we shall see, the main problem comes from the fact that, when one has singular (e.g. Coulombic) interactions, the eigenfunctions $u_j(x, y)$ of $Q(x)$ are only C^2 with respect to the x-variables (Combes-Seiler [5]).

Nevertheless, one can overcome this difficulty by introducing some x-dependent changes in the y- variables that will regularize the u_j's. Since these changes can be done only locally in x, one then has to glue them together (at least in a compact region) and to construct a kind of semiclassical pseudodifferential calculus, adapted to these changes.

This will be enough to get the complete asymptotic expansion of the discrete spectrum of P. In this case, we can even construct complete WKB expansions for the eigenfunctions of P, which allows us to estimate very precisely the tunneling in some standard situations.

However, this technique is not sufficient when one wants to study the continuous spectrum of P, or, a fortiori, its resonances, the reason for this being that the classically allowed region (with respect to x) becomes unbounded. Thus, one has first to make another change of variables whose purpose is to localize in a compact region the x-dependent singularities with respect to y in the interactions. This can be done for diatomic molecules ($n = 1$). After that, the previous ideas can be adapted to any complex dilation of P. The study of this can be reduced to the one of a matrix of smooth h-pseudodifferential operators on $L^2(\mathbf{R}^3_x)$.

2. – The Feshbach Method

This is a well-known method of reduction, introduced by Feshbach in [6], also used in [4], and that we present here in a slightly different way.

For the sake of simplicity, we take into account only the first electronic level $\lambda_1(x)$, and we call $u_1(x) = u_1(x, y)$ the first eigenfunction of $Q(x)$, normalized in $L^2(\mathbf{R}^{3p}_y)$ for any x in $\mathbf{R}^{3n}$. We also assume that $\lambda_1(x)$ is separated by a constant gap $\delta > 0$ from the rest of the spectrum of $Q(x)$.

For $\lambda \in \mathbf{C}$, $\mathrm{Re}\lambda < \mathrm{Inf}(\sigma(Q(x))\backslash\{\lambda_1(x)\})$, consider the matrix-operator

$$\mathbf{P}_\lambda = \begin{pmatrix} P - \lambda & u_1(x) \\ \langle \cdot, u_1(x)\rangle_{L^2(\mathbf{R}^{3p})} & 0 \end{pmatrix} \tag{2.1}$$

acting on $L^2(\mathbf{R}^{3(n+p)}) \oplus L^2(\mathbf{R}^{3n})$.

It is then straighforward to see that $\mathbf{P}_\lambda$ is invertible as an operator from $H^2(\mathbf{R}^{3(n+p)}) \oplus L^2(\mathbf{R}^{3n})$ into $L^2(\mathbf{R}^{3(n+p)}) \oplus H^2(\mathbf{R}^{3n})$, and if we denote

$$\mathbf{P}_\lambda^{-1} = \begin{pmatrix} E(\lambda) & E^+(\lambda) \\ E^-(\lambda) & E^{-+}(\lambda) \end{pmatrix}$$

we have the formal identities

$$\begin{aligned} (P-\lambda)^{-1} &= E(\lambda) - E^+(\lambda)[E^{-+}(\lambda)]^{-1}E^-(\lambda) \\ (E^{-+}(\lambda))^{-1} &= -\langle (P-\lambda)^{-1}(\cdot u_1), u_1\rangle_{L^2(\mathbf{R}^{3p})} \end{aligned} \tag{2.2}$$

where $(\cdot u_1)$ denotes the operator of multiplication by u_1.

In particular, we get the reduction result :

Proposition 2.1. — $\lambda \in \sigma(P) \iff 0 \in \sigma(E^{-+}(\lambda))$.

Moreover, one can compute directly that :

$$\lambda - E^{-+}(\lambda) = -h^2\Delta_x + \lambda_1(x) + h^4\Pi\Delta_x\hat{\Pi}(\hat{P}-\lambda)^{-1}\hat{\Pi}\Delta_x\Pi \tag{2.3}$$

where $\Pi = \Pi(x)$ is the orthogonal projection of $L^2(\mathbf{R}^{3p})$ on $u_1(x)$, $\hat{\Pi} = 1 - \Pi$, and $\hat{P}$ is the restriction of $\hat{\Pi}$ to $\{v \in H^2(\mathbf{R}^{3n+3p})\,;\, \hat{\Pi}v = v\}$.

Formula (2.3) and Proposition (2.1) show that the spectral study of P on $L^2(\mathbf{R}^{3(n+p)})$ is close to the one of $-h^2\Delta_x + \lambda_1(x)$ on $L^2(\mathbf{R}^{3n})$ (at least modulo $\mathcal{O}(h^2)$).

3. – The regular case

If one wants to get high order asymptotics, the problem comes from the remainder term appearing in (2.3), where the abstract resolvent $(\hat{P}-\lambda)^{-1}$ has no obvious asymptotics with respect to h.

However, when the interaction $V(x,y)$ is regular, in the sense that :

$$V \in C^\infty(\mathbf{R}_x^{3n}\ ;\ \mathcal{L}(H^2(\mathbf{R}^{3p}), L^2(\mathbf{R}^{3p})))$$

there is a simple way to prove that $E^{-+}(\lambda)$ is actually a smooth h-pseudodifferential operator. Indeed, $\mathbf{P}_\lambda$ can be considered as a h-pseudodifferential operator in x, associated to the operator-valued symbol

$$p_\lambda(x,\xi) = \begin{pmatrix} \xi^2 + Q(x) - \lambda & u_1(x) \\ \langle \cdot, u_1(x)\rangle_{L^2(\mathbf{R}^{3p})} & 0 \end{pmatrix}$$

which is a function in $C^\infty(T^*\mathbf{R}^{3n}\ ;\ \mathcal{L}(H^2\oplus L^2, L^2\oplus L^2))$.

Since the usual pseudofifferential calculus can be generalized to such symbols (Balazard-Konlein [1], Gérard-Martinez-Sjöstrand [7]), and $p_\lambda(x,\xi)$ is invertible for any $(x,\xi) \in T^*\mathbf{R}^{3n}$, we get that $\mathbf{P}_\lambda^{-1}$ also is a similar h-pseudodifferential operator. As a consequence, so is $E^{-+}(\lambda)$, with principal symbol $\lambda - \xi^2 - \lambda_1(x)$.

4. – The Polyatomic Case

When one wants to apply the pseudodifferential calculus as in the previous paragraph, one needs to differentiate $V(x,y)$ to all order with respect to x. But now, if one assumes that the interactions between particles may have Coulomb-type singularities, then these singularities become worth with each differentiation, and everything falls down.

Then, the trick consists in regularizing the interactions, but not by modifying them. Instead, in a certain sense, it will be the notion of C^∞-regularity itself that will be changed.

To explain this here, we only consider a sum of interactions of the type $\frac{\pm 1}{|y-x_j|}, \frac{+1}{|x_j-x_k|}$, where y, x_j are in $\mathbf{R}^3$.

For ellipticity reasons (due to the fact that the nuclei-nuclei interaction is repulsive), one can assume that x is away from the collision set :

$$\mathcal{C} = \{x = (x_1, \ldots, x_n) \in \mathbf{R}^{3n} \ ; \ \exists j \neq k, \ x_j = x_k\}$$

For $x_0 = (x_1^0, \ldots, x_1^n) \in \mathbf{R}^{3n} \backslash \mathcal{C}$ and x close enough to x_0, we then consider the x-dependant change of the y-variables $y \mapsto y'$, defined by

$$y = F(x,y') \stackrel{\text{déf}}{=} y' + \sum_{j=1}^{n} (x_j - x_j^0)\, f_j(y') \tag{4.1}$$

where $f_j \in C_0^\infty(\mathbf{R}^3)$, $f_j(x_0^k) = \delta_{jk}$.

(This type of transformations has already been used by Hunzikeer for the study of resonances [11]).

We associate to it the (unitary with respect to y) operator

$$\mathcal{U}_{x_0} \varphi(x,y) = \varphi(x, F(x,y'))|\det \partial_{y'} F(x,y')|^{1/2} \tag{4.2}$$

Since by construction we have

$$F(x, x_j^0) = x_j$$

it is then easy to see that for any j :

$$\mathcal{U}_{x_0}\left(\frac{1}{|y-x_j|}\right) \in C^\infty(\Omega_{x_0}, \mathcal{L}(H^2(\mathbf{R}^{3p}), L^2(\mathbf{R}^{3p}))) \tag{4.3}$$

where Ω_{x_0} is a small neighborhood of x_0.

Then, fixing an energy level $\lambda_0 < \mathrm{Inf}(\sigma(Q(x)) \backslash \lambda_1(x))$, ans assuming that $\lambda_1^{-1}(]-\infty, \lambda_0])$ is compact, one can construct a finite number of transformations $(F_\ell)_{1 \le \ell \le N}$ like (4.1), associated to transformations U_ℓ satisfying (4.3), each one defined for x in some Ω_ℓ, such that the Ω_ℓ's cover $\lambda_1^{-1}(]-\infty,\lambda_0])$.

Denoting $W = \mathbf{R}^{3n} \setminus \bigcup_{\ell=1}^{N} \Omega_\ell$, and using the ellipticity of $P-\lambda$ in W (for $\lambda \leq \lambda_0$), one can moreover modify P in a neighborhood of W, in such a way that the modified operator $\widetilde{P}$ is smooth with respect to x in W, and (thanks to some Agmon estimates) $\sigma(\widetilde{P})$ differs from $\sigma(P)$ on $]-\infty,\lambda_0]$ by exponentially small error terms.

Denoting also $\Omega_0 = \mathbf{R}^{3n} \setminus \lambda_1^{-1}(]-\infty,\lambda_0])$, and using a partition of unity $(\chi_\ell^4)_{0\leq\ell\leq N}$ adapted to $(\Omega_\ell)_{0\leq\ell\leq N}$, one can then see by (4.3) that the matrix operator $\widetilde{\mathbf{P}}_\lambda$ associated to $\widetilde{P}$ as in (2.1) can be written :

$$\widetilde{\mathbf{P}}_\lambda = \sum_{\ell=0}^{N} \chi_\ell(x)\mathcal{U}_\ell\, \chi_\ell(x)\, \check{\mathbf{P}}_\ell(\lambda)\, \chi_\ell(x)\mathcal{U}_\ell^{-1}\, \chi_\ell(x) \tag{4.4}$$

where the $\check{\mathbf{P}}_\ell(\lambda)$'s are smooth pseudodifferential operators, with operator-valued symbol, and where $\mathcal{U}_0$ is just the identity.

Then, the technical difficulty consists in showing there exists a nice calculus for operators of type (4.4). Once this is done, we get that $\widetilde{\mathbf{P}}_\lambda^{-1}$ is also of this type, and thus, writing :

$$\widetilde{\mathbf{P}}_\lambda^{-1} = \begin{pmatrix} \tilde{E}(\lambda) & \tilde{E}^{+}(\lambda) \\ \tilde{E}^{-}(\lambda) & \tilde{E}^{-+}(\lambda) \end{pmatrix}$$

and using the fact that the transformations $\mathcal{U}_\ell$ only act on the y-variables, we get :

THEOREM 4.1. — *$\tilde{E}^{-+}(\lambda)$ is a smooth h-pseudodifferential operator, with principal symbol* $\lambda - (\xi^2 + \lambda_1(x))$.

Using finally the analyticity with respect to ξ of the symbol, one can then get by standard arguments (cf. Helffer-Sjöstrand [9]) the WKB-type expansions for the eigenfunctions of P, in the case where the well $\lambda_1(]-\infty,\lambda_0])$ is a subvariety of $\mathbf{R}^{3n}$ and is non degenerate.

Moreover, when the interactions are rotationnally invariant, the group $\mathcal{O}(3)$ acts on the well, and this action becomes free for non planar molecules. In this case, the well is necessarily non connected, and a double-well problem appear. Then, using the WKB expansions, estimates on the tunneling effect can also be obtained.

All the details of these results, as well as some generalizations, can be found in [12].

5. – Resonances

We now describe the results of [15], which concern the case of diatomic molecules ($n = 1$).

Assuming that the interactions are dilation analytic in the sense of Balslev-Combes [2], one can work with the distorded hamiltonian P_μ ($\mu \in \mathbf{C}$ small enough, $\mathrm{Im}\mu > 0$) obtained by the distortion (cf. [11]) :

$$\begin{aligned} &x \mapsto x + \mu v(x) \\ &y_j \mapsto y_j + \mu v(y_j) \qquad (1 \leq k \leq p) \end{aligned}$$

where v is a smooth vector field of $\mathbf{R}^3$, equal to the identity outside a compact set.

Under this distortion, the singularities of the potential are not changed, and one can try to apply to P_μ the same techniques as before. But to get resonances, we can no longer assume that $\lambda_1(]-\infty, \lambda_0])$ is compact. For this reason, the constructions of the previous paragraph cannot be generalized directly.

A first attempt to localize the x-dependent singularities in the y-variables, could be to make the change of variables : $y \mapsto y' = \dfrac{1}{|x|} y$ (at least for $x \notin C$). But then, the operator $\dfrac{\partial}{\partial x}$ is changed to $\dfrac{\partial}{\partial x} + \dfrac{x}{|x|}\left(y' \dfrac{\partial}{\partial y'}\right)$, so that now the domain of the operator with respect to y depends on x. To avoid this problem, we instead use a change of the type :

$$y \mapsto y' = \theta_{|x|}(y)$$

where

$$\begin{aligned} &\theta_{|x|}(y) = \frac{1}{|x|} y \qquad \text{for} \quad |y| \leq |x| \\ &\theta_{|x|}(y) = y \qquad \text{for} \quad |y| \geq 2|x| \end{aligned}$$

and $\theta_{|x|}(y)$ depends smoothly on y and $|x|$.

After such a change, the singularities of the interactions become localized at $y'_j = \dfrac{\pm x}{|x|}$ for some j.

Since these singularities stay in a compact set with respect to y', one can then use the techniques of Section 4 to conclude (denoting again $\widetilde{P}_\mu$ the regularization of P_μ in the elliptic region) :

THEOREM 5.1. — *For any $\lambda \in \mathbf{C}$ with $\mathrm{Re}\lambda < \mathrm{Inf}(\sigma(Q(x)) \backslash \{\lambda_1(x)\})$, and $\mu \in \mathbf{C}$ small enough ($\mathrm{Im}\mu > 0$), there exists a smooth h-pseudodifferential operator $\tilde{E}_\mu^{-+}(\lambda)$ on $\mathbf{R}^3$ such that :*

$$\lambda \in \sigma(\tilde{P}_\mu) \Longleftrightarrow 0 \in \sigma(\tilde{E}_\mu^{-+}(\lambda)).$$

In fact, one can also prove that $\tilde{E}_\mu^{-+}(\lambda)$ is nothing but some regularization of the distortion of $E^{-+}(\lambda)$ defined in Section 2. In particular, its principal symbol on $T^* \lambda_1^{-1}(]-\infty, \lambda_0])$ is given by the distortion of $\lambda - (\xi^2 + \lambda_1(x))$.

Theorem 5.1 can also be generalized to several electronic levels, in which case $\tilde{E}_\mu^{-+}(\lambda)$ becomes a matrix of pseudodifferential operators.

Since the spectrum of $\widetilde{P}_\mu$ is exponentially close to the spectrum of P_μ, this allows us to give a rather precise localization of the resonances of P, in some standard situations (see e.g. [10], [14]).

References

[1] **A. Balazard-Konlein** : *Calcul fonctionel pour des opérateurs h-admissibles à symbole opérateurs et applications.* Thèse 3ème cycle, University of Nantes (1985).

[2] **E. Balslev, J.M. Combes** : *Spectral Properties of Many-body Schrödinger Operators with Dilation-analytic Interactions.* Comm. Math. Phys. 22, 280-294 (1971).

[3] **M. Born, R. Oppenheimer** : *Zur quantentheorie der Molekeln.* Annalen der Physik 84, p. 457 (1927).

[4] **J.M. Combes, P. Duclos, R. Seiler** : *The Born-Oppenheimer Approximation.* Rigorous atomic and molecular physics, (eds. G. Velo, A. Wightman), p. 185-212, Plenum New-York (1981).

[5] **J.M. Combes, R. Seiler** : *Regularity and Asymptotic Properties of the Discrete Spectrum of Electronic Hamiltonians.* Int. J. Quant. Chem., Vol. XIV, p. 213-229, (1978).

[6] **H. Feshbach** : *Unified Theory of Nuclear Reactions.* I and II, Ann. Phys. 5, 363 (1958), and 19, 287 (1962).

[7] **C. Gérard, A. Martinez, J. Sjöstrand** : *A Mathematical Approach to the Effective Hamiltonian in Perturbed Periodic Problems*, to appear in Comm. Math. Phys.

[8] **G.A. Hagedorn** : *High Order Corrections to the Time-Independent Born-Oppenheimer Approximation. I - Smooth Potentials.* Ann. Inst. H. Poincaré 47, p. 1-16 (1987); and *II - Diatomic Coulomb Systems.* Comm. Math. Phys. 116, p. 23-44 (1988).

[9] **B. Helffer, J. Sjöstrand** : *Puits multiples en mécanique semi-classique VI.* Ann. Inst. Henri Poincaré 46, p. 353-372 (1987).

[10] **B. Helffer, J. Sjöstrand** : *Résonances en limite semi-classique.* Mémoires S.M.F., Tome 114, Fasc. 3 (1986).

[11] **W. Hunziker** : *Distorsion Analyticity and Molecular Resonance Curves.* Ann. Inst. Henri Poincaré 45, p. 339-358 (1986).

[12] **M. Klein, A. Martinez, R. Seiler, X.P. Wang** : *On the Born-Oppenheimer Expansion for Polyatomic Molecules.* Preprint Univ. Paris-Nord (1990), and to appear in Comm. Math. Phys.

[13] **A. Martinez** : *Developpements asymptotiques et effet tunnel dans l'approximation de Born-Oppenheimer.* Ann. Inst. Henri Poincaré 49, p. 239-257 (1989).

[14] **A. Martinez** : *Résonances dans l'approximation de Born-Oppenheimer.* I - J. Diff. Eq. 91 n°2 (1991) and II - Comm. Math. Phys. 135 (1991).

[15] **A. Martinez, B. Messirdi** : *Resonances of Diatomic Molecules in the Born-Oppenheimer Approximation*, in preparation.

Asymptotic Neutrality of Polyatomic Molecules

Mary Beth Ruskai [1], Jan Philip Solovej [2]

[1]Department of Mathematics, University of Massachusetts • Lowell, Lowell, Massachusetts 01854
[2]Department of Mathematics, Princeton University, Princeton, N.J. 08544

Abstract: Polyatomic molecules are studied in the limit as the total charge Z becomes infinite with the number of nuclei and their charge ratios fixed. It is shown that, in the Born-Oppenheimer approximation, if such a system has a stable bound state then it is asymptotically neutral in the sense that it satisfies the inequality $|Z - N| < C_1 Z^{1-\epsilon}$ where N denotes the number of electrons, C_1 is a positive constant and we can choose $\varepsilon = 1/7$. The proof uses comparisons with Thomas-Fermi theory, as in the proof in [23] of a similar result for diatomic molecules. A critical element in the proof is a demonstration that, for a stable configuration, all internuclear distances are bounded below by $C_2 Z^{(-1/3)(1-\epsilon)}$ for some positive constant C_2

1 Introduction

In the last 10 years there has been considerable progress in studying the stability of multi–particle systems. It is now well–established that sufficiently negative ions do not have bound states both for atomic (see, e.g., [3,5,10,11,13,14,18,21,22]) and for molecular [10,16] systems. For diatomic molecules with dynamic nuclei, Ruskai [15,17] showed that extremely positive systems have no bound states. Vugal'ter [26] recently announced an extension of this result to polyatomic molecules in which *all* of the nuclei have sufficiently large charges. Solovej [23] proved that, if the Born–Oppenheimer approximation is used, then stable diatomic molecules are asymptotically neutral. Duclos and Hogreve [4] showed that if the nuclear charges in a homonuclear diatomic molecule are sufficiently large, then the molecule is unstable in the sense that all Born-Oppenheimer energy curves are monotonically decreasing.

The purpose of this paper is to extend the result of [23] to polyatomic molecules in which the number of nuclei and their charge ratios are fixed. The correct asymptotics of the excess charge for polyatomic molecules as the number of nuclei tends to infinity seems to be an open question, whose answer may depend upon the shape of the molecular system.

We study a molecule with N electrons (charge $e = 1$) and M nuclei with charges $Z_1, \ldots, Z_M > 0$ where $Z_k = Zz_k$ so that Z is the total charge, i.e., $\sum_k z_k = 1$. We will use the Born-Oppenheimer approximation so that the locations of the nuclei $\mathbf{R}_1, \ldots, \mathbf{R}_M$ can be regarded as parameters in $\mathbb{R}^3$. The Hamiltonian for a molecule in the Born-Oppenheimer approximation is

$$H_{\mathrm{mol}}(N, \underline{Z}, \underline{\mathbf{R}}) = \sum_{i=1}^{N} [-\Delta_i - V(x_i)] + \sum_{1 \le i < j \le N} \frac{1}{|x_i - x_j|} + U \quad , \tag{1}$$

where $\underline{Z} = (Z_1, \ldots, Z_M)$ and $\underline{\mathbf{R}} = (\mathbf{R}_1, \ldots, \mathbf{R}_M)$, the attraction between an electron at $x \in \mathbb{R}^3$ and the nuclei is described by the potential $-V$ with

$$V(x) = \sum_{k=1}^{M} Z_k |x - \mathbf{R}_k|^{-1} \quad ,$$

and U describes the nuclear-nuclear repulsion

$$U = \sum_{1 \le \ell < k \le M} Z_\ell Z_k |\mathbf{R}_\ell - \mathbf{R}_k|^{-1} \quad .$$

In the atomic case, $M = 1$, and the term U is absent. The operator H_{mol} acts on a dense domain of suitable (core corresponding to H^1) functions in the antisymmetric space

$$\mathcal{H} = \bigwedge^{N} L^2(\mathbb{R}^3; \mathbb{C}^2) \quad .$$

Let

$$E^0_{\mathrm{mol}}(N, \underline{Z}, \underline{\mathbf{R}}) = \inf \operatorname{spec}_{\mathcal{H}} H_{\mathrm{mol}}(N, \underline{Z}, \underline{\mathbf{R}}) \quad ,$$

and define the Born-Oppenheimer **ground state energy** of the molecule to be

$$E^0_{\mathrm{mol}}(N, \underline{Z}) = \inf_{\underline{\mathbf{R}}} E^0_{\mathrm{mol}}(N, \underline{Z}, \underline{\mathbf{R}}).$$

We say that the molecule given by $(N, \underline{Z})$ has a **stable bound state** if *both* of the following conditions are satisfied.

(a) There is at least one configuration $\underline{\mathbf{R}}_0$ such that

$$E^0_{\mathrm{mol}}(N, \underline{Z}) = E^0_{\mathrm{mol}}(N, \underline{Z}, \underline{\mathbf{R}}_0),$$

and $E^0_{\rm mol}(N,\underline{Z})$ is a discrete eigenvalue (i.e., an eigenvalue strictly below the continuum) for $H_{\rm mol}(N,\underline{Z},\underline{\mathbf{R}}_0)$

(b) $E^0_{\rm mol}(N,\underline{Z}) < \lim_{\lambda\to\infty}\inf\{E^0_{\rm mol}(N,\underline{Z},\underline{\mathbf{R}})|\max_{k\neq\ell}|\mathbf{R}_k-\mathbf{R}_\ell|>\lambda\}$

We call a configuration $\underline{\mathbf{R}}_0$ such that (a) is satisfied a **stable configuration.** If $\psi\in\mathcal{H}$ denotes the ground state of $H_{\rm mol}$ with eigenvalue $E^0_{\rm mol}$ then the **ground state electronic density** is

$$\rho(x) = N\int \|\psi(x,x_2,\ldots,x_N)\|^2_{\wedge\mathbb{C}^2}\, d^3x_2\ldots\, d^3x_N \quad ,$$

and the **ground state two point correlation** is

$$\rho^{(2)}(x,y) = N(N-1)\int \|\psi(x,y,x_3,\ldots,x_N)\|^2_{\wedge\mathbb{C}^2}\, d^3x_3\ldots\, d^3x_N \quad ,$$

where $\|\ \|^2_{\wedge\mathbb{C}^2}$ denotes summation over spin.

Condition (b) implies that the ground state energy $E^0_{\rm mol}(N,\underline{Z})$ is strictly less than the ground state energy of any decomposition into two or more non–interacting subsystems. To be more precise if $1\leq L\leq M$ let $\alpha = \{\alpha_1,\ldots,\alpha_L\}$ be a decomposition of $\{1,\ldots,M\}$ into L disjoint subsets α_n i.e., an L–cluster decomposition of $\{1,\ldots,M\}$. Then for all corresponding decompositions of the number of electrons $(N_1,\ldots,N_L)$ with $\sum_n^L N_n = N$

$$E^0_{\rm mol}(N,\underline{Z}) < \sum_{n=1}^{L} E^0_{\rm mol}(N_n,(Z_k;k\in\alpha_n)) \quad . \tag{2}$$

In the case where a cluster only contains one nucleus we will usually write $E^0_{\rm at}$ instead of $E^0_{\rm mol}$, reflecting that such a cluster, in fact, describes an atom.

We shall study such molecules in the limit as $Z\to\infty$ with M and $z_1,\ldots,z_M$ fixed, and show that the existence of a stable bound state implies that the molecule is asymptotically neutral. This can be stated more precisely as follows.

Theorem 1 (Asymptotic Neutrality) *Given $z_1,\ldots,z_M$ then in the sense of convergence of sets*

$$\lim_{Z\to\infty}\{\frac{N}{Z}|\text{The molecule } (N,\underline{Z}) \text{ has a stable bound state}\} = \{1\} \quad . \tag{3}$$

In fact, there exist $C_1, C_2 > 0$ depending only on M and $z_1,\ldots,z_M$ such that

$$|Z-N| < C_1 Z^{1-\varepsilon} \quad , \tag{4}$$

with $\varepsilon = 1/7$. For any stable configuration $\underline{\mathbf{R}}$

$$R_{\rm min} \equiv \min_{k\neq\ell}|\mathbf{R}_k-\mathbf{R}_\ell| > C_2 Z^{-(1/3)(1-\varepsilon)} \quad . \tag{5}$$

The proof here follows the strategy of [23]. Indeed, in Sect. 3 we shall show that the bound (5) can be proved by reducing to the diatomic case.

It should be remarked that without the Pauli exclusion principle, i.e., the restriction to the antisymmetric space, the above theorem would be false (see [23]).

As in [23] the main idea is to compare with Thomas–Fermi (TF) theory. For an excellent introduction to the mathematical aspects of TF theory we refer to Lieb's review article [9]. (The recent review of Spruch [24] may also be consulted.) In the TF model it is a consequence of the celebrated Teller's theorem that a molecule will split into M independent atoms. The exact break up is studied in Sect. 4. To prove Theorem 1 we shall hence compare the energy of the molecule with a break up into M independent atoms. This is certainly not the most energetically favorable break up, but it is simple and sufficient for our purposes. The essential ingredient is the bound (5) which implies that $R_{\min} \gg Z^{-1/3}$ which is the atomic scale. This bound, which is proven in Sect. 3, allows one to localize in clusters of independent atoms. In Sect. 2 we summarize some standard results from TF theory, as well as some technical results from [23], that we shall use in proving (4). The proof of (4) requires two bounds. The first is that the excess positive charge is negligible this is proved in Sect. 5; the second is that the excess negative charge is negligible. Since the proof of the second bound is exactly as in the diatomic case, we refer the reader to Sect. 8 of [23] rather than repeating the argument here.

2 Preliminary Results

We first remark that for all choices of M and $z_1, \ldots, z_M$ there are constants A_1 and A_2 depending on $z_1, \ldots, z_M$ and M such that if H_{mol} has a stable bound state, then

$$A_1 Z \leq N \leq A_2 Z \quad . \tag{6}$$

The upper bound was proved in the Born–Oppenheimer approximation by Lieb [10] who shows that $N \leq 2Z + M$; a weaker upper bound was established for dynamic nuclei by Ruskai [16]. A lower bound of the form $N \geq C(Z z_{\min})$ (in which case $A_1 = C z_{\min}$) was recently announced by Vugal'ter [26] for dynamic nuclei. Since details of his proof are not yet available, and since a proof of the lower bound of (6) in the Born–Oppenheimer case can be obtained by a simple modification of the method of Sect. 5, we present this in an appendix.

Secondly we state some estimates from TF theory that we shall need in our subsequent analysis. The energy functional in TF theory is defined to be

$$\mathcal{E}^{\text{TF}}(\rho, \underline{Z}, \underline{\mathbf{R}}) = \int \rho^{5/3} - \int V\rho + \tfrac{1}{2} \iint \rho(x)|x-y|^{-1}\rho(y)\, d^3x\, d^3y + U \quad , \tag{7}$$

where the potentials U and V are as in (1). The TF energy is given by minimizing this functional:

$$E^{\mathrm{TF}}(N,\underline{Z},\underline{\mathbf{R}}) = \inf\{\mathcal{E}^{\mathrm{TF}}(\rho,\underline{Z},\underline{\mathbf{R}})|\rho \in L^1(\mathbb{R}^3), \int \rho = N\} \quad . \tag{8}$$

The energy E^{TF} is, for fixed $\underline{Z}$ and $\underline{\mathbf{R}}$, a non-increasing convex function of N. The chemical potential is defined as $\mu(N,\underline{Z},\underline{\mathbf{R}}) = -\frac{\partial E^{\mathrm{TF}}}{\partial N}$. To distinguish the atomic ($M = 1$) and molecular ($M > 1$) cases we shall as, before, write $E_{\mathrm{at}}^{\mathrm{TF}}$ and $E_{\mathrm{mol}}^{\mathrm{TF}}$ respectively. Likewise we write $\mu_{\mathrm{at}}(N,Z)$ and $\mu_{\mathrm{mol}}(N,\underline{Z},\underline{\mathbf{R}})$. If $N \leq Z$ (i.e., the molecule is neutral or positively charged), then there exists a unique minimizing ρ in (8). We shall denote this minimizer as $\rho_{\mathrm{mol}}^{\mathrm{TF}}(x,N) = \rho_{\mathrm{mol}}^{\mathrm{TF}}(x,N,\underline{Z},\underline{\mathbf{R}})$ or $\rho_{\mathrm{at}}^{\mathrm{TF}}(x,N)$ respectively.

We shall need the following two basic theorems from TF theory; proofs can be found in the review [9] or in the original paper [12].

Theorem 2 *If $R_{\min} > cZ^{-1/3}$ and $N > AZ$ for some constants $c, A > 0$ then as $Z \to \infty$ the TF energy E^{TF} behaves as $Z^{7/3}$ and approximates the quantum energy according to*

$$E^{\mathrm{TF}}(N,\underline{Z},\underline{\mathbf{R}}) - E^0(N,\underline{Z},\underline{\mathbf{R}}) = O(Z^{7(1-\varepsilon)/3}) \quad , \tag{9}$$

for some $\varepsilon > 0$. Furthermore we have the following operator inequality

$$\sum_{i=1}^{N}\Big(- \Delta_i - V(x_i) + \rho_{\mathrm{mol}}^{\mathrm{TF}} * |x_i|^{-1}\Big) + U \geq E^0(N,\underline{Z},\underline{\mathbf{R}})$$
$$+ \tfrac{1}{2}\iint \rho_{\mathrm{mol}}^{\mathrm{TF}}(x)|x-y|^{-1}\rho_{\mathrm{mol}}^{\mathrm{TF}} - CZ^{7(1-\varepsilon)/3} \quad . \tag{10}$$

The error term in (9) and the constant in (10) depend on $z_1,\ldots,z_M$, M, c and A but are independent of $\underline{\mathbf{R}}$.

It is worth noting that Ivrii and Sigal [8] recently showed that the next term in an asymptotic expansion of the energy has the cZ^2 form conjectured by Scott, which implies that one can actually *choose $\varepsilon = 1/7$ above, as well as in all other theorems of this paper.* Their work extends to molecules a result established earlier for atoms by Siedentop and Weikard [19, 20] and also partially by Hughes [7]; more recently, further refinements in the energy asymptotics have been proved by Fefferman and Seco [6] in the atomic case. For a review of much of this work see the joint contribution by Fefferman, Ivrii, Seco and Sigal in the present volume.

Theorem 3 (Teller's No Binding Theorem) *For any cluster decomposition of $\{1,\ldots,M\}$ as described in the paragraph preceding (2) there exists a sequence of (not necessarily integer valued) numbers $N_1,\ldots,N_L$ (where L*

as before denotes the number of clusters) with $\sum_{n=1}^{L} N_n = N$ *such that the TF energies satisfy*

$$E_{\text{mol}}^{\text{TF}}(N, \underline{Z}) \geq \sum_{n=1}^{L} E_{\text{mol}}^{\text{TF}}(N_n, (Z_k; k \in \alpha_n)) \quad . \tag{11}$$

Comparing with (2) we see that the physical consequence of no binding in TF theory follows if (11) holds for even one decomposition. However, for our analysis below it is important that it actually holds for *all* decompositions.

Next, we present estimates that allow us to compare the quantum density and 2-point correlation function with the corresponding quantities in TF theory. The results given here are almost identical to those given in [23] Sects. 4 and 6. But since the results there were not stated completely adequately, we shall here give the proof of a more precise formulation.

Let g_1 be the multiple of the characteristic function of the unit ball on $\mathbb{R}^3$ which satisfies $\int g_1 = 1$, and let $g_Z(x) = Z^3 g_1(Zx)$ denote the indicated rescaled function. Then g_Z is supported on the ball of radius Z^{-1} and satisfies $\int g_Z = 1$. Define the function $K_N : \mathbb{R}^{3N} \to \mathbb{R}$ by

$$K_N(\underline{x}) = \tfrac{1}{2} \iint R_N(\underline{x}; x)|x - y|^{-1} R_N(\underline{x}; y)\, d^3x\, d^3y \quad , \tag{12}$$

where $\underline{x} = (x_1, \ldots, x_N)$ and

$$R_N(\underline{x}; x) = \sum_{i=1}^{N} g_Z(x - x_i) - \rho_{\text{mol}}^{\text{TF}}(x) \quad . \tag{13}$$

The TF theory implies the following operator bound for the molecular Hamiltonian involving the non-negative function K_N considered as a multiplication operator on $\mathcal{H}$.

Lemma 4 *If* $A_1 Z \leq N \leq A_2 Z$ *and* $R_{\text{min}} > cZ^{-1/3}$ *we have*

$$H_{\text{mol}}(N, \underline{Z}, \underline{\mathbf{R}}) \geq E^0(N, \underline{Z}, \underline{\mathbf{R}}) + K_N - CZ^{7(1-\varepsilon)/3} \quad , \tag{14}$$

with $\varepsilon = 1/7$ *and where the constant* C *depends on* $z_1, \ldots, z_M$, M, A_1, A_2 *and* c.

Remark. We shall see in Lemma 7 below that the condition $R_{\text{min}} > cZ^{-1/3}$ for some $c > 0$ follows if H_{mol} has a stable bound state. In view of (6) we can thus conclude that all assumptions of Lemma 4 are satisfied if H_{mol} has a stable bound state.

Proof. By smearing the point charges we obtain the inequality

$$\sum_{i<j}|x_i-x_j|^{-1} \geq \tfrac{1}{2}\iint\left(\sum_{i=1}^{N} g_Z(x-x_i)\right)|x-y|^{-1}\left(\sum_{i=1}^{N} g_Z(y-x_i)\right)d^3x\,d^3y$$
$$-\tfrac{1}{2}N\iint g_Z(x)|x-y|^{-1}g_Z(y)\,d^3x\,d^3y$$
$$\geq K_N(\underline{x})+\sum_{i=1}^{N}\rho_{\mathrm{mol}}^{\mathrm{TF}}*g_Z*|x_i|^{-1}$$
$$-\tfrac{1}{2}\iint \rho_{\mathrm{mol}}^{\mathrm{TF}}(x)|x-y|^{-1}\rho_{\mathrm{mol}}^{\mathrm{TF}}(y)\,d^3x\,d^3y-(\mathrm{const.})NZ \quad , \tag{15}$$

where the constant in front of the last term is exactly

$$\tfrac{1}{2}\iint g_1(x)|x-y|^{-1}g_1(y)\,d^3x\,d^3y \quad .$$

Since g_Z is spherically symmetric, it follows from Newton's theorem (in which g_Z is regarded as the distribution of a charge density) that $g_Z*|y|^{-1} = |y|^{-1}$ for $y \notin \mathrm{supp}(g_Z)$; therefore

$$\rho_{\mathrm{mol}}^{\mathrm{TF}}*g_Z*|x|^{-1}-\rho_{\mathrm{mol}}^{\mathrm{TF}}*|x|^{-1} \geq -\int_{y\in \mathrm{supp}(g_Z)}\rho_{\mathrm{mol}}^{\mathrm{TF}}(x-y)|y|^{-1}\,d^3y \quad .$$

Combining this with the following well-known (see [9]) inequality from TF theory $\rho_{\mathrm{mol}}^{\mathrm{TF}}(x) \leq (\sum_k Z_k|x-\mathbf{R}_k|^{-1})^{3/2}$ and the inequality $\int_{|y|\leq Z^{-1}}|x-y|^{-3/2}|y|^{-1}\,d^3y \leq (\mathrm{const.})Z^{-1/2}$ we obtain that

$$\rho_{\mathrm{mol}}^{\mathrm{TF}}*g_Z*|x|^{-1} \geq \rho_{\mathrm{mol}}^{\mathrm{TF}}*|x|^{-1}-(\mathrm{const.})Z \quad .$$

Thus from (15)

$$\sum_{i<j}|x_i-x_j|^{-1} \geq K_N(\underline{x})+\sum_{i=1}^{N}\rho_{\mathrm{mol}}^{\mathrm{TF}}*|x_i|^{-1}$$
$$-\tfrac{1}{2}\iint \rho_{\mathrm{mol}}^{\mathrm{TF}}(x)|x-y|^{-1}\rho_{\mathrm{mol}}^{\mathrm{TF}}(y)\,d^3x\,d^3y-(\mathrm{const.})NZ \quad , \tag{16}$$

Since $N \leq A_2 Z$ it follows that the error term in (16) is $O(Z^2) = O(Z^{7(1-\varepsilon)/3})$ with $\varepsilon = 1/7$. Inserting the bound (16) into (1) and using (10) gives (14). □

Lemma 5 *For and fixed k with $1 \leq k \leq M$ let $\lambda(x)$ and $\eta(x)$ denote functions satisfying*

(a) $\lambda, \eta \in C^\infty(\mathbb{R}^3)$ *with* $0 \leq \lambda(x) \leq 1$ *and* $0 \leq \eta(x) \leq 1$

(b) $\mathrm{supp}(\lambda) \subset \{x| \quad |x - \mathbf{R}_k| < R(Z)\}$ *with* $R(Z) = \beta Z^{-1/3(1-\epsilon)}$
(c) $\mathrm{supp}(\eta) \subset \{x| \quad |x - \mathbf{R}_k| \ > \ R_\eta\}$ *for some* $R_\eta > 2R(Z)$
(d) $\|\nabla\lambda\|_\infty < c_1 R(Z)^{-1}$

If $A_1 Z < N < A_2 Z$ *and* $R_{\min} > cZ^{-1/3}$ *we have the following estimates on the ground state density and 2-point correlation for* Z *large enough*

$$|\int [\rho(x) - \rho^{\mathrm{TF}}_{\mathrm{mol}}(x, N)]\lambda(x)\, d^3x| \leq C' Z^{(1-\epsilon)} \quad , \tag{17}$$

and

$$\left|\iint \left[\rho^{(2)}(x,y) - \rho^{\mathrm{TF}}_{\mathrm{mol}}(y,N)\rho(x)\right] |x-y|^{-1}\eta(x)^2\lambda(y)\, d^3x\, d^3y\right| \leq$$
$$C'\left(Z^{(1-\epsilon)} R_\eta^{-1} \int \rho(x)\eta(x)^2\, d^3x + Z^{(5+\epsilon)/6}\|\nabla\eta\|_{L^\infty} R_\eta^{-1}\right) \tag{18}$$

where the constant C' *depends on the same quantities as in Lemma 4 as well as* β *and* c_1.

These two estimates are easily proved using the main technical lemma from [23], i.e.,

Lemma 6 *If* $A_1 Z \leq N \leq A_2 Z$ *and* $R_{\min} > cZ^{-1/3}$ *then given non-negative functions* $\chi \in C^\infty(\mathbb{R}^3 \times \mathbb{R}^3)$ *and* $\eta \in C^\infty(\mathbb{R}^3)$ *we have*

$$\left|\iint \left[\rho^{(2)}(x,y) - \rho^{\mathrm{TF}}_{\mathrm{mol}}(y, N-1)\rho(x)\right]\eta(x)^2\chi(x,y)\, d^3x\, d^3y\right| \leq$$
$$C\int \rho(x)\eta(x)^2 dx \left\{ Z^{(7/6)(1-\epsilon)} \sup_x \|\nabla_y \chi_x\|_{L^2(\mathbb{R}^3)} + \|\nabla_y\chi\|_{L^\infty(\mathbb{R}^3\times\mathbb{R}^3)}\right\}$$
$$+ CZ^{1/2}\|\nabla\eta\|_{L^\infty(\mathbb{R}^3)} \sup_x \|\nabla_y\chi_x\|_{L^2(\mathbb{R}^3)} \left(\int \rho(x)\eta(x)^2\, d^3x\right)^{1/2} \quad . \tag{19}$$

The constant C *depends on the same quantities as the constant in Lemma 4.*

Proof. First of all we shall use the simple estimate

$$|\chi_x * g_Z - \chi_x| \leq Z^{-1}\|\nabla_y\chi_x\|_{L^\infty(\mathbb{R}^3\times\mathbb{R}^3)} \quad ,$$

to replace χ_x by $\chi_x * g_Z$. We obtain that

$$\left|\iint \left[\rho^{(2)}(x,y) - \rho^{\mathrm{TF}}_{\mathrm{mol}}(y, N-1)\rho(x)\right]\eta(x)^2\chi(x,y)\, d^3x\, d^3y\right|$$
$$\leq \left|\iint \left[\rho^{(2)}(x,y)\chi_x * g_Z(y) - \rho^{\mathrm{TF}}_{\mathrm{mol}}(y,N-1)\rho(x)\chi_x(y)\right]\eta(x)^2\, d^3x\, d^3y\right|$$
$$+ \int \rho(x)\eta(x)^2\, d^3x (N-1)Z^{-1}\|\nabla_y\chi\|_{L^\infty(\mathbb{R}^3\times\mathbb{R}^3)} \quad .$$

We can now rewrite the first term above in the following way

$$\iint \left[\rho^{(2)}(x,y)\chi_x * g_Z(y) - \rho^{\mathrm{TF}}_{\mathrm{mol}}(y,N-1)\rho(x)\chi_x(y)\right]\eta(x)^2\,d^3x\,d^3y =$$

$$\sum_{i=1}^{N}\iint \|\psi(\underline{x})\|^2_{\wedge\mathbb{C}^2} R_{N-1}(x_1,\ldots,\hat{x}_i,\ldots,x_N,y)\chi_{x_i}(y)\eta(x_i)^2\,d^3y\,d^{3N}\underline{x} \quad ,$$

where R_{N-1} was defined in (13) and the notation $\hat{x}_i$ denotes that this variable is absent.

Using Fourier transform it is easy to see that

$$\begin{aligned}&\left|\int R_{N-1}(x_2,\ldots,x_N;y)\chi_{x_1}(y)\,d^3y\right|\\ &\le (\text{const.})K_{N-1}(x_2,\ldots,x_N)^{1/2}\|\nabla_y\chi_{x_1}\|_{L^2(\mathbb{R}^3)}.\end{aligned}$$

By applying the Cauchy-Schwarz inequality and then combining the estimates above we find that

$$\begin{aligned}&\left|\iint \left[\rho^{(2)}(x,y) - \rho^{\mathrm{TF}}_{\mathrm{mol}}(y,N-1)\rho(x)\right]\eta(x)^2\chi(x,y)\,d^3x\,d^3y\right|\\ &\le \alpha^{1/2}\left(\int \rho(x)\eta(x)^2\,d^3x\right)^{1/2}\sup_x\|\nabla_y\chi_x\|_{L^2(\mathbb{R}^3)}\\ &\quad+\int \rho(x)\eta(x)^2\,d^3x(N-1)Z^{-1}\|\nabla_y\chi\|_{L^\infty} \quad ,\end{aligned}$$

where

$$\alpha = \sum_{i=1}^{N}\iint \|\psi(\underline{x})\|^2_{\wedge\mathbb{C}^2}K_{N-1}(x_1,\ldots,\hat{x}_i,\ldots,x_N)\eta(x_i)^2\,d^3x_1\cdots d^3x_N \quad .$$

We can get an upper bound on α by separating off one electron from the Hamiltonian in the following way

$$\begin{aligned}&E^0(N,\underline{Z},\mathbf{R})\int \rho(x)\eta(x)^2\,d^3x = \sum_{i=1}^{N}\langle\psi|\,\eta(x_i)^2H_{\mathrm{mol}}\,|\psi\rangle\\ &\ge \sum_{i=1}^{N}\langle\psi|\,\eta(x_i)H^{(i)}_{\mathrm{mol}}\eta(x_i) + \eta(x_i)(-\Delta_i - V(x_i))\eta(x_i) - (\nabla\eta(x_i))^2\,|\psi\rangle \quad .\end{aligned}$$

Here $H^{(i)}_{\mathrm{mol}}$ denotes the molecular Hamiltonian with particle i removed. Since $\eta(x_i)\psi$ is antisymmetric in all variables except x_i we can use Lemma 4 for the term above involving $H^{(i)}_{\mathrm{mol}}$. If we furthermore use that

$$-\Delta_i - V(x_i) = \sum_{k=1}^{M}(-M^{-1}\Delta_i - Z_k|x-\mathbf{R}_k|^{-1}) \ge -(\text{const.})Z^2 \quad ,$$

we get with $\varepsilon = 1/7$

$$[E^0(N,\underline{Z},\underline{\mathbf{R}}) - E^0(N-1,\underline{Z},\underline{\mathbf{R}})] \int \rho(x)\eta(x)^2\, d^3x$$
$$\geq -(\text{const.}) \left(Z^{7(1-\varepsilon)/3} \int \rho(x)\eta(x)^2\, d^3x + N\|(\nabla\eta)^2\|_{L^\infty} \right) + \alpha \quad .$$

Since $E^0(N,\underline{Z},\underline{\mathbf{R}}) \leq E^0(N-1,\underline{Z},\underline{\mathbf{R}})$ we have

$$\alpha \leq (\text{const.}) \left(Z^{7(1-\varepsilon)/3} \int \rho(x)\eta(x)^2\, d^3x + N\|(\nabla\eta)^2\|_{L^\infty} \right)$$

from which the lemma follows. □

Remark. For simplicity we have replaced $\rho^{\mathrm{TF}}_{\mathrm{mol}}(x, N-1)$ by $\rho^{\mathrm{TF}}_{\mathrm{mol}}(x, N)$ in Lemma 5. It is not difficult to see that the lemma is still correct with this replacement.

3 Estimates on $R_{\min}$

In order to prove (5), we first need to establish a similar, but weaker, result which holds without any assumptions on the relationship between N and Z, in particular, without assuming that the lower bound holds in (6).

Lemma 7 *If H_{mol} has a stable bound state, then for any stable configuration $R_{\min} \geq cN^{-1/3}$ where c depends only on $z_1, \ldots, z_M$.*

Proof. The well-known 'united atom' bound (see, e.g., [25]) states that

$$E^0_{\mathrm{mol}}(N,\underline{Z}) \geq E^0_{\mathrm{at}}(N,Z) + \sum_{k<\ell} \frac{Z_k Z_\ell}{|\mathbf{R}_\ell - \mathbf{R}_k|} \quad .$$

However, by stability the molecular energy is smaller than the energy corresponding to the decomposition where all electrons are in one cluster together with the nucleus Z_{ℓ_0}. Thus

$$E^0_{\mathrm{mol}}(N,\underline{Z}) \leq E^0_{\mathrm{at}}(N, Z_{\ell_0}).$$

Combining these yields

$$\sum_{\{k:k\neq\ell_0\}} \frac{Z_k Z_{\ell_0}}{|\mathbf{R}_{\ell_0} - \mathbf{R}_k|} \leq \sum_{k<\ell} \frac{Z_k Z_\ell}{|\mathbf{R}_\ell - \mathbf{R}_k|} \leq E^0_{\mathrm{at}}(N, Z_{\ell_0}) - E^0_{\mathrm{at}}(N,Z)$$
$$\leq \gamma Z_{\ell_0}(Z - Z_{\ell_0})N^{1/3} \quad ,$$

where $\gamma > 0$ is a constant and the last inequality was proved by Ruskai [15]. Now choose k_0 and ℓ_0 so that $R_{\min} = |\mathbf{R}_{\ell_0} - \mathbf{R}_{k_0}|$. Then

$$\frac{z_{k_0}}{1 - z_{\ell_0}} \frac{1}{R_{\min}} \leq \sum_{\{k:k \neq \ell_0\}} \left(\frac{Z_k}{Z - Z_{\ell_0}} \right) \frac{1}{|\mathbf{R}_{\ell_0} - \mathbf{R}_k|} \leq \gamma N^{1/3} \quad . \tag{20}$$

Now it follows that $R_{\min} \geq N^{-1/3} z_{\min}/\gamma(1 - z_{\min})$ where $z_{\min}$ has the obvious meaning. □

Using the upper bound in (6), it follows immediately that $R_{\min} \geq cZ^{-1/3}$. We next show that this result can actually be improved to $R_{\min} \geq cZ^{-(1-\epsilon)/3}$. The proof of the following theorem relies heavily on the estimates from TF theory given in the previous section.

Theorem 8 *If H_{mol} has a stable bound state, then for any stable configuration $R_{\min} \geq cZ^{-1/3(1-\epsilon)}$ where c depends only on $z_1, \dots, z_M$, and M.*

Proof. By the last remark, one can apply Theorem 2 so that there exist a_1 and a_2 such that for any choices of positive integers N_k, $k = 1, \dots, M$, with $\sum_k N_k = N$

$$\begin{aligned} a_1 Z^{7/3(1-\epsilon)} + \sum_k E_{\text{at}}^{\text{TF}}(N_k, Z_k) &\geq \sum_k E_{\text{at}}^0(N_k, Z_k) > E_{\text{mol}}^0(N, \underline{Z}, \underline{\mathbf{R}}) \\ &\geq E_{\text{mol}}^{\text{TF}}(N, \underline{Z}, \underline{\mathbf{R}}) - a_2 Z^{7/3(1-\epsilon)} \quad , \end{aligned}$$

where the second inequality follows from the stability condition (2) since $\underline{\mathbf{R}}$ is a stable configuration. Thus

$$E_{\text{mol}}^{\text{TF}}(N, \underline{Z}, \underline{\mathbf{R}}) - \sum_k E_{\text{at}}^{\text{TF}}(N_k, Z_k) \leq a Z^{7/3(1-\epsilon)} \quad . \tag{21}$$

We will denote the supremum of the left side of (21) over the set of all decompositions $\{(N_1, \dots, N_M) | \sum_k N_k = N\}$ by $\Gamma(N, \underline{Z}, \underline{\mathbf{R}})$. Teller's Theorem implies $\Gamma(N, \underline{Z}, \underline{\mathbf{R}}) > 0$. We will now argue that for any pair $(\mathbf{R}_\ell, \mathbf{R}_m)$ from $\underline{\mathbf{R}}$, there exists a decomposition $(N_1, \dots, N_M)$ with $\sum_k N_k = N$ such that

$$\Gamma(N, \underline{Z}, \underline{\mathbf{R}}) \geq \Gamma(N_\ell + N_m, (Z_\ell, Z_m), (\mathbf{R}_\ell, \mathbf{R}_m)) \quad . \tag{22}$$

To see this let $N' = N - N_\ell - N_m$ and let $\underline{Z}'$ and $\underline{\mathbf{R}}'$ denote the $M - 2$ component vectors corresponding to a molecule with the ℓ-th and m-th nuclei omitted. [Thus, for example, in the simple case $\ell = 1, m = 2$, we have $\underline{Z}' = (Z_3, \dots, Z_M)$, and $\underline{\mathbf{R}}' = (\mathbf{R}_3, \dots, \mathbf{R}_M)$]. Then, temporarily altering our notation so that the subscript *mol* is replaced by one denoting the number of nuclei,

$$E_{\mathrm{M}}^{\mathrm{TF}}(N,\underline{Z},\underline{\mathbf{R}})-\sum_k E_{\mathrm{at}}^{\mathrm{TF}}(N_k,Z_k)=\Big[E_{\mathrm{M}}^{\mathrm{TF}}(N,\underline{Z},\underline{\mathbf{R}})$$
$$-E_{\mathrm{M-2}}^{\mathrm{TF}}(N',\underline{Z}',\underline{\mathbf{R}}')-E_2^{\mathrm{TF}}(N_\ell+N_m,(Z_\ell,Z_m),(\mathbf{R}_\ell,\mathbf{R}_m))\Big]$$
$$+\Big[E_{\mathrm{M-2}}^{\mathrm{TF}}(N',\underline{Z}',\underline{\mathbf{R}}')-\sum_{k\neq m,\ell}E_{\mathrm{at}}^{\mathrm{TF}}(N_k,Z_k)\Big]$$
$$+\Big[E_2^{\mathrm{TF}}(N_\ell+N_m,(Z_\ell,Z_m),(\mathbf{R}_\ell,\mathbf{R}_m))-E_{\mathrm{at}}^{\mathrm{TF}}(N_\ell,Z_\ell)-E_{\mathrm{at}}^{\mathrm{TF}}(N_m,Z_m)\Big]$$

By Teller's Theorem we can choose $N_\ell+N_m$ so that the term in the first set of square brackets [ ] above, which represent a break up of the molecule into a diatomic molecule and a molecule with $M-2$ nuclei, is positive. Next, we can consider a full atomic decomposition $(N_k : 1\leq k\leq M, k\neq m,\ell)$ with $\sum_{k\neq m,\ell}N_k=N'$ of the latter so that Teller's Theorem again implies that the term in the second set of [ ] is positive. Finally we can choose N_ℓ and N_m such that the last term in [ ] is exactly $\Gamma(N_\ell+N_m,(Z_\ell,Z_m),(\mathbf{R}_\ell,\mathbf{R}_m))$. We hence obtain (22). From (21) we now conclude that

$$\Gamma(N_\ell+N_m,(Z_\ell,Z_m),(\mathbf{R}_\ell,\mathbf{R}_m))\leq aZ^{7/3(1-\epsilon)}\quad . \tag{23}$$

It was proved in [23] (see paragraph following equation (32) in [23]) using results from Benguria [1] and Brezis and Lieb [2] that if $|\mathbf{R}_\ell-\mathbf{R}_m|>cZ^{-1/3}$ then

$$\Gamma(N_\ell+N_m,(Z_\ell,Z_m),(\mathbf{R}_\ell,\mathbf{R}_m))\geq C|\mathbf{R}_\ell-\mathbf{R}_m|^{-7}\quad .$$

The previous lemma established that we can apply this result to complete the proof. □

4 Break up in Thomas-Fermi theory

In this section we shall describe how molecules split up in Thomas-Fermi theory. As was already described in Sect. 3 a molecule with M nuclei will in TF theory split into M independent atoms. The question is how will the electrons distribute themselves among the atoms and how close will the actual molecular density be to a sum of atomic densities. We shall assume in this section that $N<Z$.

It is clear that the break up will result in M atoms with electron numbers $N_1^{\mathrm{TF}},\ldots,N_M^{\mathrm{TF}}$ (not necessarily integers) that minimize $E_{\mathrm{at}}^{\mathrm{TF}}(N_1^{\mathrm{TF}},Z_1)+\ldots+E_{\mathrm{at}}^{\mathrm{TF}}(N_M^{\mathrm{TF}},Z_M)$ under the constraint $N_1^{\mathrm{TF}}+\ldots+N_M^{\mathrm{TF}}=N$. We arrive at the well-known condition that all the chemical potentials $\mu_{\mathrm{at}}(N_k^{\mathrm{TF}},Z_k)$ for the atoms must be identical

$$\mu_{\mathrm{at}}(N_k^{\mathrm{TF}},Z_k)=\mu_{\mathrm{mol}}(N,\underline{Z},\infty),\quad k=1,\ldots M$$

where, for reasons which will become clear below, $\mu_{\rm mol}(N, \underline{Z}, \infty)$ denotes the common value of these chemical potentials. The main results of this section are the following estimates on the molecular TF density $\rho_{\rm mol}^{\rm TF}(x, N)$.

Lemma 9 *If $cZ^{-1/3} < R' < R_{\rm min}/2$ then we get for all $k = 1, \ldots, M$*

$$\int_{|x-\mathbf{R}_k|<R'} \rho_{\rm mol}^{\rm TF}(x, N)\, d^3x = N_k^{\rm TF} + O(R'^{-3}) \tag{24}$$

and if $|x - \mathbf{R}_k| > 3R_{\rm min}/4$

$$\int_{|y-\mathbf{R}_k|<R'} \rho_{\rm mol}^{\rm TF}(y, N)|x-y|^{-1}\, d^3y = (N_k^{\rm TF} + O(R'^{-3}))\, |x - \mathbf{R}_k|^{-1} \quad . \tag{25}$$

Proof. To study the molecular TF density we introduce the TF potential for the molecule as the function

$$\varphi_{\rm mol}^{\rm TF}(x) = \sum_{k=1}^{M} Z_k |x - \mathbf{R}_k|^{-1} - \rho_{\rm mol}^{\rm TF} * |x|^{-1} \quad .$$

The functions $\rho_{\rm mol}^{\rm TF}$ and $\varphi_{\rm mol}^{\rm TF}$ satisfy the TF equation, i.e., the Euler-Lagrange equation for the minimization problem in (8)

$$\rho_{\rm mol}^{\rm TF}(x)^{2/3} = [\varphi_{\rm mol}^{\rm TF}(x) - \mu_{\rm mol}(N, \underline{Z}, \underline{\mathbf{R}})]_+ \quad , \tag{26}$$

where as before $\mu_{\rm mol}(N, \underline{Z}, \underline{\mathbf{R}})$ is the chemical potential for the molecule and $[\cdot]_+$ means $\max\{0, \cdot\}$. We now define $N_k^{\rm TF}(\underline{\mathbf{R}})$ to be the number of electrons in a TF atom with nuclear charge Z_k and chemical potential equal to the molecular chemical potential $\mu_{\rm mol}(N, \underline{Z}, \underline{\mathbf{R}})$. Correspondingly we denote the TF density and potential function for such an atom centered at $\mathbf{R}_k$ by $\rho_k^{\rm TF}$ and $\varphi_k^{\rm TF}$. These functions satisfy a TF equation similar to (26).

The monotonicity of TF theory implies that

$$\max\{\varphi_1^{\rm TF}(x), \ldots, \varphi_M^{\rm TF}(x)\} \leq \varphi_{\rm mol}^{\rm TF}(x) \leq \sum_{k=1}^{M} \varphi_k^{\rm TF}(x) \quad . \tag{27}$$

Using the TF equations and the estimate $\varphi_k^{\rm TF}(x) \leq c|x - \mathbf{R}_k|^{-4}$ we can translate (27) into the following estimate for the densities,

$$\rho_k^{\rm TF}(x) < \rho_{\rm mol}^{\rm TF}(x) < \rho_k^{\rm TF}(x) + c|x - \mathbf{R}_k|^{-2} R'^{-4} \quad \text{for } |x - \mathbf{R}_k| < R' \quad . \tag{28}$$

From this estimate we find that

$$\begin{aligned}\int_{|x-\mathbf{R}_k|<R'} \rho_{\rm mol}^{\rm TF}(x)\, d^3x &= \int_{|x-\mathbf{R}_k|<R'} \rho_k^{\rm TF}(x)\, d^3x + O(R'^{-3}) \\ &= N_k^{\rm TF}(\underline{\mathbf{R}}) + O(R'^{-3}) \quad .\end{aligned}$$

The $\underline{\mathbf{R}}$ dependent electron number $N_k^{\mathrm{TF}}(\underline{\mathbf{R}})$ can be compared with N_k^{TF} as follows. First of all it is easy again using (27) and (28) to see that

$$N - O(R'^{-3}) < \sum_{k=1}^{M} N_k^{\mathrm{TF}}(\underline{\mathbf{R}}) < N \quad .$$

Since $\mu_{\mathrm{at}}(N_k^{\mathrm{TF}}, Z_k)/\mu_{\mathrm{at}}(N_k^{\mathrm{TF}}(\underline{\mathbf{R}}), Z_k)$ is independent of k for $k = 1, \ldots, M$ we conclude (see also [23]) from well-known properties of TF theory (see [9]) that

$$0 < \kappa_1 < \left|(N_\ell^{\mathrm{TF}}(\underline{\mathbf{R}}) - N_\ell^{\mathrm{TF}})/(N_k^{\mathrm{TF}}(\underline{\mathbf{R}}) - N_k^{\mathrm{TF}})\right| < \kappa_2 \quad .$$

for constants κ_1 and κ_2 depending on $z_1, \ldots, z_M$. Since $N = \sum_{k=1}^{M} N_k^{\mathrm{TF}}$ we arrive at (24); then (25) follows since ρ_k^{TF} is spherically symmetric. □

Combining Lemmas 5 and 9 we can obtain estimates on the true density and two-point correlation functions.

We shall also need the following estimate on the ratio between ionization charges in a break up in TF theory.

Proposition 10 *If $\mu_{\mathrm{mol}}(N, \underline{Z}, \infty) > 0$ then the ionization ratios are bounded above and below by positive constants κ and κ' depending on $z_1, \ldots, z_M$, i.e.,*

$$\kappa < (Z_k - N_k^{\mathrm{TF}})/(Z_\ell - N_\ell^{\mathrm{TF}}) < \kappa' \quad .$$

If $\mu_{\mathrm{mol}}(N, \underline{Z}, \infty) = 0$ then $Z_k = N_k^{\mathrm{TF}}$

Proof. The particle number in TF theory is uniquely given if we know the chemical potential. Thus we shall prove the estimate by controlling the chemical potential. From the scaling of TF theory

$$\mu_{\mathrm{at}}(N', Z') = Z'^{4/3}\mu_{\mathrm{at}}(N'/Z', 1) \quad .$$

If $\mu_{\mathrm{at}}(N'/Z', 1)$ is bounded away from zero then N'/Z' is bounded above by a number strictly less than 1. The only problem is if $\mu_{\mathrm{at}}(N_k/Z_k, 1)$ and $\mu_{\mathrm{at}}(N_\ell/Z_\ell, 1)$ approach zero. Recall that $\mu_{\mathrm{at}}(N_k/Z_k, 1)/\mu_{\mathrm{at}}(N_\ell/Z_\ell, 1) = z_k^{4/3}/z_\ell^{4/3}$ and thus they approach zero at the same time. The estimate on the ionization ratios follows from the asymptotic relation $\mu_{\mathrm{at}}(N'/Z', 1) \sim (\mathrm{const.})(1 - N'/Z')^{4/3}$ when $N'/Z' \sim 1$. (see [9]). □

In the next section, we shall also use the facts which follow from Proposition 10 that $Z_k - N_k^{\mathrm{TF}} > 0$ for all $k = 1, \ldots, M$ when $Z > N$ and $Z_k = N_k^{\mathrm{TF}}$ when $Z \leq N$

5 Bound on the Excess Positive Charge

Bounding the excess positive charge means giving a lower bound on the number of electrons N for which the molecule is stable. We thus assume that the molecule is stable and that the nuclei are in a stable configuration $\underline{\mathbf{R}} = \underline{\mathbf{R}}_0$ as defined by stability condition (a). We shall furthermore assume $N < Z$.

We first choose localizing functions $0 \leq \theta_k \in C^\infty(\mathbb{R}^3)$, $k = 0, \ldots, M$ having the following properties.

1. For $k \geq 1$ we have $\theta_k(x) = \theta(|x - \mathbf{R}_k|/R_{\text{min}})$, where θ is a function in $C^\infty(\overline{\mathbb{R}_+})$ satisfying $0 \leq \theta \leq 1$ with $\theta(t) = 1$ if $t < 1/5$ and $\theta(t) = 0$ if $t > 1/4$.
2. $\sum_{k=0}^{M} \theta_k(x)^2 = 1$ (which defines θ_0).

These properties imply

3. $|\nabla\theta_k(x)| \leq CR_{\text{min}}^{-1}$ for all k

We shall now use these localizing functions to compare the energy of the molecule with the energy of M independent atoms. Let again ψ be the ground state of H_{mol} with eigenvalue E_{mol}^0. Then

$$\begin{aligned} E_{\text{mol}}^0 &= \langle\psi| \prod_{i=1}^{N} \left(\sum_{k=0}^{M} \theta_k(x_i)^2 \right) H_{\text{mol}} |\psi\rangle \\ &= \sum_{\alpha} \langle\psi| \Theta_\alpha H_{\text{mol}} \Theta_\alpha - \sum_{i=1}^{N} (\nabla_i \Theta_\alpha)^2 |\psi\rangle \quad , \end{aligned} \qquad (29)$$

where the sum is over all $(M+1)$-cluster decompositions $\alpha = \{\alpha_0, \ldots, \alpha_M\}$ of $\{1, \ldots, N\}$ and $\Theta_\alpha(x_1, \ldots, x_N) = \prod_{k=0}^{M} \prod_{i \in \alpha_k} \theta_k(x_i)$. The last term in (29) is referred to as the *localization error* (see, e.g. [3]). We interpret α as a decomposition of the N electrons into $M+1$ clusters corresponding to the M nuclei, plus an additional cluster α_0 corresponding to the outer electrons. It follows from the properties of θ_k that electrons in α_k are localized near the k-th nucleus and are a distance of at least $3R_{\text{min}}/4$ from all other nuclei, while those in α_0 are at least $R_{\text{min}}/5$ from all the nuclei.

For any given cluster decomposition α we can write the Hamiltonian as

$$H_{\text{mol}} = \sum_{k=0}^{M} H_{\alpha_k} + I_\alpha \quad ,$$

where the cluster Hamiltonians are

$$H_{\alpha_k} = \sum_{i\in\alpha_k}\Big(-\Delta_i - Z_k|x_i-\mathbf{R}_k|^{-1}\Big) + \sum_{\substack{i,j\in\alpha_k\\ i<j}} |x_i-x_j|^{-1} \quad \text{for} \quad k\geq 1 \quad ,$$

$$H_{\alpha_0} = \sum_{i\in\alpha_0} -\Delta_i + \sum_{\substack{i,j\in\alpha_0\\ i<j}} |x_i-x_j|^{-1} \quad .$$

Notice that the Hamiltonian for the outer electrons satisfies $H_{\alpha_0}\geq 0$ and the atomic Hamiltonians for $k\geq 1$ satisfy $H_{\alpha_k}\geq E^0_{\rm at}(N_k(\alpha),Z_k)$ where $N_k(\alpha)$ *is the number of elements in the cluster* α_k. The intercluster interaction is

$$\begin{aligned} I_\alpha =& \sum_{k=1}^{M}\sum_{i\notin\alpha_k} -Z_k|x_i-\mathbf{R}_k|^{-1} + \sum_{k<\ell}\sum_{i\in\alpha_k}\sum_{j\in\alpha_\ell}|x_i-x_j|^{-1} \\ &+\sum_{\ell<k} Z_\ell Z_k|\mathbf{R}_\ell-\mathbf{R}_k|^{-1} \quad , \end{aligned}$$

We can also write this as

$$\begin{aligned} I_\alpha =& \sum_{\ell<k}\Bigg(Z_kZ_\ell|\mathbf{R}_k-\mathbf{R}_\ell|^{-1} - Z_k\sum_{i\in\alpha_\ell}|x_i-\mathbf{R}_k|^{-1} - Z_\ell\sum_{i\in\alpha_k}|x_i-\mathbf{R}_\ell|^{-1} \\ &+\sum_{\substack{i\in\alpha_\ell\\ j\in\alpha_k}}|x_i-x_j|^{-1}\Bigg) \\ &+\sum_{k=1}^{M}\sum_{i\in\alpha_0}\Bigg(-Z_k|x_i-\mathbf{R}_k|^{-1}+\sum_{j\in\alpha_k}|x_i-x_j|^{-1}\Bigg) \quad . \end{aligned} \tag{30}$$

We then get from (29) and the *stability condition* (2) (notice that $\Theta_\alpha\psi$ is antisymmetric with respect to interchanging electrons within the same cluster)

$$\begin{aligned} E^0_{\rm mol} &\geq \sum_\alpha \langle\psi|\,\Theta^2_\alpha\left(\sum_{k=0}^{M}E^0_{\rm at}(N_k(\alpha),Z_k)+I_\alpha\right) - \sum_{i=1}^{N}|\nabla_i\Theta_\alpha|^2\,|\psi\rangle \\ &\geq E^0_{\rm mol}+\sum_\alpha\langle\psi|\,\Theta^2_\alpha I_\alpha - \sum_{i=1}^{N}|\nabla_i\Theta_\alpha|^2\,|\psi\rangle \quad . \end{aligned} \tag{31}$$

Using property 2. of the localizing functions θ_k and (30) we can write $\sum_\alpha\Theta^2_\alpha I_\alpha = \sum_{0\leq k<\ell\leq M} I_{k\ell}$ where for $k,\ell\geq 1$, $I_{k\ell}$ is the effective interaction between the k-th and ℓ-th screened nuclei.

$$\begin{aligned} I_{k\ell} =& Z_kZ_\ell|\mathbf{R}_k-\mathbf{R}_\ell|^{-1} - Z_k\sum_{i=1}^{N}\theta_\ell(x_i)^2|x_i-\mathbf{R}_k|^{-1} \\ &- Z_\ell\sum_{i=1}^{N}\theta_k(x_i)^2|x_i-\mathbf{R}_\ell|^{-1} + \sum_{\substack{i,j\\ i\neq j}}\theta_k(x_i)^2\theta_\ell(x_j)^2|x_i-x_j|^{-1} \end{aligned} \tag{32}$$

and for $\ell \geq 1$, $I_{0\ell}$ is the interaction of all the outer electrons with all the screened nuclei

$$I_{0\ell} = \sum_{i=1}^{N} \theta_0(x_i)^2 \left(-Z_\ell |x_i - \mathbf{R}_\ell|^{-1} + \sum_{j=1}^{N} \theta_\ell(x_j)^2 |x_i - x_j|^{-1} \right) \quad . \tag{33}$$

For the second (or localization error) term in (31) we get the bound (recall that constants depend on M)

$$\begin{aligned} \sum_\alpha \sum_{i=1}^{N} |\nabla_i \Theta_\alpha|^2 &= \sum_{k=0}^{M} \sum_{i=1}^{N} \left(\nabla_i \theta_k(x_i) \right)^2 \leq (\text{const.}) N R_{\text{min}}^{-2} \\ &\leq (\text{const.}) Z^{1+(1/3)(1-\varepsilon)} R_{\text{min}}^{-1} \leq (\text{const.}) Z^{2(1-\varepsilon)} R_{\text{min}}^{-1} \quad , \end{aligned} \tag{34}$$

where we have used property 3. of the functions θ_k. We shall now estimate the contributions from the interactions $I_{k\ell}$ using Lemmas 5 and 9. We begin with the interaction between two screened nuclei, i.e., $k, \ell \geq 1$. Then

$$\begin{aligned} \langle \psi | \, I_{k\ell} \, | \psi \rangle &= Z_k Z_\ell |\mathbf{R}_k - \mathbf{R}_\ell|^{-1} \\ &- Z_\ell \int \rho(x) \theta_k(x)^2 |x - \mathbf{R}_\ell|^{-1} \, d^3x - Z_k \int \rho(x) \theta_\ell(x)^2 |x - \mathbf{R}_k|^{-1} \, d^3x \\ &+ \iint \rho^{(2)}(x,y) \theta_k(y)^2 \theta_\ell(x)^2 |x - y|^{-1} \, d^3x \, d^3y \quad . \end{aligned} \tag{35}$$

Unfortunately we cannot use Lemma 5 on (35) directly. The problem is that θ_k is not defined with support in $\{|x - \mathbf{R}_k| < R(Z)\}$ (as required by (b) in Lemma 5) but rather with support in $\{|x - \mathbf{R}_k| < (1/4) R_{\text{min}}\}$. Appealing to Theorem 8 we can assume that β (from (b) in Lemma 5) has been chosen such that $R_{\text{min}} > 3R(Z)$. If we define $\tilde{\theta}_k(x) = \theta(|x - \mathbf{R}_k|/R(Z))$ for $k \geq 1$ we can use Lemma 5 with

$$\lambda(x) = \tilde{\theta}_k(x)^2 |\mathbf{R}_k - \mathbf{R}_\ell| (4|x - \mathbf{R}_\ell|)^{-1}$$

(the factor $|\mathbf{R}_k - \mathbf{R}_\ell|/4$ is included only to guarantee that $\lambda(x) \leq 1$) and (25) of Lemma 9 to conclude that

$$\int \rho(x) \tilde{\theta}_k(x)^2 |x - \mathbf{R}_\ell|^{-1} \, d^3x = \left(N_k^{\text{TF}} + O(Z^{1-\varepsilon}) \right) |\mathbf{R}_k - \mathbf{R}_\ell|^{-1} \quad . \tag{36}$$

We also have

$$\int \rho(x) \tilde{\theta}_k(x)^2 \, d^3x = N_k^{\text{TF}} + O(Z^{1-\varepsilon}) \quad . \tag{37}$$

Thus since $N_1^{\text{TF}} + \ldots + N_M^{\text{TF}} = N$ we conclude from (37) that

$$\sum_{k=1}^{M} \int \rho(x) (\theta_k(x)^2 - \tilde{\theta}_k(x)^2) \, d^3x = \int \rho(x) (1 - \sum_{k=1}^{M} \tilde{\theta}_k(x)^2) \, d^3x = O(Z^{(1-\varepsilon)}) \quad .$$

We can use this to control the error in replacing θ_k by $\tilde{\theta}_k$ in (35). We find that

$$\int \rho(x)\theta_k(x)^2|x-\mathbf{R}_\ell|^{-1}\,d^3x = \int \rho(x)\tilde{\theta}_k(x)^2|x-\mathbf{R}_\ell|^{-1}\,d^3x + \int \rho(x)[\theta_k(x)^2-\tilde{\theta}_k(x)^2]|x-\mathbf{R}_\ell|^{-1}\,d^3x = |\mathbf{R}_k-\mathbf{R}_\ell|^{-1}(N_k^{\mathrm{TF}}+O(Z^{(1-\varepsilon)})) \quad . \qquad (38)$$

We estimate the last term in (35) again using Lemmas 5 and 9. Thus

$$\begin{aligned}\iint \rho^{(2)}(x,y)\theta_k(y)^2\theta_\ell(x)^2|x-y|^{-1}\,d^3x\,d^3y \geq\\ &\iint \rho^{(2)}(x,y)\tilde{\theta}_k(y)^2\theta_\ell(x)^2|x-y|^{-1}\\ &\geq \iint \rho_{\mathrm{mol}}^{\mathrm{TF}}(y,N)\rho(x)|x-y|^{-1}\tilde{\theta}_k(y)^2\theta_\ell(x)^2\,d^3x\,d^3y\\ &\quad -(\mathrm{const.})\left(Z^{(1-\varepsilon)}\int \rho(x)\theta_\ell(x)^2\,d^3x + Z^{(5+\varepsilon)/6}R_{\min}^{-1}\right)|\mathbf{R}_k-\mathbf{R}_\ell|^{-1}\\ &\geq \left(N_k^{\mathrm{TF}}+O(Z^{(1-\varepsilon)})\right)\int \rho(x)\theta_\ell(x)^2|x-\mathbf{R}_k|^{-1}\,d^3x\\ &\quad -(\mathrm{const.})\left(\int \rho(x)\theta_\ell(x)^2\,d^3x Z^{(1-\varepsilon)} + Z^{(5+\varepsilon)/6}R_{\min}^{-1}\right)|\mathbf{R}_k-\mathbf{R}_\ell|^{-1}\ .\end{aligned}$$

We can now use (38) to arrive at

$$\begin{aligned}&\iint \rho^{(2)}(x,y)\theta_k(y)^2\theta_\ell(x)^2|x-y|^{-1}\,d^3x\,d^3y \geq\\ &\geq \left[\left(N_k^{\mathrm{TF}}+O(Z^{(1-\varepsilon)})\right)\left(N_\ell^{\mathrm{TF}}+O(Z^{(1-\varepsilon)})\right)-O(Z^{2(1-\varepsilon)})\right]|\mathbf{R}_k-\mathbf{R}_\ell|^{-1}\ .\end{aligned}$$

Thus we obtain the following estimate for the interaction of two screened nuclei

$$\begin{aligned}\langle\psi|\,I_{k\ell}\,|\psi\rangle \geq &\Big[\left(Z_k-N_k^{\mathrm{TF}}+O(Z^{(1-\varepsilon)})\right)\left(Z_\ell-N_\ell^{\mathrm{TF}}+O(Z^{(1-\varepsilon)})\right)\\ &+O(Z^{2(1-\varepsilon)})\Big]|\mathbf{R}_k-\mathbf{R}_\ell|^{-1} \quad . \qquad (39)\end{aligned}$$

The terms $I_{0\ell}$ are treated the same way and we get

$$\begin{aligned}\langle\psi|\,I_{0\ell}\,|\psi\rangle \geq &\Big[-\left(Z_\ell-N_\ell^{\mathrm{TF}}+O(Z^{(1-\varepsilon)})\right)\\ &O(Z^{(1-\varepsilon)})+O(Z^{2(1-\varepsilon)})\Big]|\mathbf{R}_k-\mathbf{R}_\ell|^{-1} \quad . \qquad (40)\end{aligned}$$

Inserting the estimates (34), (35), (39) and (40) into (31) gives

$$0 \geq \sum_{1\leq k<\ell\leq M} \left(Z_k - N_k^{\mathrm{TF}} + O(Z^{(1-\epsilon)})\right)\left(Z_\ell - N_\ell^{\mathrm{TF}} + O(Z^{(1-\epsilon)})\right)|\mathbf{R}_k - \mathbf{R}_\ell|^{-1}$$
$$- O(Z^{2(1-\epsilon)})R_{\min}^{-1} \quad .$$

Thus if say k_0 and ℓ_0 represent the two closest nuclei we get

$$\begin{aligned}(Z_{k_0} - N_{k_0}^{\mathrm{TF}})(Z_{\ell_0} - N_{\ell_0}^{\mathrm{TF}})R_{\min}^{-1} &\leq \\ &\leq \sum_{1\leq k<\ell\leq M} (Z_k - N_k^{\mathrm{TF}})(Z_\ell - N_\ell^{\mathrm{TF}})|\mathbf{R}_k - \mathbf{R}_\ell|^{-1} \\ &\leq O(Z^{(1-\epsilon)})\sum_{\ell=1}^{M}(Z_\ell - N_\ell^{\mathrm{TF}})R_{\min}^{-1} + O(Z^{2(1-\epsilon)})R_{\min}^{-1}\end{aligned} \quad .$$

If we divide this inequality by $Z_{\ell_0} - N_{\ell_0}^{\mathrm{TF}}$ and apply Proposition 10 we find that $Z_{k_0} - N_{k_0}^{\mathrm{TF}} \leq O(Z^{(1-\epsilon)})$. Using Proposition 10 again, we can conclude that for *all* k $Z_k - N_k^{\mathrm{TF}} \leq O(Z^{(1-\epsilon)})$. If we sum this inequality over k we finally obtain the desired bound on the excess positive charge

$$Z - N \leq C_1 Z^{(1-\epsilon)} \quad .$$

Acknowledgment

Portions of this paper were written while the authors were visiting various institutions. In particular, while MBR was visiting the Institute for Theoretical Atomic and Molecular Physics at the Harvard-Smithsonian Center for Astrophysics (and thereby also partially supported by an NSF grant to that institute), while both authors (at different times) visited the University of Michigan, and while both authors were visiting the workshop on Schrödinger operators at Aarhus University in July, 1991. The authors are grateful to these institutions for their hospitality and support.

MBR was partially supported by NSF grants DMS–8908125 and HRD–9103315. JPS was partially supported by NSF grant DMS–9002416.

Appendix: A Lower Bound on N/Z

We now give a simple argument to show that the lower bound in (6) holds, i.e. that stability implies $N \geq A_1 Z$ for some constant A_1 depending only on M and the charge ratios of the nuclei. Although this is weaker than our main theorem, the proof of Theorem 8 requires an independent proof of this lower bound. Using the notation of Sect. 5, and proceeding as in (31), we can conclude that stability implies

$$0 \geq \sum_{\alpha} \langle\psi|\, \Theta_\alpha^2 I_\alpha - \sum_{i=1}^{N} |\nabla_i \Theta_\alpha|^2 \,|\psi\rangle \geq \sum_{\alpha} \langle\psi|\, \Theta_\alpha^2 \left(I_\alpha - CNR_{\min}^{-2},\right) |\psi\rangle \quad .$$

Now it follows easily from the properties of θ_k, (30) and (32) that

$$\begin{aligned} &\Theta_\alpha^2 \left(I_\alpha - CNR_{\min}^{-2}\right) \geq \\ &\geq \Theta_\alpha^2 \left(\sum_k Z_k \left[\sum_{\ell \neq k} \left(\frac{1}{2} Z_\ell - \frac{4}{3} N_\ell\right) |\mathbf{R}_k - \mathbf{R}_\ell|^{-1} - 5N_0 R_{\min}^{-1} \right] - CNR_{\min}^{-2} \right) \\ &\geq \Theta_\alpha^2 R_{\min}^{-1} \left[Z^2 z_{k_0} z_{\ell_0} - 5NZ - CN^{4/3} \right] \quad . \end{aligned} \tag{41}$$

where z_{k_0} and z_{ℓ_0} are as in (20) and $z_{k_0} z_{\ell_0} \geq z_{\min}^2$. Note that Lemma 7, which does not require any conditions on the size of N relative to Z, was used to bound one of the $R_{\min}$'s in the term $CNR_{\min}^{-2}$. It now follows easily that there is a positive constant B such that $Z > BN$ implies that the right hand side of (41) is positive, contradicting the stability condition given above. Thus, stability implies $N > A_1 Z$ with $A_1 \geq B^{-1}$. □

It should be evident that the argument presented here is very rough; however, these crude estimates suffice for the purposes of this paper. Nevertheless, we indicate briefly how these estimates can be improved to give an asymptotic estimate of $A_1 \geq 2z_{\min}^{-1}$. (As discussed by Ruskai [15,17] the dependence on $z_{\min}$ cannot be eliminated without taking into account differences in threshold energies.) Since the term $CNR_{\min}^{-2}$, which arises from the localization error, is asymptotically negligible compared to terms of order Z^2, it will be ignored in the rest of this discussion. Now, modify the definition of θ so that $\theta(t) = 1$ if $t < 1/2$ and $\theta(t) = 0$ if $t > 1/2 + a$ where $a > 0$. Then $\sum_{k=1}^{M} \theta_k(x)^2 > 1$ which eliminates the need to consider a subcluster of "outer" electrons corresponding to θ_0 as was done previously. Now let α denote an M-cluster (rather than $M+1$) and let Θ_α be as before, but without any contributions from α_0. Since $\sum_\alpha \Theta_\alpha^2 > 1$, using standard arguments (see, e.g. [3, 11, 14, 21, 22]), one can replace Θ_α by $\Theta_\alpha \left[\sum_\alpha \Theta_\alpha^2\right]^{-1/2}$ without increasing the localization error or altering the support of Θ_α. With these modifications, one finds

$$\Theta_\alpha^2 I_\alpha \geq \Theta_\alpha^2 \sum_{k<\ell} |\mathbf{R}_k - \mathbf{R}_\ell|^{-1} \left[Z_k Z_\ell - \frac{2}{1-2a} (Z_k N_\ell + Z_\ell N_k) \right]$$

It is easy to see that the quantity in square brackets [ ] will be positive for all k, ℓ if $Z z_{\min} \geq 2/(1-2a)$ and that a can be chosen arbitrarily close to 0. (More precisely, one can let $a \to 0$ as $Z \to \infty$ while still keeping the localization error negligible.)

References

[1] R. Benguria: Dependence of the Thomas-Fermi energy on the nuclear coordinates, *Commun. Math. Phys.* **81** 419–428 (1981)

[2] H. Brezis and E. H. Lieb: Long range potentials in Thomas-Fermi theory, *Commun. Math. Phys.* **65** 231–246 (1979)

[3] H.L. Cycon, R.G. Froese, W. Kirsch and B. Simon: *Schrödinger Operators* Springer-Verlag, Berlin Heidelberg New York, (1987)

[4] P. Duclos and H. Hogreve: On the Stability of Positive Diatomic Molecular Ions, in *Rigorous Results in Quantum Dynamics* (Proceedings of the Liblice Conference), World Scientific, Singapore, Eds. J. Dittrich, P. Exner pp. 63–70 (1991)

[5] C. L. Fefferman and L. A. Seco: Asymptotic neutrality of large ions, *Commun. Math. Phys.* **128** 109–130 (1990)

[6] C. L. Fefferman and L. A. Seco: On the energy of a large atom, *Bull. Amer. Math. Soc.* **23** 525–530 (1990)

[7] W. Hughes: An atomic energy lower bound that agrees with Scott's correction, *Adv. Math.* **79** 213–270 (1990)

[8] V. Ivrii and I. M. Sigal: Asymptotics of the ground state energies of large Coulomb systems, *Annals of Math.* to appear.

[9] E. H. Lieb: Thomas-Fermi and related theories of atoms and molecules, *Rev. Mod. Phys.* **53** 603–604 (1981)

[10] E. H. Lieb: Bound on the maximum negative ionization of atoms and molecules, *Phys. Rev. A* **29** 3018–3028 (1984)

[11] E. H. Lieb, I. M. Sigal, B. Simon, and W. Thirring: Asymptotic neutrality of large-Z ions, *Commun. Math. Phys.* **116** 635–644 (1988)

[12] E. H. Lieb and B. Simon: The Thomas-Fermi theory of atoms, molecules, and solids, *Adv. Math.* **23** 22–116 (1977)

[13] M. B. Ruskai: Absence of discrete spectrum in highly negative ions, *Commun. Math. Phys.* **82** 457–469 (1982)

[14] M. B. Ruskai: Absence of discrete spectrum in highly negative ions II: extension to fermions, *Commun. Math. Phys.* **85** 325–327 (1982)

[15] M. B. Ruskai: Limits on stability of positive molecular ions, *Lett. Math. Phys.* **18** 121–132 (1989)

[16] M. B. Ruskai: Limit on the excess negative charge of a dynamic diatomic molecule, *Ann. Inst. H. Poincaré A: Physique Théorique* **52** 397-414 (1990)

[17] M. B. Ruskai: Absence of bound states in extremely asymmetric diatomic molecules, *Commun. Math. Phys.* **137** 553-566 (1991)

[18] L. A. Seco, I. M. Sigal, and J. P. Solovej: Bound on the ionization energy of large atoms, *Commun. Math. Phys.* **131** 307–315 (1990)

[19] H. Siedentop and R. Weikard: On the leading energy correction og the statistical model of thr atom: Interacting case, *Commun. Math. Phys.* **112** 471–490 (1987)

[20] H. Siedentop and R. Weikard: On the leading correction of the Thomas-Fermi model: Lower bound – with an appendix by A. M. K. Müller, *Invent. Math.* **97** 159–193 (1989)

[21] I. M. Sigal: Geometric methods in the quantum many-body problem. nonexistence of very negative ions, *Comm. Math. Phys.* **85** 309–324 (1982)

[22] I. M. Sigal: How many electrons can a nucleus bind? *Ann Phys.* **157** 307–320 (1984)

[23] J. P. Solovej: Asymptotic neutrality of diatomic molecules, *Commun. Math. Phys.* **130** 185–204 (1990)

[24] L. Spruch: Pedagogic notes on Thomas-Fermi theory (and on some improvements): atoms, stars, and the stability of bulk matter, *Rev. Mod. Phys.* **63** 151–209 (1991)

[25] W. E. Thirring: *A Course in Mathematical Physics*, volume III, Springer Verlag, New York, (1981)

[26] S. A. Vugal'ter: On the stability of positive molecular ions: in *Rigorous Results in Quantum Dynamics* (Proceedings of the Liblice Conference), World Scientific, Singapore, Eds. J. Dittrich, P. Exner pp. 71–74 (1991)

TIME-DELAY IN SHORT RANGE POTENTIAL SCATTERING

Kalyan B. Sinha*
Indian Statistical Institute
New Delhi 110016, India

and

A. Mahapatra†
Department of Mathematics
Sambalpur University
Orissa 768019, India

1 Introduction

This paper is divided into two parts. In the first part, we give a modified definition of time-delay as the limit as $r \to \infty$ of the difference of the sojourn times of a scattering state and of the associated free state in a 'fuzzy' ball of radius r in $\mathbb{R}$. The potential W is assumed to be smooth and behave like $|x|^{-\alpha}(\alpha > 1)$ at infinity. For earlier studies on this problem, the reader is referred to [1]-[6]. Here we content ourselves with giving the results, the details appearing in [7].

The first author came to know of the very recent work of Jensen and Nakamura [8] on the configuration space pro. perties of the S-operator during his visit to Aarhus and the second part of this paper uses this result to simplify some of the results in the first part. Here we retain the original definition of Jauch as given

*† The first author thanks Prof. E. Balslev for his hospitality at the university of Aarhus and the second author acknowledges the financial support of C.S.I.R., India

in [1] in terms of sharp localization function, viz. that by the characteristic function of the ball of radius r in $\mathbb{R}^n$.

2 Time-delay for short range potentials

We consider Schroödinger operators $H_0 = -\Delta$ and $H = H_0 + W$ on $L^2(\mathbb{R}^n)$ for any $n \geq 1$, where W is a potential satisfying the following assumption.

Assumption A2.1 : (i) $W \in C^5(\mathbb{R}^n)$, (ii) $|D^m W(x)| \leq K(1+|x|)^{-|m|-\alpha}$ for any multi-index m with $|m| \leq 5$ and some $\alpha > 1$.

It is well known (see [9] and [10]) that if W satisfies A2.1, then H is selfadjoint, bounded below, does not have any positive singular spectrum and also the wave operators $\Omega_\pm \equiv \text{s-lim}_{t\to\pm\infty} e^{itH} e^{-itH_0}$ exist and are complete. Thus the scattering operator $S \equiv \Omega_+^* \Omega_-$ is unitary and commutes with H_0. Furthermore, S is decomposable as $\{S(\lambda)\}_{\lambda\in[0,\infty)}$ in the spectral representation w.r.t. H_0 with $S(\lambda)$ unitary for a.a.λ. In this context, we state a theorem of Jensen [6] without proof.

Proposition 2.2 : Let W satisfy A2.1. Then the map $\lambda \to S(\lambda)$ is five times continuously norm differentiable in $(0,\infty)$.

We shall use the following notations.

$$\begin{aligned}
U_t &= \exp(-iH_0 t),\, V_t = \exp(-iHt),\, t \in \mathbb{R} \\
|Q| &= (\sum_{j=1}^{n} Q_j^2)^{1/2},\, < Q >= (1+|Q|^2)^{1/2}, \\
|P| &= (\sum_{j=1}^{n} P_j^2)^{1/2},\, A_0 = \frac{1}{2}(H_0^{-1} P.Q + Q.P H_0^{-1}),
\end{aligned}$$

where Q and P are the standard position and momentum operators in $L^2(\mathbb{R}^n)$.

We shall also set for $\mu \geq 0$,

$$\mathcal{D}_\mu = \{f \in L^2(\mathbb{R}^n) | f \in D(< Q >^\mu) \text{ and the Fouraier transform } \tilde{f}$$

$$\text{of } f \text{ has compact support in } \mathbb{R}^n \backslash \{0\}\} .$$

Then it is clear that $\mathcal{D}_\mu$ is dense, $\mathcal{D}_\mu \subseteq \mathcal{D}_\nu$ if $\mu \geq \nu$ and that A_0 is symmetric on $\mathcal{D}_1$.

For each $r > 1$, consider the function $\phi_r \in C^1[0, \infty)$ such that

$$\left.\begin{array}{llll} \text{(i)} & 0 \leq \phi_r(u) & \leq & 1 \text{ for all } u \in [0, \infty) \\ \text{(ii)} & \phi_r(u) & = & \begin{cases} 1 & \text{for } \ 0 \leq u \leq 1 \\ 0 & \text{for } \ u > 1 + cr^{-\delta} \text{ with} \\ & \qquad c > 0, 0 < \delta < 1. \end{cases} \\ \text{(iii)} & |\phi_r'(u)| & \leq & kr^\delta \text{ for some } k > 0 \text{ and all } u, \end{array}\right\} \tag{2.1}$$

and let P_r be the multiplication operator in $L^2(\mathbb{R}^n)$ by the function $\phi_r(\frac{|x|}{r})$. In the first part of this paper , P_r replaces F_r (the multiplication operator by the characteristic function χ_r of the ball of radius r) in the definition of time delay and unlike a uniformly scaled C_0^∞ function, its tail grows at the rate $r^{1-\delta}$ while the main support grows linearly with r.

The time-delay $\sigma_r(f)$ for each $r > 1$ is defined as :

$$\sigma_r(f) = \int_{-\infty}^{\infty} ||P_r V_t \Omega_- f||^2 dt - \int_{-\infty}^{\infty} ||P_r U_t f||^2 dt. \tag{2.2}$$

It is easy to see that by the local smoothness results ([10] and [11]) $\sigma_r(f) < \infty$ for every $f \in \mathcal{D}_0$ and $1 \leq r < \infty$. The main result of [7] is given in the next theorem.

Theorem 2.3 : Let W satisfy A2.1 and let P_r be as above with $\delta < (\alpha - 1)/2(\alpha + 2)$. Then for each $f \in \mathcal{D}_{4+\eta} (\eta > 0)$, $\lim_{r \to \infty} \sigma_r(f) \equiv \sigma(f)$ exists and equals $\int_0^\infty d\lambda (f_\lambda, \{-iS^*(\lambda) \frac{dS(\lambda)}{d\lambda}\} f_\lambda)$.

Next, we give the basic steps in the proof of the above theorem, referring for the details to [7]. The first result is essentially due to Martin [3], with the necessary modifications forced upon us by the slower decrease of the potential. We set

$$\left.\begin{aligned} X_t^{\pm}(P) &= \pm\textstyle\int_t^{\pm\infty} W(2sP)ds, \\ Y_t^{\pm} &= \exp(-iX_t^{\pm}) \text{ and} \\ T_t^{\pm} &= U_t Y_t^{\pm}, \end{aligned}\right\} \tag{2.3}$$

where the sign $\pm$ is chosen according as $t \gtrless 0$. $T_t^{\pm}$ is a kind of modified free evolution similar to the one used in the theory of long range scattering.

Lemma 2.4 : Let $f \in \mathcal{D}_0$ be such that

(i) $\|V_t\Omega_- f - T_t^- f\| \in L^1(-\infty, 0)$, and

(ii) $\|V_t\Omega_- f - T_t^+ S f\| \in L^1(0, \infty)$. Then

$$\begin{aligned} \sigma(f) &= \lim_{r\to\infty}\Big[\int_0^\infty (\|P_r T_t^+ Sf\|^2 - \|P_r T_t^+ f\|^2)dt \\ &\quad + \int_0^\infty (\|P_r T_t^+ f\|^2 - \|P_r U_t f\|^2)dt \\ &\quad + \int_{-\infty}^0 (\|P_r T_t^- f\|^2 - \|P_r U_t f\|^2)dt\Big]. \end{aligned}$$

The next lemma lists the properties of the modifier $Y_t^{\pm}$ that are needed in the rest of this paper.

Lemma 2.5 : Let W satisfy A2.1 and let $X_t^{\pm}, Y_t^{\pm}, T_t^{\pm}$ be as in equation (2.3). Suppose furthermore that $\psi \in C_0^\infty(0, \infty)$. Then

(i) $\text{s-lim}_{t\to\pm\infty} Y_t^{\pm} = I$, $\quad \text{s-lim}_{t\to\pm\infty} V_t^* T_t^{\pm} = \Omega_{\pm}$,

(ii) for $0 \le \mu \le 4$, $\quad \| < Q >^\mu Y_t^{\pm}\psi(H_0) < Q >^{-\mu} \| \le K$, uniformly in $t \in I\!R$,

(iii) for $0 \le \mu \le 3, j = 1, 2, ..., n$, $\quad \| < Q >^\mu \partial_j X_t^{\pm}\psi(H_0) < Q >^{-\mu} \| \le K$, uniformly in $t \in I\!R$,

(iv) for $t \neq 0$, $\quad \| < Q >^2 (Y_t^{\pm} - I)\psi(H_0) < Q >^{-2} \| \le K(|t|^{1-\alpha} + |t|^{2-2\alpha})$, and $\|(Y_t^{\pm} - I)\psi(H_0)\| \le K|t|^{1-\alpha}$, where K is independent of t.

This lemma combines the lemmas 3.1 and 3.2 of [7].

Lemma 2.6 : Let P_r be as defined in (2.1) with $\delta < \min(\alpha-1, \beta/2)$ with $0 < \beta < 1$. Then for $f \in \mathcal{D}_{1+\beta}$ one has the following asymptotic behaviour as $r \to \infty$.

(i) $\int_{\mathbb{R}_\pm} ||P_r U_t f||^2 dt = (\frac{r}{2}\int_0^\infty \phi_r(u)^2 du)(f, H_0^{-\frac{1}{2}} f)$

$$\mp\frac{1}{2}(f, A_0 f) + O(r^{\delta-\frac{\beta}{2}}),$$

(ii) $\int_{\mathbb{R}_\pm} ||P_r T_t^\pm f||^2 dt = (\frac{r}{2}\int_0^\infty \phi_r(u)^2 du)(f, H_0^{-\frac{1}{2}} f)$

$$\mp\frac{1}{2}(f, A_0 f) + O(r^{\delta-\min(\alpha-1,\frac{\beta}{2})}),$$

where $\mathbb{R}_+ = [0,\infty)$ and $\mathbb{R}_- = (-\infty, 0]$.

Lemma 2.7 : Let W satisfy A2.1. Then for every $\eta > 0, S$ maps $\mathcal{D}_{3+\eta}$ into $\mathcal{D}_{1+\beta}$ with $\beta \in (0, \frac{\alpha-1}{\alpha+2})$. The lemmas 2.6 and 2.7 are respectively theorems 4.1 and A.10 of [7].

Sketch of the proof of theorem 2.3 : Assume for the moment the hypotheses (i) and (ii) of lemma 2.4. Then by lemma 2.6, the second and the third terms on the R.H.S. of the conclusion of lemma 2.4 converges to 0 as $r \to \infty$, while the first term by lemma 2.7 converges to $-\frac{1}{2}(f, S^*[A_0, S]f)$. Now it is easy to verify from the definition of A_0 that in the spectral representation w.r.t. H_0, A_0 has the representation as $2i\frac{d}{d\lambda}$ and this leads to the Eisenbud-Wigner expression for time-delay as stated in the theorem.

The verification of the hypothesis (i) of lemma 2.4 is straightforward for $f \in \mathcal{D}_{2+\eta}(\eta > 0)$ on expanding $V_t\Omega_- f - T_t^- f$ as is standard for long-range scattering (see for example Chapter 13 of [9]). However, the verification of the hypothesis (ii) of lemma 2.4 is much more difficult because of the appearance of Sf instead of f in the expansion of $V_t\Omega_- f - T_t^+ Sf$. This is achieved for $f \in \mathcal{D}_{4+\eta}$ in theorems A.14

and A.15 of [7] and it is this part of the proof which is simplified by virtue of the recent result of Jensen and Nakamura [8].

3 Time-delay : improvement of the result in section 2

In this section, we assume that the potential W satisfy

Assumption A3.1

$W \in C^\infty(\mathbb{R}^n)$ and there exists $\alpha > 1$ such that for all multi-indices m

$$|D^m w(x)| \leq K_m(1+|x|)^{-\alpha-|m|}.$$

Then we have the following proposition of Jensen and Nakamura [8].

Proposition 3.1 : Let W satisfy A3.1 and let $\psi \in C_0^\infty(0,\infty)$. Then for every $\mu_1 > \mu_2 \geq 0, \| < Q >^{\mu_2} S\psi(H_0) < Q >^{-\mu_1} \| \leq K$.

We shall retain the original definition of time-delay by Jauch in terms of the sharp localization and write

$$\tau_r(f) = \int_{-\infty}^{\infty} \|F_r U_t \Omega_- f\|^2 dt - \int_{-\infty}^{\infty} \|F_r U_t f\|^2 dt, \tag{3.1}$$

which, as before, is well-defined for $f \in \mathcal{D}_0$ by the local smoothness results. The main result is given below:

Theorem 3.2 : Let w satisfy A3.1 and let $f \in \mathcal{D}_{3+\eta}$. Then $\lim_{r\to\infty} \tau_r(f) = \tau(f)$ exists and

$$\tau(f) = \int_0^\infty d\lambda (f_\lambda, \left\{-iS^*(\lambda)\frac{dS(\lambda)}{d\lambda}\right\} f_\lambda).$$

For the proof of theorem 3.2, we first note that the lemma 2.4 remains unchanged with P_r replaced by F_r. However, the proof of lemma 2.6 or that of the main theorem on page 484 of [1] has to be modified because of the appearance of $T_t^\pm$

instead of U_t. We use the notation of [7] and set $Z_t = \exp(iQ^2t), s = r/2|t|, \nu = \frac{1}{2r}$ so that $|t| = (4\nu s)^{-1}$ and adopt the convention that $Y^{\pm}_{\pm\infty} = I$.

Theorem 3.3 : Suppose $f \in \mathcal{D}_2$. Then

$$\lim_{\nu\to 0+} \int_0^\infty \frac{ds}{4\nu s^2}[(Y^{\pm}_{\pm\rho}f, Z^*_{\pm\nu s}\chi(\frac{|P|}{s})Z_{\pm\nu s}Y^{\pm}_{\pm\rho}f) - (f, \chi(\frac{|P|}{s})f)]$$

$$\pm\frac{1}{2}(f, A_o f) = 0. \tag{3.2}$$

where ρ is either ∞ or $(4\nu s)^{-1}$. We proceed via a few lemmas.

Lemma 3.4 : If $f \in \mathcal{D}_2$, then

$$-\frac{i}{4}\int_0^\infty \frac{ds}{s}\{(Q^2f, \chi(\frac{|P|}{s})f) - (f, \chi(\frac{|P|}{s})Q^2f)\} = -\frac{1}{2}(f, A_0 f).$$

This lemma has been proven in page 487 of [1].

Lemma 3.5 : Let $f \in \mathcal{D}_0$ and let $N_0 = \sup\{|k| \mid \underline{k} \in$ the support of $\tilde{f}\}$. Set $\chi^\perp = 1 - \chi$. Then $\chi^\perp(\frac{|P|}{s})f = 0$ for $s > N_0$.

This result is easy to see since $\tilde{f}(\underline{k}) = 0$ for $|k| > N_0$ while $\chi^\perp(\frac{|k|}{s}) = 0$ for $|k| < s$.

Proof of Theorem 3.3 (for +ve sign) : Fix $f \in \mathcal{D}_2$ and set $f_\rho = Y^+_\rho f$. As in [7], we can rewrite the L.H.S. of (3.2) by using lemma 3.4 as $\lim_{\nu\to 0+}\int_0^\infty ds\mathcal{I}(\nu, s)$, where

$$\mathcal{I}(\nu, s) = \frac{1}{4\nu s^2}[(f_\rho, Z^*_{\nu s}\chi(\frac{|P|}{s})Z_{r s}f_\rho) - (f_\rho, \chi(\frac{|P|}{s})f_\rho)$$

$$+i\nu s\{(Q^2f, \chi(\frac{|P|}{s})f) - (f, \chi(\frac{|P|}{s}Q^2f)\}]$$

$$\equiv \frac{1}{4\nu s^2}[\mathcal{I}^{(1)}(\nu, s) + \mathcal{I}^{(2)}(\nu, s)]$$

$$\equiv \frac{-1}{4\nu s^2}[\mathcal{I}^{(3)}(\nu, s) + \mathcal{I}^{(4)}(\nu, s)]. \tag{3.3}$$

In (3.3),

$$\begin{aligned}
\mathcal{I}^{(1)}(\nu,s) &= ((Z_{\nu s}-I)f_\rho, \chi(\frac{|P|}{s}(Z_{\nu s}-I)f_\rho),\\
\mathcal{I}^{(2)}(\nu,s) &= 2\,\mathrm{Re}\{((Z_{\nu s}-I)f_\rho, \chi(\frac{|P|}{s})f_\rho)\\
&\quad +i\nu s(Q^2 f, \chi(\frac{|P|}{s})f)\}, \qquad (3.4)
\end{aligned}$$

and $\mathcal{I}^{(3)}$ and $\mathcal{I}^{(4)}$ have expressions exactly the same as for $\mathcal{I}^{(1)}$ and $\mathcal{I}^{(2)}$ respectively with χ repalced by $\chi^\perp$. Thus it suffices to show that

$$\lim_{\nu\to 0+}\int_0^{N_0}\frac{ds}{4\nu s^2}[\mathcal{I}^{(1)}(\nu,s)+\mathcal{I}^{(2)}(\nu,s)]=0$$

and

$$\lim_{\nu\to 0+}\int_{N_0}^{\infty}\frac{ds}{4\nu s^2}[\mathcal{I}^{(3)}(\nu,s)+\mathcal{I}^{(4)}(\nu,s)]=0. \qquad (3.5)$$

First we note that by lemma 3.5, $\mathcal{I}^{(4)}(\nu,s)=0$ identically for $s>N_0$. Also since $[P_j, Z_{\nu s}]=2\nu s\, Q_j Z_{\nu s}$ and since $||(Z_{\nu s}-I)f|| \le K|\nu s|\,||Q^2 f||$, we have

$$\begin{aligned}
|\mathcal{I}^{(3)}(\nu,s)| &\le \frac{1}{s^2}|(\frac{s}{|P|})^2\chi^\perp(\frac{|P|}{s})||\sum_{j=1}^{n}||P_j(Z_{\nu s}-J)f_\rho||^2\\
&\le Ks^{-2}(\nu s)^2\{\sum_{j=1}^{n}||Q^2P_jf_\rho||^2+\langle Q\rangle f_\rho||^2\}\\
&\le K'\nu^2
\end{aligned}$$

by lemma 2.5 (ii) and hence we have verified the second equation of (3.5).

An identical line of reasoning as above also gives

$$|\int_0^{N_0}\frac{ds}{4\nu s^2}\mathcal{I}^{(1)}(\nu,s)| \le KN_0||Q^2f_\rho||\nu \to 0 \text{ as } \nu\to 0+.$$

Finally we rewrite $\mathcal{I}^{(2)}$ as $\mathcal{I}^{(2)}(\nu,s)=2Re[\mathcal{I}_1^{(2)}(\nu,s)-\mathcal{I}_2^{(2)}(\nu,s)]$, with

$$(\nu s)^{-1}\mathcal{I}_1^{(2)}(\nu,s)=((\frac{Z_{\nu s}-I}{\nu s}-iQ^2)f_\rho, \chi(\frac{|P|}{s})f_\rho)$$

and

$$(\nu s)^{-1}\mathcal{I}_2^{(2)}(\nu,s)=i[(Q^2f_\rho,\chi(\frac{|P|}{s})f_\rho)-(Q^2f,\chi(\frac{|P|}{s})f)].$$

Thus

$$\begin{aligned}|\frac{1}{\nu s^2}\mathcal{I}_1^{(2)}(\nu,s)| &\le \|(\frac{Z_{\nu s}-I}{\nu s}-iQ^2)f_\rho\|\,\|\frac{|P|}{s}\chi(\frac{|P|}{s})\|\,\|\frac{1}{|P|}f_\rho\| \\ &\le K\|\frac{1}{|P|}f\|\,\|Q^2 f_\rho\|\end{aligned}$$

and hence by lemma 2.5 (ii) and the dominated convergence theorem

$$\lim_{\nu\to 0+}\int_0^{N_0}\frac{ds}{4\nu s^2}\mathcal{I}_1^{(2)}(\nu,s)=0.$$

If we note that for $\rho=\infty, \mathcal{I}_2^{(2)}$ is identically zero, then the proof of theorem 3.3 is complete in this case. For the case $\rho=(4\nu s)^{-1})$ choose $\psi\in C_0^\infty(0,\infty)$ such that $\psi(H_0)f=f$ and we have by lemma 2.5 (iv) that

$$\begin{aligned}|\frac{\mathcal{I}_2^{(2)}(\nu,s)}{\nu s}| &\le \{\|Q^2(Y_\rho^+-I)\psi(H_0)<Q>^{-2}\|\,\|<Q>^2 f\|\,\|f\| \\ &\quad +\|Q^2 f\|\,\|(Y_\rho^+-I)f\| \\ &\le K[(\nu s)^{a-1}+(\nu s)^{2a-2}],\end{aligned}$$

which leads to the verification of the first equation in (3.5) and completes the proof.

∎

Proof of Theorem 3.2 : As in the proof of theorem 2.3, we can use theorem 3.3 to observe that the last two terms in the conclusion of lemma 2.4 converge to 0 as $r\to\infty$ while the first term converges to the required Eisenbud-Wigner expression. It only remains to verify the two integrability hypotheses of lemma 2.4.

The verification of hypothesis (i) of lemma 2.4 is the same as in the case of the proof of theorem 2.3. For that of the hypothesis (ii) we use the proposition 3.1. Expanding $V_t\Omega_- f - T_t^+ Sf = (V_t\Omega_+ - T_t^+)Sf$ as in the case of long range scattering (see page 532 of [9]), we have

$$||V_t\Omega_- f - T_t^+ Sf|| \leq \int_t^\infty ds \int_0^1 d\rho ||\nabla W(\rho Q) U_{s/\rho} \cdot QY_s^+ Sf|| +$$
$$\int_t^\infty s ds \int_0^1 d\rho ||\nabla W(\rho Q) U_{s/\rho} Y_s^+ Sf||. \qquad (3.6)$$

That $\int_1^\infty ||V_t \Omega f - T_t^+ Sf|| dt < \infty$ follows from (3.6) by applying lemma 2.5 (ii), (iii) and proposition 3.1, and by using the time decay estimates for the free evolution group (e.g. lemma 13.3 in [8] or [12]). ∎

References

[1] W.O. Amrein and M.B. Cibils, Helv. Phys. Acta, 60 , 1987, p. 481.

[2] W.O. Amrein, M.B. Cibils and K.B. Sinha, Ann. Inst. Henri. Poincare, 47 , 1987, p. 367.

[3] Ph. martin, Comm. Math. Phys., 47 , 1976, p. 221.

[4] X.P. Wang, Helv. Phys. Acta, 60, 1987, p. 501.

[5] S. Nakamura, Comm. Math. Phys., 109, 1987, p. 397.

[6] A. Jensen, Comm. Math. Phys. 82, 1981 , p. 435.

[7] A. Mahapatra, K.B. Sinha and W.O. Amrein, "Configuration space properties of scattering operator and time delay for potentials decaying like $|x|^{-\alpha}, \alpha > 1$", to appear in Ann. Inst. Henri Poincare.

[8] A. Jensen & S. Nakamura, "Mapping properties of Wave and Scattering Operators for two-body Schrodinger Operators", preprint, Aarhus University, 1991.

[9] W.O. Amrein, J.M. Jauch and K.B. Sinha, "Scattering theory in Quantum Mechanics", W.A. Benjamin, Reading, Mass, 1977.

[10] M. Reed and B. Simon, "Methods of Modern Mathematical Physics, IV. Analysis of Operators, Academic Press, New York, 1978.

[11] R. Lavine, J. Funct. Anal. 12 , 1973, p. 30.

[12] P. Perry; Comm. Math. Phys. 81, 1981, p. 243.

On smoothness of the N-body S-matrix

Erik Skibsted Matematisk Institut Aarhus Universitet 8000 Aarhus C Denmark

Section 1 Introduction and results

In this note we give an account on some recent results on the smoothness of quantum mechanical N–body scattering amplitudes [S1]. We have results for the 2–cluster—2–cluster and 2–cluster—N–cluster amplitudes under a short range condition on the potential and in addition under a discreteness assumption on the 2–cluster channel energies. This gives a rather complete picture for $N = 3$ while a number of interesting cases remain to be treated for $N > 3$. To explain our results and what could be expected for other scattering amplitudes we state the following conjecture. (All notations are given in precise terms below.)

We consider an incoming 2–cluster channel α and an outgoing channel β that is not 2–cluster. The variable $\theta_a'(\in S_a)$ should be understood as the (normalized) incoming intercluster velocity vector. Similarly the variable θ_b labels all possible outgoing intercluster velocity vectors defined for the cluster decomposition of β. The scattering amplitude $\hat{S}_{\beta\alpha}\left(\lambda, \theta_b, \theta_a'\right)$ takes incoming data to outgoing data at total energy λ. We denote by $I_{\beta\alpha}\backslash\mathcal{F}$ the set of possible energies excluding the thresholds. The set $Z_b \cap S_b$ consists of those θ_b for which all of the corresponding intercluster velocity vectors are non-vanishing. The conjecture is

$$\hat{S}_{\beta\alpha}\left(\lambda, \theta_b, \theta_a'\right) \in C^\infty(I_{\beta\alpha}\backslash\mathcal{F} \times Z_b \cap S_b \times S_a). \tag{1.1}$$

We prove (1.1) for the outgoing channel being free (i.e. for the N–cluster channel). The result is optimal in the sense that singularities can appear in the complement of $Z_b \cap S_b$ in S_b as shown recently by Isozaki [I] for $N = 3$ (under a stronger decay assumption on the potential). This depends on possible existence of zero energy bound states or zero energy resonance states. Another motivation for the conjecture (1.1) is our second type of results namely for the outgoing channel also being 2–cluster. In that case a similar result holds except for possible singularities at the diagonal. (See Theorem 1.4.) Finally the method of proof consisting of pseudodifferential localization in external coordinates (cf. [I-K]) together with strong estimates on the resolvent (by the commutator method cf. [M1–2],[J],[J-M-P]) gives some further indication. As we see it the main obstacle for (1.1) for an arbitrary outgoing channel is some good propagation estimates. We shall give a brief outline of the proofs in Section 2.

For some previous work on the N–body S–matrix we refer to [A-P-S], [C-T], [D1], [E-S] and the above mentioned paper [I].

To state our results precisely we need various notations.

We consider $N(\geq 3)$ 3–dimensional particles labelled $1, \cdots, N$ in the center of mass configuration space

$$X = \left\{ x = (x_1, \cdots, x_N) | x_i \in \mathbf{R}^3, \sum_{i=1}^{N} m_i x_i = 0 \right\}.$$

Here x_i and m_i denote the position vectors and the masses, respectively.

We introduce for any cluster decomposition $a = (C_1, \cdots, C_{\#a})$ the subspace of external motion

$$X_a = \{x \in X | x_i = x_j \text{ if } i, j \in C \text{ for some } C \in a\},$$

and the subspace of internal motion

$$X^a = X_a^{\perp} = X^{C_1} \oplus \cdots \oplus X^{C_{\#a}},$$

where the metric on X is given by

$$x \cdot y = \sum_{i=1}^{N} 2m_i x_i \cdot y_i,$$

and

$$X^{C_j} = \{x \in X | x_i = 0 \text{ if } i \notin C_j\}.$$

Given a subset C of $\{1, \cdots, N\}$ we may identify C with the cluster decomposition given by C and one-particle clusters.

We use the notation $x = x^a \oplus x_a \in X^a \oplus X_a$ for $x \in X$, and similarly for the momentum operator $p = -i\nabla$ we write $p = p^a \oplus p_a$. The unit-sphere in X_a is denoted by S_a. Put $\hat{p} = p/|p|$.

Given two cluster decompositions a and b we write $a \subset b$, if all clusters in a are subsets of clusters in b.

With the notation Δ used for Laplacians we consider the Hamiltonians

$$H^a = -\Delta + \sum_{(ij)\subset a} v_{ij}(x_i - x_j) \text{ on } L^2(X^a)$$

under the following assumptions on the pair potentials.

Assumption 1.1 $\exists \epsilon_0 > 1 \forall i, j \leq N, i \neq j : v_{ij}(y) = v_{ij}^{(1)}(y) + v_{ij}^{(2)}(y)$, *where*

(1) $v_{ij}^{(1)}(y)$ *is smooth and for all multiindices* α

$$\partial_y^\alpha v_{ij}^{(1)}(y) = O\left(|y|^{-|\alpha|-\epsilon_0}\right) \textit{ for } y \to \infty \textit{ in } \mathbf{R}_y^3,$$

(2) $v_{ij}^{(2)}(y)$ *is compactly supported and*

$$v_{ij}^{(2)}(-\Delta_y + 1)^{-1} \textit{ is compact on } L^2(\mathbf{R}_y^\nu).$$

By definition the full Hamiltonian is given by $H = H^a, a = (1\cdots N)$, corresponding to the trivial cluster decomposition, while the sub Hamiltonian corresponding to the cluster decomposition into N clusters, $a = 0 := (1)\cdots(N)$, is given by $H^0 = 0$ on $L^2(X^0) := \mathbf{C}$.

By a channel $\alpha = (a, \phi_\alpha, E_\alpha)$ we mean a triple consisting of a cluster decomposition $a \neq (1\cdots N)$ and a bound state ϕ_α for H^a with energy E_α, i.e. $(H^a - E_\alpha)\phi_\alpha = 0$ and $||\phi_\alpha|| = 1$. We refer to $\alpha = ((1)\cdots(N), 1, 0) =: 0$ as the free channel. For each channel we introduce the identification operator $J_\alpha : L^2(X_a) \to L^2(X)$ given by $(J_\alpha\psi_a)(x) = \phi_\alpha(x^a)\psi_a(x_a)$. The wave operators given in terms of channels α are

$$W_\alpha^{\stackrel{+}{(-)}} = s - \lim_{t\to\stackrel{+}{(-)}} e^{itH}e^{-itH_a}J_\alpha,$$

where we use the notation $H_a = H^a \otimes I + I \otimes p_a^2 = H - I_a; I_a(x) = \sum\limits_{(ij)\not\subset a} v_{ij}(x_i - x_j)$, $I_a^{(1)}(x) = \sum\limits_{(ij)\not\subset a} v_{ij}^{(1)}(x_i - x_j)$. Under Assumption 1.1 the completeness of the wave operators follows from [G]. This statement reads

$$\mathcal{H}_c(H) = \sum_\alpha \oplus \mathcal{R}\left(W_\alpha^{\stackrel{+}{(-)}}\right),$$

i.e. that the continuous subspace of H is decomposed orthogonally into the ranges of the wave operators.

Given an incoming channel α the total S–matrix $\sum\limits_\beta \oplus S_{\beta\alpha}, S_{\beta\alpha} = \left(W_\beta^+\right)^* W_\alpha^- :$ $L^2(X_a) \to L^2(X_b)$, is isometric. From the point of view of physics the most interesting case is α being 2–cluster, i.e. $\#a = 2$.

We diagonalize $p_a^2 + E_\alpha$ by the modified Fourier transform $F_\alpha : L^2(X_a) \to L^2((E_\alpha, \infty), L^2(S_a))$ given with $\sigma = \dim X_a$ and $C_\sigma = 2^{-1/2}(2\pi)^{-\sigma/2}$ by

$$(F_\alpha\psi_a)(\lambda, \theta_a) = C_\sigma(\lambda - E_\alpha)^{\frac{\sigma-2}{4}} \int_{X_a} e^{-i\sqrt{\lambda - E_\alpha}\theta_a\cdot x_a}\psi_a(x_a)\, dx_a\, ; \lambda > E_\alpha, \theta_a \in S_a. \tag{1.2}$$

Clearly $F_\alpha(p_a^2 + E_\alpha)F_\alpha^{-1}$ is just multiplication by λ.

Given channels β and α we shall use the notation $\hat{S}_{\beta\alpha} = F_\beta S_{\beta\alpha} F_\alpha^{-1}$. By well known abstract theory the following fibered representation holds:

$$\hat{S}_{\beta\alpha} = \int_{I_{\beta\alpha}} \oplus \hat{S}_{\beta\alpha}(\lambda)\, d\lambda\, ; I_{\beta\alpha} = (\max(E_\beta, E_\alpha), \infty).$$

Here the fibers $\hat{S}_{\beta\alpha}(\lambda)$ are defined only almost everywhere as bounded linear operators from $L^2(S_a)$ to $L^2(S_b)$. The issue here is more detailed pointwise properties (including properties of the kernels).

Given a cluster decomposition a the intercluster potential $I_a(x)$ (defined above) "goes to zero" on $X\setminus\bigcup_{b\not\subset a} X_b$. This motivate the notations $Y_a = X\setminus\bigcup_{b\not\subset a} X_b$ and $Z_a = X_a\setminus\bigcup_{b\not\subset a} X_b$ (cf. [D2]). Clearly if a is 2–cluster, then $Z_a = X_a\setminus\{0\}$.

The set of thresholds is given by

$$\mathcal{F} = \bigcup_{a\neq(1\cdots N)} \sigma_{pp}(H^a),$$

i.e. the union of the pure point spectra of H^a. It is known to closed and countable. The notations $\sigma(\cdot)$ and $\sigma_d(\cdot)$ stand for spectrum and discrete spectrum, respectively.

We impose the following condition on both incoming and outgoing channels.

Condition 1.2 *With* $\alpha = (a, \phi_\alpha, E_\alpha)$

$$E_\alpha \in \sigma_d(H^a).$$

By the HVZ theorem this condition is equivalent to the requirement $E_\alpha < \inf_{b\subsetneq a} \sigma(H^b)$; in particular for $N = 3$ to $E_\alpha < 0$.

We can now state the our results. The first deals with a 2–cluster incoming channel and the free as outgoing channel. By $\mathbf{R}_+$ we mean the set of positive reals.

Theorem 1.3 *Suppose Assumption 1.1 and that α is a 2–cluster channel obeying Condition 1.2. Then the operators $\hat{S}_{0\alpha}(\lambda)$ have kernels*

$$\hat{S}_{0\alpha}\left(\lambda, \theta_0, \theta_a'\right) \in C^\infty(\mathbf{R}_+ \times Z_0 \cap S_0 \times S_a).$$

The next result deals with the incoming as well as the outgoing channels being 2–cluster.

Theorem 1.4 *Suppose Assumption 1.1. Let α and β be 2–cluster channels both obeying Condition 1.2. Then*

(1) If $b \neq a$ or $E_\beta \neq E_\alpha$ the operators $\hat{S}_{\beta\alpha}(\lambda)$ have kernels

$$\hat{S}_{\beta\alpha}\left(\lambda, \theta_b, \theta_a'\right) \in C^\infty(I_{\beta\alpha}\setminus\mathcal{F} \times S_b \times S_a).$$

(2) If $b = a$, $E_\beta = E_\alpha$ and $\beta \neq \alpha$ the operators $\hat{S}_{\beta\alpha}(\lambda)$ have kernels

$$\hat{S}_{\beta\alpha}\left(\lambda, \theta_b, \theta_a'\right) \in C^\infty(I_{\beta\alpha}\setminus\mathcal{F} \times (S_b \times S_a)\setminus D_a);$$

$$D_a = \left\{\left(\theta_b, \theta_a'\right) \in S_a \times S_a | \theta_b = \theta_a'\right\}.$$

Moreover with $\psi(x_a) = \left(\nabla_x I_a^{(1)}\right)(x_a)$ and $\hat{\psi}(\theta_a)$ its Fourier transform

$$|\hat{S}_{\beta\alpha}\left(\lambda, \theta_b, \theta_a'\right) + i\pi(2\pi)^{-\frac{3}{2}}\sqrt{\lambda - E_\alpha} < \phi_\beta, x^a\phi_\alpha > \cdot\hat{\psi}\left(\sqrt{\lambda - E_\alpha}\left(\theta_b - \theta_a'\right)\right)|$$
$$\leq C(\lambda)|\theta_b - \theta_a'|^{-\kappa};$$
$$\kappa = \tilde{\kappa} := 3 - \epsilon_0 - \min(2, \epsilon_0) \textit{ if } \tilde{\kappa} > 0,$$
$$\kappa > 0 \textit{ if } \tilde{\kappa} = 0, \kappa = 0 \textit{ if } \tilde{\kappa} < 0,$$

and $C(\lambda)$ locally bounded on $I_{\beta\alpha}\setminus\mathcal{F}$.

(3) If $\beta = \alpha$ the operators $\hat{T}_{\beta\alpha}(\lambda) := \hat{S}_{\beta\alpha}(\lambda) - I$ have kernels

$$\hat{T}_{\beta\alpha}\left(\lambda, \theta_b, \theta_a'\right) \in C^\infty(I_{\beta\alpha}\backslash\mathcal{F} \times (S_b \times S_a)\backslash D_a).$$

Moreover with $\eta(x_a) = I_a^{(1)}(x_a)$ and $\hat{\eta}(\theta_a)$ its Fourier transform

$$|\hat{T}_{\beta\alpha}\left(\lambda, \theta_b, \theta_a'\right) + i\pi(2\pi)^{-\frac{3}{2}}\sqrt{\lambda - E_\alpha}\hat{\eta}\left(\sqrt{\lambda - E_\alpha}\left(\theta_b - \theta_a'\right)\right)|$$
$$\leq C(\lambda)|\theta_b - \theta_a'|^{-\kappa};$$
$$\kappa = \tilde{\kappa} := 3 - \epsilon_0 - \min(1, \epsilon_0 - 1) \textit{ if } \tilde{\kappa} > 0,$$
$$\kappa > 0 \textit{ if } \tilde{\kappa} = 0, \kappa = 0 \textit{ if } \tilde{\kappa} < 0,$$
and $C(\lambda)$ locally bounded on $I_{\beta\alpha}\backslash\mathcal{F}$.

Section 2 Proofs

We shall give a brief outline of the proof of Theorem 1.3. The proof of Theorem 1.4 is similar but in some sense simpler. It is noticed that the singularities at the diagonal in Theorem 1.4 (2) and (3) come from the appearance of a "Born term" similar to the one appearing in the 2–body problem [I-K]. In fact our proofs go along the same scheme as in [I-K], however with some modifications especially for the one of Theorem 1.3.

First we localize in momentum space. For any channel $\beta = (b, \phi_\beta, E_\beta)$ we define for $\lambda > E_\beta$ and $\delta > \frac{1}{2}$ the operator $F_\beta(\lambda) \in \mathcal{B}\left(\langle x_b\rangle^{-\delta}L^2(X_b), L^2(S_b)\right)$ by (cf. (1.2)) $F_\beta(\lambda)\psi_b = (F_\beta\psi_b)(\lambda, \theta_b);, \psi_b \in \langle x_b\rangle^{-\delta}L^2(X_b)$.

Let $\tilde{N} \in \mathbf{N}, \theta \in Z_0 \cap S_0$ and $\tilde{\lambda} \in \mathbf{R}_+$. Then we put

$$n = \tilde{N} + \dim X \text{ and } m = 2n.$$

We want to find open neighborhoods N_θ of θ and $N_{\tilde{\lambda}}$ of $\tilde{\lambda}$ in $Z_0 \cap S_0$ and $\mathbf{R}_+$, respectively, such that

$$\forall\chi \in C_0^\infty(N_\theta)\forall f \in C_0^\infty(N_{\tilde{\lambda}}):$$
$$F_0 f(p^2)\chi(\hat{p})S_{0\alpha}F_\alpha^{-1} = \int_{\mathbf{R}_+} \oplus\, \hat{S}_{f,\chi}(\lambda)\, d\lambda,$$

where

$$\hat{S}_{f,\chi}(\lambda) = F_0(\lambda)\langle x\rangle^{-n}B(\lambda)J_\alpha\langle x_a\rangle^{-n}F_\alpha(\lambda)^*;$$
$$B(\cdot) \in C^{\tilde{N}}(\mathbf{R}_+, \mathcal{B}(L^2(X))). \tag{2.1}$$

Given (2.1) we get by inspection that the kernel $\hat{S}_{f,\chi}\left(\lambda, \theta_0, \theta_a'\right)$ of $\hat{S}_{f,\chi}(\lambda)$ satisfies

$$\hat{S}_{f,\chi}(\cdot,\cdot,\cdot) \in C^{\tilde{N}}(\mathbf{R}_+ \times S_0 \times S_a). \tag{2.2}$$

By a covering argument and (2.2) we obtain the statement of Theorem 1.3 with ∞ replaced by $\tilde{N}$. But since $\tilde{N}$ is arbitrary the theorem follows.

The basic strategy for proving (2.1) is to write for some f_1 supported near $\tilde{\lambda}$ (as f)

$$
\begin{aligned}
f(p^2)\chi(\hat{p})S_{0\alpha} &= f_1(p^2)\left(\tilde{W}_0^+\right)^* \tilde{W}_\alpha^-;\\
\tilde{W}_0^+ &= s - \lim_{t\to+\infty} e^{itH} B_3 e^{-itH_0},\\
\tilde{W}_\alpha^- &= s - \lim_{t\to-\infty} e^{itH} B_3' e^{-itH_a} J_\alpha,
\end{aligned}
\tag{2.3}
$$

where the commutators

$$
\begin{aligned}
i(HB_3 - B_3H_0) &= B_1 + B_2,\\
i\left(HB_3' - B_3'H_a\right) &= B_1' + B_2'
\end{aligned}
\tag{2.4}
$$

localize to forbidden regions in phase space. Explicitly this mean that $B_1 = P_-(x,p)$ is a bounded Ps.D.Op. (i.e. pseudodifferential operator) on X with the symbol $p_-(x,\xi)$ supported in

$$
x\cdot\xi \le (1-\mu_0)|x||\xi| \tag{2.5}
$$

for some $\mu_0 > 0$ depending on 0 and the N–body geometry. Similarly B_1' is a finite sum of operators of the product type $|\phi_\alpha^{(k)}><\phi_\alpha^{(l)}| \otimes P_+(x_a,p_a)$ where the first rank-one operator is built up of vectors in the null-space of $H^a - E_\alpha$, E_α being the channel energy, and the second factor is a bounded Ps.D.Op. on X_a with the symbol $p_+(x_a,\xi_a)$ supported in

$$
x_a\cdot\xi_a \ge (1-\epsilon_a)|x_a||\xi_a| \tag{2.6}
$$

for some $\epsilon_a > 0$ that can be chosen as small as one would like. As for the remainder terms B_2 and B_2' these take the forms $B_2 = \langle x\rangle^{-m} B\langle x\rangle^{-m}$ and $B_2' = \langle x\rangle^{-m} B'\langle x_a\rangle^{-m}$, respectively. The operator B_3 is a Ps.D.Op. on X. The operator B_3' is also a finite sum of operators of the tensor product type. The internal parts maps to functions with strong decay, while the external parts are Ps.D.Op.'s.

In the construction of the above "channel localization operators" we make use of Condition 1.2. (It is not quite clear to the author whether this condition is essential or rather just technically convenient.)

By using (2.3) and (2.4) we obtain the following formal expression for $\hat{S}_{f,\chi}(\lambda)$.

$$
\begin{aligned}
\hat{S}_{f,\chi}(\lambda) &= D_1(\lambda) + \lim_{\epsilon\downarrow 0} D_{2,\epsilon}(\lambda);\\
D_1(\lambda) &= -2\pi f_1(\lambda)F_0(\lambda)B_3^*\left(B_1' + B_2'\right)J_\alpha F_\alpha(\lambda)^*,\\
D_{2,\epsilon}(\lambda) &= 2\pi i f_1(\lambda)F_0(\lambda)(B_1+B_2)^*\\
&\quad (H-(\lambda+i\epsilon))^{-1}\left(B_1' + B_2'\right)J_\alpha F_\alpha(\lambda)^*.
\end{aligned}
\tag{2.7}
$$

We justify (2.7) (and in fact prove (2.1)) by expanding the products and making sense to all terms. The difficult ones come from the second term with the resolvent in the middle. For these we use strong resolvent estimates (cf. [M1–2],[J],[J-M-P]) together with the indicated properties of the B's. We use as conjugate operator A (this means an operator with a positive localized commutator with the Hamiltonian) an operator that roughly has the symbol

$$a(x,\xi) \simeq x \cdot \xi - \kappa\langle x\rangle,$$

where $0 < \kappa < \sqrt{\lambda}$ is adjusted to make B_1 localize to the negative part of A, while B_1' localizes to the positive part of A. This adjustment is possible due to (2.5), (2.6) and energy localization.

References

[A-P-S] Amrein,W.O., Pearson,D.B., Sinha,K.B.: Bounds on the total scattering cross-section for N-body systems. Nuovo Cimento **52** A–1, 115–131 (1979)

[C-T] Combes,J.M., Tip,A.: Properties of the scattering amplitude for electron-atom collisions. Ann.Inst.Henri Poincaré **40**–2, 117–139 (1984)

[D1] Derezinski,J.: Existence and analyticity of many body scattering amplitudes at low energy. J.Math.Phys. **28**, 1080–1088 (1987)

[D2] Derezinski,J.: A new proof of the propagation theorem for N–body quantum systems. Commun.Math.Phys. **122**, 203–231 (1989)

[E-S] Enss,V., Simon,B.: Finite total cross sections in non-relativistic quantum mechanics. Commun.Math.Phys. **76**, 177–209 (1980)

[F-H] Froese,R.G., Herbst,I.: Exponential bounds and absence of positive eigenvalues for N–body Schödinger operators. Commun.Math.Phys. **87**, 429–447 (1982)

[Gr1] Graf,G.M.: Asymptotic completeness for N-body short-range quantum systems:a new proof. Commun.Math.Phys. **132**, 73–101 (1990)

[I] Isozaki,H.: Structure of S–matrices for three body Schrödinger operators. Preprint 1991

[I-K] Isozaki,H., Kitada,H.: Scattering matrices for two-body Schrödinger operators. Scientific papers of the college of arts and sciences, Tokyo Univ. **35**, 81–107 (1985)

[J] Jensen,A.: Propagation estimates for Schrödinger-type operators. Transactions of the Am.Math.Soc. **291**–1, 129–144 (1985)

[J-M-P] Jensen,A., Mourre,E., Perry,P.: Multiple commutator estimates and resolvent smoothness in quantum scattering theory. Ann.Inst.Henri Poincaré **41**–2, 2o7–225 (1984)

[M1] Mourre,E.: Operateurs conjugues et proprietes de propagation. Commun.Math.Phys. **91**, 279–300 (1983)

[M2] Mourre,E.: Absence of singular continuous spectrum for certain selfadjoint operators. **78**, 391–408 (1981)

[S1] Skibsted,E.: Smoothness of N–body scattering amplitudes. Preprint 1992

[S2] Skibsted,E.: Propagation estimates for N–body Schrödinger operators. Commun.Math.Phys. **142**–1, 67–98 (1991)

Semiclassical Approximation for Schrödinger Operators at High Energy

Lawrence E. Thomas [1], Stephen R. Wassell [2]

[1]Department of Mathematics, University of Virginia, Charlottesville, VA 22903, USA
[2]Department of Mathematical Sciences, Sweet Briar College, Sweet Briar, VA 24595, USA

Abstract: Let $H_\hbar = -\hbar^2\partial_\varphi^2/2 + U(\varphi)$ be a Schrödinger operator acting in $L^2(\mathbb{T})$ with $\mathbb{T}$ an ℓ-dimensional torus and U an analytic periodic function on $\mathbb{T}$. Approximate semiclassical expansions for the eigenfunctions and eigenvalues of $H_\hbar$ are developed which are asymptotic in inverse powers of the classical action variables for the corresponding classical Hamiltonian. The leading term in the eigenvalue expansion is the energy associated with a KAM torus; the fact that KAM tori are abundant at high energy is exploited to show that the rank of the approximate eigenfunctions with energy $\leq E$ approaches the rank of the true eigenvalues of $H_\hbar$, for E large.

1 Introduction

Let $H_\hbar$ be the Schrödinger operator, acting in $L^2(\mathbb{T})$ where $\mathbb{T}$ is an ℓ-dimensional torus,

$$H_\hbar\psi = -\frac{\hbar^2}{2}\frac{\partial^2}{\partial\varphi^2}\psi(\varphi) + U(\varphi)\psi(\varphi). \tag{1}$$

Here, ∂_φ^2 is the ℓ-dimensional Laplacian, and the potential U is periodic in φ and real analytic in the sense that if $U(\varphi) = \tilde{U}(z)$ with $z_j = e^{i\varphi_j}$, $j = 1,\ldots,\ell$, then $\tilde{U}(z)$ extends to a $\mathbb{C}^\ell$-neighborhood W of the torus,

$$W = \{z \mid e^{-\delta} < |z_j| < e^{\delta},\ j = 1,\ldots,\ell\}, \tag{2}$$

for some $\delta > 0$. Functions ψ in the domain of $H_\hbar$ will be assumed to satisfy the self-adjoint α-periodicity conditions

$$\psi(\varphi + 2\pi\hat{\jmath}) = e^{2\pi i\alpha_j}\psi(\varphi), \quad j = 1, \ldots, \ell, \tag{3}$$

with $\hat{\jmath}$ a unit vector in the φ_j-direction, and $0 \le \alpha_j \le 1$. (Physically, $H_\hbar$ with α-periodicity conditions can be regarded as the reduced Hamiltonian for a particle moving through an infinite crystal in $\mathbb{R}^\ell$ with periodic potential U, and having quasi-momentum α.) For fixed α, $H_\hbar$ has discrete spectrum, and in particular has a compact resolvent (cf. [Thomas]).

The intent of this article is to provide asymptotic expansions for eigenfunctions of $H_\hbar$ and their eigenvalues at high energy. *A priori,* the expansions are in powers of Planck's constant $\hbar$, in the manner of WKB or semiclassical expansions. Our expansions have the property that if the pth partial sum (including terms up to $O(\hbar^p)$) has energy expectation E, then the variance of the energy is essentially $O(E^{-p-2+\varepsilon})$ for arbitrarily small ε (see (56) for the exact exponent). We also discuss the (approximate) orthogonality of these partial sums and estimate the dimension of the space spanned by them, the conclusion being that the dimension of the partial sums with energy expectation $\le E$, divided by the dimension of the actual eigenfunctions of $H_\hbar$ with energy $\le E$, approaches 1, for $E \to \infty$, almost surely in α.

Recall that in the WKB expansion one assumes that the eigenfunction is of the form

$$\psi = \exp(iS/\hbar)(W_0 + \hbar W_1 + \cdots), \tag{4}$$

where, in particular, the action $S = S(A^\infty, \varphi)$ is a function of the angle variables φ satisfying the Hamilton-Jacobi equation (A^∞ are ℓ auxilliary action variables to be adjusted so that ψ satifies the boundary conditions). But in general, the Hamilton-Jacobi equation does not admit a global (in φ) solution depending continuously and non-trivially on the A^∞, since if it did the classical Hamiltonian corresponding to $H_\hbar$ would be integrable (with the A^∞ as constants of motion). On the other hand, KAM theory (see particularly [Gallavotti]) does provide a C^∞-function $S(A^\infty, \varphi)$ which is a solution to the Hamilton-Jacobi equation, at least for A^∞ in a Cantor set V^∞ (which depends on the potential U). Thus, one obtains an approximate eigenfunction, i.e. the eikonal approximation $\exp(iS/\hbar)$, provided that A^∞ is in this set V^∞ and is such that $\exp(iS/\hbar)$ satisfies the appropriate periodicity conditions.

Fortunately, at least at high energy, V^∞ has large fractional measure. To be precise, let $N > 0$, $a_j > 0$, $j = 1, \ldots, \ell$, and set

$$V_N = \{A^\infty | \, a_j N \le |A_j^\infty| \le N, \; j = 1, \ldots, \ell\} \tag{5}$$

$$V_N^\infty = V^\infty \bigcap V_N. \tag{6}$$

It is shown in [Thomas and Wassell] that $\mathrm{vol} V_N^\infty / \mathrm{vol} V_N \to 1$ as $N \to \infty$. This allows one to construct the present eikonal approximation with ***large probability*** (this notion is made precise in section 4).

Having constructed the eikonal approximation, one then proceeds to recursively solve for the coefficients, $W_0, W_1, \ldots$, of ψ, which are also functions of A^∞, φ. The W_n are solutions to first order partial differential equations and are to be periodic in φ; it is not obvious that such solutions exist (conceivably, V^∞ might need to be further restricted to obtain solutions). But by a change of variable $\varphi \mapsto \varphi^\infty$ (with φ^∞ canonically conjugate to A^∞) one can rewrite the differential equations in a form where one readily solves for and estimates the W_n. In fact, in the φ^∞ representation one does encounter small denominator problems, but it turns out that they have already been dealt with in the KAM construction of the generator S; in particular no further restriction of V^∞ is required.

We note an interesting connection between this work and [Feldman, Knörrer and Trubowitz]. These authors consider the Schrödinger operators (1) with $\hbar = 1$, although with possibly more general quadratic kinetic energies $\partial_\varphi^2 = \sum m_j \partial_{\varphi_j}^2$, and again U periodic (although not necessarily analytic) and normalized with average value zero over the torus. They show that, in 2 and 3 dimensions, most eigenvalues of $H_\hbar$ are stable in the sense that for $\varepsilon > 0$ fixed, an eigenvalue λ of $H_\hbar$ is within a distance $\lambda^{\varepsilon - 1}$ of an eigenvalue of the kinetic energy only ($U = 0$); in particular they show that the fraction of stable eigenvalues $\leq E$ approaches 1, $E \to \infty$, for generic α. These authors estimate the zeros of the regularized determinant $\det M(z)$, where $M(z) = (-\hbar^2 \partial_\varphi^2/2 - z\mathbb{1})^{-1}(H_\hbar - z\mathbb{1})$. Writing $M(z)$ in the basis of eigenvectors of the Laplacian $-\hbar^2\partial_\varphi^2/2$ and considering a neighborhood of one of its eigenvalues λ_0, they "block" $M(z)$ into pieces, one block of which involves matrix elements having small denominators, $(\hbar^2 q^2/2 - z)^{-1}\hat{U}(q - q')$, with $\hbar^2 q^2/2$ and z near λ_0 (here $\hat{U}$ is the Fourier transform of U). The idea is to generically control this block.

On the other hand, our results hold in higher dimensions, and provide explicit (asymptotic) formulae for the eigenvalues and eigenvectors. Moreover we believe our results shed some light on the eigenvalue stability question; in particular we find that if $E_0(A^\infty)$ is the leading term in the asymptotic expansion for an eigenvalue, then the difference between $E_0(A^\infty)$ and $(A^\infty)^2/2$ (the corresponding eigenvalue of $-\hbar^2\partial_\varphi^2/2$) is $O((A^\infty)^{-2+\varepsilon})$, for some $\varepsilon > 0$, which is essentially the stability result of [Feldman et al.] in 2 and 3 dimensions. But $E_0(A^\infty)$ is the *classical energy*, equal (for $A^\infty \in V^\infty$) to the classical Hamiltonian corresponding to $H_\hbar$, but written in terms of the conserved A^∞ (it is also the energy appearing in the Hamilton-Jacobi equation). Thus, the energy shift is primarily a *classical effect*; higher order terms in the expansion for the eigenvalue contribute corrections only of order $O((A^\infty)^{-4+\varepsilon})$.

Finally, we note the connection of our work with [Graffi and Paul]. These authors consider the problem of the semiclassical limit of perturbation theory for a system of quantum harmonic oscillators (as opposed to the anisochronous rotator system which we consider) with an analytic interaction εU. By transforming the corresponding Schrödinger operator into the Bargmann representation (which is closely related to the action-angle representation of the classical Hamiltonian), these authors show that the Rayleigh-Schrödinger perturbation expansion for an eigenvalue λ converges to the classical Birkhoff expansion, $\hbar \to 0$. In particular, they show that the coefficient of ε^n, $\lambda_n(\hbar)$, has a convergent series in $\hbar$ about its classical counterpart, and that the series for λ, in powers of ε, is asymptotic. By way of contrast, again, our asymptotic series for the eigenfunctions and eigenvalues are in inverse powers of the classical actions for a potential of fixed strength. We also give probabilistic estimates for these actions to belong to the Cantor set V_N^∞ and to satisfy some additional quantization (periodicity) conditions; these estimates are used in turn to estimate the rank of the approximate eigenfunctions.

The outline of the paper is as follows: Sect. 2 outlines the expansion scheme for the approximate eigenfunctions and eigenvalues, and establishes the existence of the coefficients in these expansions. Sect. 3 discusses the linear independence of the partial sums of these approximate eigenfunctions and indeed the sense in which they are approximate. Sect. 4 contains probability estimates on the abundance of the approximate eigenfunctions. Estimates on the coefficients in the expansions are sketched in the appendix.

2 Approximate Eigenfunctions and Eigenvalues

In this section, we obtain asymptotic expansions for eigenfunctions and eigenvalues of $H_\hbar$; in particular, we show how to solve for the coefficients in the following expansions:

$$\psi_\hbar(A^\infty,\varphi) = (W_0(A^\infty,\varphi) + \hbar W_1(A^\infty,\varphi) + \cdots)\exp(\frac{i}{\hbar}S(A^\infty,\varphi)) \quad (7)$$

$$E_\hbar(A^\infty) = E_0(A^\infty) + \hbar E_1(A^\infty) + \cdots. \quad (8)$$

(Note that we regard (7) and the differential equations below as local expressions in φ; later we incorporate the periodicity conditions, whereby, e.g., (7) is defined on the torus $\mathbb{T}$.) Substituting these expansions into the Schrödinger equation,

$$H_\hbar\psi_\hbar(A^\infty,\varphi) = E_\hbar(A^\infty)\psi_\hbar(A^\infty,\varphi), \quad (9)$$

we obtain a sequence of differential equations for S and the $\{W_n\}$ by matching terms of order $\hbar^0, \hbar^1, \ldots, \hbar^{n+1}, \ldots$, respectively:

$$h(\frac{\partial S}{\partial\varphi},\varphi) = E_0(A^\infty) \quad (10)$$

$$\frac{\partial S}{\partial\varphi}\cdot\frac{\partial W_0}{\partial\varphi} + \frac{1}{2}\frac{\partial^2 S}{\partial\varphi^2}W_0 = 0 \quad (11)$$

$$\vdots$$

$$-\frac{1}{2}\frac{\partial^2 W_{n-1}}{\partial\varphi^2} - \mathrm{i}\frac{\partial S}{\partial\varphi}\cdot\frac{\partial W_n}{\partial\varphi} - \frac{\mathrm{i}}{2}\frac{\partial^2 S}{\partial\varphi^2}W_n = E_2 W_{n-1} + \cdots + E_{n+1}W_0, \quad (12)$$

$$\vdots$$

the last equation being valid for $n \geq 1$. Here h is the classical Hamiltonian,

$$h(A,\varphi) = \frac{1}{2}A^2 + U(\varphi), \tag{13}$$

so that (10) is just the Hamilton-Jacobi equation. Note that $E_1 = 0$; we will comment on this further below.

We remind the reader that (11) is the time-independent transport equation which can be written

$$\frac{\partial}{\partial\varphi}\cdot(W_0^2 A) = 0, \tag{14}$$

where

$$A = \frac{\partial S}{\partial\varphi}(A^\infty,\varphi). \tag{15}$$

We seek solutions to these differential equations so that $\psi_\hbar$ satisfies the α-periodicity conditions

$$\psi_\hbar(A^\infty,\varphi + 2\pi\hat{\jmath}) = \mathrm{e}^{2\pi\mathrm{i}\alpha_j}\psi_\hbar(A^\infty,\varphi), \quad j = 1,\ldots,\ell, \tag{16}$$

with $\hat{\jmath}$ a unit vector in the φ_j-direction. Indeed, $\psi_\hbar$ will satisfy the α-periodicity conditions provided that A^∞ is chosen so that

(i) S can be written

$$S = A^\infty\cdot\varphi + \hat{S}(A^\infty,\varphi) \tag{17}$$

with

$$A_j^\infty = (\alpha_j + m_j)\hbar, \quad j = 1,\ldots,\ell \tag{18}$$

for some integer m_j, and $\hat{S}(A^\infty,\varphi)$ is periodic in φ;

(ii) The $W_n(A^\infty,\varphi)$ are periodic functions in φ.

We proceed to the question of existence of solutions for the differential equations (10) through (12). First, as regards the Hamilton-Jacobi equation (10), we have

Theorem 1 *[Thomas and Wassell] There exists a C^∞-function $\hat{S}(A^\infty,\varphi)$, defined on $V_N\times\mathbb{T}$ (in particular, periodic in φ), a C^∞-function $E_0(A^\infty)$ defined on V_N, and a closed subset $V_N^\infty \subset V_N$ such that for $A^\infty \in V_N^\infty$, $S(A^\infty,\varphi) = A^\infty\cdot\varphi + \hat{S}(A^\infty,\varphi)$ is a solution to the Hamilton-Jacobi equation (10). Moreover*

$$\lim_{N\to\infty}\frac{|V_N^\infty|}{|V_N|} = 1; \tag{19}$$

here $|\cdot|$ *denotes* ℓ*-dimensional Lebesgue measure. For* $0 < \nu < 1$, $j \geq 1$, $0 \leq n \leq j$, *and* $A^\infty \in V_N^\infty$, *we have*

$$\left(\frac{\partial}{\partial A^\infty}\right)^n \left(\frac{\partial}{\partial \varphi}\right)^{j-n} \hat{S} = O(N^{(1-\nu)j-\nu-n}). \tag{20}$$

Note. The proof of this theorem is contained in [Thomas and Wassell]; see Theorem 1.1 and Sect. 4 there. See also Proposition 7 of the appendix for remarks on the estimates of $\hat{S}$.

We turn now to the transport equation (11). Let $S(A^\infty, \varphi)$ be the generating function of the previous theorem.

Proposition 2 *The* C^∞*-function on* $V_N \times \mathbb{T}$ *defined by*

$$W_0(A^\infty, \varphi) = \left[\det\left(\frac{\partial^2 S}{\partial A^\infty \partial \varphi}(A^\infty, \varphi)\right)\right]^{\frac{1}{2}} \tag{21}$$

is a solution to the transport equation (11) for $A^\infty \in V_N^\infty$. *Here,*

$$\frac{\partial^2 S}{\partial A^\infty \partial \varphi}(A^\infty, \varphi) = \mathbb{1} + \frac{\partial^2 \hat{S}}{\partial A^\infty \partial \varphi}(A^\infty, \varphi), \tag{22}$$

with the second term of (22) being $O(N^{1-3\nu})$, ν *chosen as in Theorem 1.*

Proof. The estimate of $\hat{S}$ is given in Theorem 1. Let $C : (A^\infty, \varphi^\infty) \to (A, \varphi)$ be the canonical transformation defined implicitly by the equations

$$A = A^\infty + \frac{\partial \hat{S}}{\partial \varphi}(A^\infty, \varphi) \tag{23}$$

$$\varphi^\infty = \varphi + \frac{\partial \hat{S}}{\partial A^\infty}(A^\infty, \varphi). \tag{24}$$

The classical time evolution generated by h yields the equations

$$\frac{d}{dt}\varphi = A = \frac{\partial S}{\partial \varphi}(A^\infty, \varphi) \tag{25}$$

$$\frac{d}{dt}A^\infty = -\frac{\partial(h \circ C)}{\partial \varphi^\infty} = -\frac{\partial(E_0(A^\infty))}{\partial \varphi^\infty} = 0 \tag{26}$$

$$\frac{d}{dt}\varphi^\infty = \frac{\partial(E_0(A^\infty))}{\partial A^\infty} \equiv \omega^\infty(A^\infty) \tag{27}$$

for $A^\infty \in V_N^\infty$. Consequently, we write

$$\frac{d}{dt} = \frac{\partial S}{\partial \varphi} \cdot \frac{\partial}{\partial \varphi} \tag{28}$$

as an operator on functions of (A^∞, φ), for $A^\infty \in V_N^\infty$.

Set $J = W_0^2 = \det(\partial_{A^\infty}\partial_\varphi S)$. Then for $A^\infty \in V_N^\infty$,

$$\begin{aligned}
\frac{\partial S}{\partial \varphi}\cdot\frac{\partial J}{\partial \varphi} &= \frac{d}{dt}J \\
&= J \operatorname{tr}\left[\left(\frac{\partial^2 S}{\partial A^\infty \partial \varphi}\right)^{-1} \frac{\partial S}{\partial \varphi}\cdot\frac{\partial}{\partial \varphi}\frac{\partial^2 S}{\partial A^\infty \partial \varphi}\right] \\
&= J \operatorname{tr}\left[\left(\frac{\partial^2 S}{\partial A^\infty \partial \varphi}\right)^{-1} \frac{\partial}{\partial \varphi}\left(\frac{\partial S}{\partial \varphi}\cdot\frac{\partial}{\partial \varphi}\frac{\partial S}{\partial A^\infty}\right)\right] \\
&\quad - J \operatorname{tr}\left[\left(\frac{\partial^2 S}{\partial A^\infty \partial \varphi}\right)^{-1} \frac{\partial^2 S}{\partial \varphi \partial \varphi}\left(\frac{\partial^2 S}{\partial A^\infty \partial \varphi}\right)\right] \\
&= -J\frac{\partial^2 S}{\partial \varphi^2},
\end{aligned} \tag{29}$$

since

$$\frac{\partial S}{\partial \varphi}\cdot\frac{\partial}{\partial \varphi}\frac{\partial S}{\partial A^\infty} = \frac{d}{dt}\varphi^\infty = \omega^\infty(A^\infty), \quad A^\infty \in V_N^\infty \tag{30}$$

is independent of φ. It follows immediately from this differential equation for J that $W_0 = J^{1/2}$ is a solution to the transport equation (11). □

(We are now able to justify $E_1 = 0$ (see (11)). Were $E_1 \neq 0$, then (11) becomes, in view of the above proposition,

$$-\mathrm{i}\frac{\partial S}{\partial \varphi}\cdot\frac{\partial}{\partial \varphi}J^{-\frac{1}{2}}W_0 = E_1 J^{-\frac{1}{2}}W_0, \tag{31}$$

which has solution

$$W_0 = J^{\frac{1}{2}}\mathrm{e}^{\mathrm{i}\boldsymbol{r}\cdot\varphi^\infty}, \quad \boldsymbol{r} \in \mathbb{Z}^\ell \tag{32}$$

and

$$E_1 = \omega^\infty\cdot\boldsymbol{r} = \frac{\partial}{\partial A^\infty}E_0\cdot\boldsymbol{r}. \tag{33}$$

Thus the subsequent expansions for $\psi_\hbar$ and $E_\hbar$ correspond to the unperturbed (U=0) state $\exp(\mathrm{i}(\boldsymbol{m}+\boldsymbol{r})\cdot\varphi)$ and energy $(\boldsymbol{m}+\boldsymbol{r})^2\hbar^2/2$, respectively. For this reason we take $\boldsymbol{r} = 0$ and $E_1 = 0$.)

It remains to establish a solution to (12), which can now be rewritten as

$$-\frac{1}{2}J^{-\frac{1}{2}}\frac{\partial^2 W_{n-1}}{\partial \varphi^2} - \mathrm{i}\frac{\partial S}{\partial \varphi}\cdot\frac{\partial}{\partial \varphi}J^{-\frac{1}{2}}W_n = J^{-\frac{1}{2}}(E_2 W_{n-1} + \cdots + E_{n+1}W_0), \tag{34}$$

again for $n \geq 1$. We shall do so for $n = 1$ (which corresponds to $\hbar^2$). The analysis of the higher order equations is analogous and hence omitted. For $n = 1$, (34) can be written as

$$-\mathrm{i}\frac{d}{dt}\left(\frac{W_1}{W_0}\right) = E_2 + \frac{1}{2}\frac{1}{W_0}\frac{\partial^2 W_0}{\partial \varphi^2}, \tag{35}$$

using (28) as before. We have the following:

Proposition 3 *There exists a* C^∞*-function* $W_1(A^\infty,\varphi)$ *defined on* $V_N \times \mathbb{T}$ *such that if* $A^\infty \in V_N^\infty$, W_1 *is a solution to (35), with* $E_2(A^\infty)$ *given by*

$$E_2(A^\infty) = \frac{1}{2}(2\pi)^{-\ell} \int \left(\frac{\partial W_0}{\partial\varphi}\right)^2 d\varphi. \tag{36}$$

For $0 < \nu < 1$, $W_1 = O(N^{1-4\nu+(1-\nu)(2\ell+3)})$ *and* $E_2 = O(N^{4-8\nu})$.

Remark. The estimates on W_1, E_2 and indeed estimates on all higher W_n, derivatives of W_n and E_n is the content of Proposition 8.

Proof. If we regard (35) as an equation in the $(A^\infty,\varphi^\infty)$ variables, then d/dt takes the form

$$\frac{d}{dt} = \omega^\infty(A^\infty)\cdot\frac{\partial}{\partial\varphi^\infty} \tag{37}$$

for $A^\infty \in V_N^\infty$. For such A^∞, (35) is solvable, with W_1 periodic, if the 0th Fourier coefficient of the right hand side is zero, i.e., if

$$(2\pi)^\ell E_2 = -\frac{1}{2}\int \frac{1}{W_0}\frac{\partial^2 W_0}{\partial\varphi^2}d\varphi^\infty = -\frac{1}{2}\int W_0\frac{\partial^2 W_0}{\partial\varphi^2}d\varphi, \tag{38}$$

by a change of variables, from which (36) follows.

With E_2 so chosen, we have that, in the $(A^\infty,\varphi^\infty)$ representation,

$$\frac{W_1}{W_0} = \frac{1}{2}\sum_{k\neq 0}\frac{\widehat{\frac{1}{W_0}\partial_\varphi^2 W_0}(k)\exp(ik\cdot\varphi^\infty)}{\omega^\infty(A^\infty)\cdot k}, \tag{39}$$

where again, by a change of variables,

$$\widehat{\frac{1}{W_0}\frac{\partial^2 W_0}{\partial\varphi^2}}(k) = \frac{1}{(2\pi)^\ell}\int W_0\frac{\partial^2 W_0}{\partial\varphi^2}\exp\left[-ik\cdot\left(\varphi+\frac{\partial\hat{S}}{\partial A^\infty}(A^\infty,\varphi)\right)\right]d\varphi. \tag{40}$$

The denominator in the right hand side of (39) satisfies the estimate (see Equation 4.41 of [Gallavotti])

$$|\omega^\infty(A^\infty)\cdot k|^{-1} \le 8C_0\|k\|^{\ell+1}, \quad A^\infty \in V_N^\infty, \tag{41}$$

whereas we have the bounds on the Fourier coefficients, from integration by parts p times,

$$\left|\widehat{\frac{1}{W_0}\frac{\partial^2 W_0}{\partial\varphi^2}}(k)\right| \le \frac{c(p)}{\|k\|^p}\sup_\varphi\left|\left(\left(\frac{\partial\varphi^\infty}{\partial\varphi}\right)^{-1}\frac{\partial}{\partial\varphi}\right)^p\frac{1}{W_0}\frac{\partial^2 W_0}{\partial\varphi^2}(A^\infty,\varphi)\right|. \tag{42}$$

These estimates show the convergence of the series (39) for $A^\infty \in V_N^\infty$. Now W_1 extends to $V_N \times \mathbb{T}$ as a C^∞-function by an argument analogous to that for $S(A^\infty, \varphi)$; the argument will not be given here. For bounds on derivatives of W_0 and the bounds on W_1 and E_2, see Proposition 8. □

We end this section by noting that the N-dependence of C_0, i.e. $C_0 = O(N^{-\nu})$, appearing in (41), is of key importance in [Thomas and Wassell] and remains so presently.

3 Rank and Variance of the Approximations

In the previous section we solved the differential equations arising from the Schrödinger equation in order to obtain aysmptotic expansions for the eigenfunctions and eigenvalues of $H_\hbar$ in powers of $\hbar$. The purposes of this section are to show that the partial sums for these expansions are approximately orthogonal (for different values of $A^\infty \in V_N^\infty$ satisfying the α-periodicity conditions) and to elaborate on the sense in which the expansions are asymptotic and indeed provide approximate eigenfunctions. We refer the reader to Proposition 8 of the appendix where estimates of the W_n and E_n are given for these purposes.

We begin by showing that the approximate eigenfunctions satisfying the α-periodicity conditions (16) with their corresponding $A^\infty \in V_N^\infty$ are linearly independent. More precisely, we have the following result.

Proposition 4 *Fix $p > 0$ and let*

$$\left\{\psi^{(p)}(A^\infty, \varphi) \equiv \left((W_0 + \hbar W_1 + \cdots + \hbar^p W_p)\exp(\frac{i}{\hbar}S)\right)(A^\infty, \varphi)\right\} \tag{43}$$

be the collection of partial sums of approximate eigenfunctions of order p, with $A^\infty \in V_N^\infty$ satisfying the α-periodicity conditions. Form the matrix

$$M = \left\langle \psi^{(p)}(A^\infty, \cdot), \psi^{(p)}(A^{\infty\prime}, \cdot)\right\rangle. \tag{44}$$

Then for sufficiently large N, M is of maximal rank, i.e., the rank of M equals the number of $A^\infty \in V_N^\infty$ satisfying the α-periodicity conditions.

Proof. It suffices to show that

$$\inf_{A^\infty} \|\psi^{(p)}(A^\infty, \cdot)\|_2^2 > \sup_{A^\infty} \sum_{A^{\infty\prime} \neq A^\infty} \left|\left\langle \psi^{(p)}(A^\infty, \cdot), \psi^{(p)}(A^{\infty\prime}, \cdot)\right\rangle\right| \tag{45}$$

where A^∞ and $A^{\infty\prime}$ of course range over V_N^∞. (This condition ensures that the Holmgren norm of the off-diagonal part of M is less than the inf of the diagonal part of M, so that M is invertible.) Here $A^\infty = (m + \alpha)\hbar$ and $A^{\infty\prime} = (m' + \alpha)\hbar$, so that

$$\left\langle\psi^{(p)}(A^{\infty},\cdot),\psi^{(p)}(A^{\infty\prime},\cdot)\right\rangle$$
$$= \int \exp\left[\frac{i}{\hbar}\left(\hbar(m-m')\cdot\varphi + \hat{S}(A^{\infty},\varphi) - \hat{S}(A^{\infty\prime},\varphi)\right)\right]\cdot$$
$$\cdot \sum_{n,\ n'}^{p} W_n(A^{\infty},\varphi)W_{n'}(A^{\infty\prime},\varphi)\hbar^{n+n'}d\varphi. \quad (46)$$

We now change variables in the integral; let

$$\varphi_j' = \varphi_j + \frac{1}{(m_j - m_j')\hbar}(\hat{S}(A^{\infty},\varphi) - \hat{S}(A^{\infty\prime},\varphi)) \quad (47)$$

for the index j with the largest $|m_j - m_j'|$, and otherwise let the components of φ' cqual those of φ. We have that

$$\sup_{\substack{A^{\infty},A^{\infty\prime}\in V_N \\ \varphi\in\mathbf{T}}} \left|\frac{\partial}{\partial\varphi_j}\left[\frac{1}{(m_j-m_j')\hbar}(\hat{S}(A^{\infty},\varphi) - \hat{S}(A^{\infty\prime},\varphi))\right]\right| \to 0, \quad (48)$$

$N \to \infty$. (Note that $(\hat{S}(A^{\infty},\varphi) - \hat{S}(A^{\infty\prime},\varphi))/\|A^{\infty} - A^{\infty\prime}\| \to 0$, along with its φ-derivatives, as $N \to \infty$ is non-trivial. The function $\hat{S}$ can be thought of as a limit of generating functions obtained from the KAM iteration scheme, each of which has small derivatives, but which is *a priori* defined on a domain filled with holes and hence not convex. See Sect. 4 of [Gallavotti] and the proof of Lemmas 2.1 and 2.2 in [Thomas and Wassell] for a proof of this estimate.) Hence, for sufficiently large N, (47) is invertible $\varphi_j = \varphi_j(\varphi')$ as a C^{∞}-function, in particular of the form

$$\varphi_j = \varphi_j' + \hat{\varphi}_j(\varphi'), \quad (49)$$

with $\hat{\varphi}_j$ periodic in the φ' and with the φ'-derivatives of $\hat{\varphi}_j$ approaching zero as $N \to \infty$. Making this substitution of variables into (46) and integrating by parts r times, we obtain that

$$\left|\left\langle\psi^{(p)}(A^{\infty},\cdot),\psi^{(p)}(A^{\infty\prime},\cdot)\right\rangle\right|$$
$$\leq \frac{c(r)}{|m_j - m_j'|^r}\sup_{\varphi\in\mathbf{T}}\left[\left(\left(\frac{\partial\varphi}{\partial\varphi'}\right)\frac{\partial}{\partial\varphi}\right)^r \sum_{n,\ n'}^{p} W_n(A^{\infty},\varphi)W_{n'}(A^{\infty\prime},\varphi)\hbar^{n+n'}\cdot\right.$$
$$\left.\cdot\left(1 + \frac{1}{(m_j-m_j')\hbar}\frac{\partial}{\partial\varphi_j}\left(\hat{S}(A^{\infty},\varphi) - \hat{S}(A^{\infty\prime},\varphi)\right)\right)^{-1}\right], \quad (50)$$

which for $r > \ell$ is summable in $m' \neq m$, and which goes to zero, $N \to \infty$, by Proposition 8 and (48).

Finally, we note that the left hand side of (45) exceeds

$$(2\pi)^\ell \inf_{A^\infty,\varphi} \left| \sum_{n=0}^{p} W_n(A^\infty,\varphi)\hbar^n \right| \to (2\pi)^\ell, \tag{51}$$

as $N \to \infty$, by Proposition 8. This observation, together with the result of the previous paragraph, completes the proof of the proposition. □

Finally we show that the partial sums for the series $\psi_\hbar(A^\infty,\varphi)$ and $E_\hbar(A^\infty)$ serve as approximate eigenfunctions and eigenvalues. For p a positive integer, set

$$E^{(p+1)}(A^\infty) = (E_0 + \hbar^2 E_2 + \cdots + \hbar^{p+1}E_{p+1})(A^\infty), \tag{52}$$

and let $\psi^{(p)}(A^\infty,\varphi)$ be the partial sum as defined in Proposition 4 (43).

Proposition 5 *If $A^\infty \in V_N^\infty$ and satisfies α-periodicity conditions (16) then*

$$H_\hbar\psi^{(p)}(A^\infty,\varphi) = \left(E^{(p+1)}(A^\infty) + k^{(p)}(A^\infty,\varphi)\right)\psi^{(p)}(A^\infty,\varphi), \tag{53}$$

where

$$k^{(p)}(A^\infty,\varphi) = O(\hbar^{p+2}N^{\gamma p+3-5\nu}), \tag{54}$$

$0<\nu<1$, *and* $\gamma = (1-\nu)(2\ell+3)-\nu$. *Moreover,*

$$\mathrm{var}_{\psi^{(p)}}(H_\hbar) \equiv \left\langle H_\hbar\psi^{(p)}, H_\hbar\psi^{(p)}\right\rangle - \left\langle \psi^{(p)}, H_\hbar\psi^{(p)}\right\rangle^2 \tag{55}$$

$$= O(\hbar^{2(p+2)}N^{2\gamma p+6-10\nu}); \tag{56}$$

where $\langle\ ,\ \rangle$ denotes the $L^2(\mathbb{T})$-inner product, and $\psi^{(p)}$ has been normalized.

Proof. The function $k^{(p)}$ is given by

$$k^{(p)}(A^\infty,\varphi) = \left(-\frac{\hbar^{p+2}}{2}\frac{\partial^2}{\partial\varphi^2}W_p - {\sum}' E_n W_{n'}\hbar^{n+n'}\right) \Big/ \sum_{n=0}^{p} W_n\hbar^n, \tag{57}$$

where the sum $\sum'$ extends over n,n' satisfying $2\le n\le p+1$, $1\le n'\le p$, and $p+2\le n+n'\le 2p+1$. The estimates then follow from Proposition 8 of the appendix. □

Remark. Proposition 5 allows us to estimate the distance between $E^{(p+1)}$ and an actual eigenvalue, as well as the L^2-distance between $\psi^{(p)}$ and the corresponding eigenfunction. First we note that if $\bar{E} \equiv \langle\psi^{(p)}, H_\hbar\psi^{(p)}\rangle$, then $|\bar{E} - E^{(p+1)}| \le \|k^{(p)}\|_\infty$ and so $E^{(p+1)}$ must be within a distance $(\mathrm{var}_{\psi^{(p)}}(H_\hbar))^{1/2} + \|k^{(p)}\|_\infty$ of some eigenvalue λ of $H_\hbar$; thus by the above proposition,

$$|\lambda - E^{(p+1)}| = O(\hbar^{p+2}N^{\gamma p+3-5\nu}). \tag{58}$$

Next, if we write $\psi^{(p)} = c\psi_\lambda + \psi_\perp$ with ψ_λ the eigenfunction corresponding to λ and $\psi_\perp$ orthogonal to ψ_λ, then we obtain that $\|\psi_\perp\|_2 \le \Delta^{-1}(\mathrm{var})^{1/2}$, where Δ is the minimal distance $\Delta = \inf_{\lambda'} |\lambda' - \bar{E}|$ with λ' ranging over the eigenvalues of $H_\hbar$ not equal to λ. Now the typical spacing between eigenvalues is $O(\hbar^\ell N^{2-\ell})$, and so if $(\text{var})^{1/2}$ is small relative to this spacing, we have that $|c| \geq (1 - \Delta^{-2}\text{var})^{1/2}$, whence

$$\|\psi_\lambda - \psi^{(p)}\|_2 \leq (1 - |c|) + \|\psi_\perp\|_2 = O(\hbar^{p+2-\ell} N^{\gamma p+1-5\nu+\ell}). \tag{59}$$

4 Probability Estimates

Let $\mathcal{N}_\hbar(\alpha, V_N^\infty)$ be the number of $A^\infty \in V_N^\infty$ satisfying the α-periodicity conditions (16), i.e.,

$$\mathcal{N}_\hbar(\alpha, V_N^\infty) = \#\{A^\infty \in V_N^\infty |\ A^\infty = (m + \alpha)\hbar \text{ for some } m \in \mathbb{Z}^\ell\}. \tag{60}$$

$\mathcal{N}_\hbar$ could be very small, or even zero, in which case our construction of approximate eigenfunctions is of limited interest. On the other hand, if we regard $\mathcal{N}_\hbar$ as a random variable over the probability space $\{\alpha |\ 0 \leq \alpha_i \leq 1\}$ with Lebesgue measure, then at least the expectation of $\mathcal{N}_\hbar$ is large in the sense that

$$\langle \mathcal{N}_\hbar(\cdot, V_N^\infty) \rangle \equiv \int d\alpha \sum_m \chi_{V_N^\infty}((m + \alpha)\hbar) = \hbar^{-\ell} |V_N^\infty|, \tag{61}$$

where $\chi_{V_N^\infty}$ is the indicator function for V_N^∞, and $|V_N^\infty|$ its Lebesgue measure. This expectation is to be compared with $\hbar^{-\ell}|V_N|$, the number of eigenvalues (for fixed α) when the potential U is identically zero; the quotient of these numbers approaches 1 as $N \to \infty$ since $|V_N^\infty|/|V_N| \to 1$.

To elaborate somewhat on these ideas, we define $N_\hbar(\alpha, E)$ to be the actual number of eigenfunctions of $H_\hbar$ satisfying α-periodicity conditions and having energy eigenvalues $\leq E$. By a considerable abuse of notation, let $\mathcal{N}_\hbar(\alpha, E)$ now be the number of approximate eigenfunctions of a given order p (see (43)) which we can construct by the above method, which satisfy the α-periodicity conditions, and which have their 0th energy eigenvalue coefficient satisfying $E_0(A^\infty) \leq E$. (Note that the linear independence of the approximate eigenfunctions of a given order p is guaranteed, for N sufficiently large, by Proposition 4.)

Proposition 6 *We have that*

$$\lim_{E\to\infty} \left\langle \left(\frac{\mathcal{N}_\hbar(\alpha, E)}{N_\hbar(\alpha, E)} \right) \right\rangle = 1. \tag{62}$$

Proof. We consider the denominator first. Under perturbations by U, an eigenvalue of $-\hbar^2\partial_\varphi^2/2$ (with fixed α-periodicity conditions) can be perturbed at most a distance $\|U\|_\infty$. Consequently,

$$c(\ell)\hbar^{-\ell}(2(E-\|U\|_\infty-1))^{\frac{\ell}{2}} \leq N_\hbar(\alpha,E) \leq c(\ell)\hbar^{-\ell}(2(E+\|U\|_\infty+1))^{\frac{\ell}{2}}, \quad (63)$$

where $c(\ell)$ is the volume of an ℓ-dimensional sphere of unit radius.

Let $\varepsilon > 0$ be given, and let a in the definition of V_N (5) be chosen so that $a_i = \varepsilon/2\ell$, $i = 1,\ldots,\ell$. For this choice of a, $N^\ell(1-\varepsilon) < |V_N| < N^\ell$. For $N > 0$, let

$$Y_N^0 \equiv \{A^\infty |\; a^2\ell N^2 \leq (A^\infty)^2 \leq N^2\} \quad (64)$$

$$Y_N \equiv Y_N^0 \cap V_N \quad (65)$$

$$Y_N^\infty \equiv Y_N^0 \cap V_N^\infty. \quad (66)$$

Then for some constant $c_1 = c(\ell)$,

$$|Y_N^0|(1-c_1\varepsilon) \leq |Y_N^0| - \varepsilon N^\ell < |Y_N| < |Y_N^0|. \quad (67)$$

Moreover, since $|V_N^\infty|/|V_N| \to 1$, $N \to \infty$, and $|Y_N|/|V_N| \geq \text{const} > 0$, it follows that $|Y_N^\infty|/|Y_N| \to 1$, $N \to \infty$, so that

$$(1-c_1\varepsilon) \leq \liminf \frac{|Y_N^\infty|}{|Y_N^0|} \leq \limsup \frac{|Y_N^\infty|}{|Y_N^0|} \leq 1. \quad (68)$$

Next, we note that for $A^\infty \in V_N^\infty$,

$$E_0(A^\infty) - \frac{1}{2}(A^\infty)^2 = O(N^{2-4\nu}), \quad (69)$$

by Proposition 8 of the appendix. Hence, if $E_0(A^\infty) \leq E$, then $|A^\infty| \leq (2E)^{1/2} + O(N^{2-4\nu}/E^{1/2})$. Consequently, we have that

$$\begin{aligned} \mathcal{N}_\hbar(\alpha,E) &\leq \frac{c(\ell)}{\hbar^\ell}\left((2E)^{\frac{1}{2}} + O\left(\frac{N^{2-4\nu}}{E^{1/2}}\right)\right)^\ell \\ &= \frac{c(\ell)}{\hbar^\ell}\left((2E)^{\frac{\ell}{2}} + o(E^{\frac{\ell}{2}-1})\right). \end{aligned} \quad (70)$$

On the other hand, taking $N = (2E)^{1/2} - E^{-1/2}$, we have that

$$\begin{aligned} \langle\mathcal{N}_\hbar(\cdot,E)\rangle &\geq \int d\alpha \sum_m \chi_{Y_N^\infty}((m+\alpha)\hbar) \\ &= \hbar^{-\ell}|Y_N^\infty| \\ &\geq \hbar^{-\ell}\frac{|Y_N^\infty|}{|Y_N^0|}c(\ell)\left((2E)^{\frac{\ell}{2}} + O(E^{\frac{\ell}{2}-1})\right)\left(1-(\|a\|\ell^{\frac{1}{2}})^\ell\right) \\ &\geq \hbar^{-\ell}(1-c_1'\varepsilon)c(\ell)\left((2E)^{\frac{\ell}{2}} + O(E^{\frac{\ell}{2}-1})\right), \end{aligned} \quad (71)$$

for some c_1', since $\|a\|$ is of order ε. Here we have used the fact that, by (64), Y_N^0 is a hollowed sphere of volume $c(\ell)N^\ell(1-(\|a\|\ell^{1/2})^\ell)$. Estimates (63), (70), and (71) combine to give the proposition. □

Appendix

Recall that $\hat{S} = \hat{S}_N(A^\infty, \varphi)$ was defined on $V_N \times \mathbb{T}$. In view of (23) and (24), bounds on the derivatives of $\hat{S}$ are estimates on how nearly the canonical transformation differs from the identity. In the following, we write, for example, that

$$f = f_N(A^\infty, \varphi) = O(N^\beta) \tag{72}$$

if

$$\limsup_{N\to\infty} \frac{1}{N^\beta} \sup_{\substack{A^\infty \in V_N^\infty \\ \varphi \in \mathbb{T}}} |f_N(A^\infty, \varphi)| < \infty. \tag{73}$$

The basic facts we need concerning $\hat{S} = \hat{S}_N$ can be summarized:

Proposition 7 *For $0 < \nu < 1$, $j \geq 1$, $0 \leq n \leq j$, and $A^\infty \in V_N^\infty$ we have*

$$\left(\frac{\partial}{\partial A^\infty}\right)^n \left(\frac{\partial}{\partial \varphi}\right)^{j-n} \hat{S} = O(N^{(1-\nu)j-\nu-n}). \tag{74}$$

Proof. We do not prove this in detail. However, [Gallavotti] provides estimates on the (A, φ) derivatives of the functions D_1, D_2 defined implicitly by

$$D_1(A, \varphi) \equiv \frac{\partial \hat{S}}{\partial A^\infty}(A^\infty, \varphi) \tag{75}$$

$$D_2(A, \varphi) \equiv \frac{\partial \hat{S}}{\partial \varphi}(A^\infty, \varphi) \tag{76}$$

in terms of the constants employed in the KAM iteration scheme (see particularly Eqs. 4.22 and 4.23 of [Gallavotti] and Eqs. 3.1–3.9 of [Thomas and Wassell]). Application of the chain rule, expressing (A^∞, φ) derivatives in terms of the (A, φ) derivatives, yields the estimate of the proposition. □

This proposition, and an estimate on a small denominator, provide estimates on the coefficients $\{W_n\}, \{E_n\}$. For the following proposition, set

$$|W_n|_j = \sup_\varphi \left\| \frac{\partial^j}{\partial \varphi^j} W_n \right\|. \tag{77}$$

Proposition 8 *For $0 < \nu < 1$, we have*

$$E_0(A^\infty) = \frac{1}{2}(A^\infty)^2 + O(N^{2-4\nu}) \tag{78}$$
$$E_n = O(N^{\gamma n+\sigma}), \quad n \geq 2 \tag{79}$$
$$|W_0 - 1|_j = O(N^{(1-\nu)j+1-3\nu}), \quad j \geq 0 \tag{80}$$
$$|W_n|_j = O(N^{\gamma n+(1-\nu)j+1-3\nu}), \quad n \geq 1,\ j \geq 0, \tag{81}$$

where

$$\sigma = 2(2-3\nu) - 2(1-\nu)(2\ell+3) \tag{82}$$
$$\gamma = (1-\nu)(2\ell+3) - \nu. \tag{83}$$

Remark. Note that $\gamma \to -1$ as $\nu \to 1$.

Proof. We note first that by the above proposition

$$W_0 = J^{\frac{1}{2}} = \det\left(\mathbb{1} + \frac{\partial^2 \hat{S}}{\partial A^\infty \partial\varphi}\right) = 1 + O(N^{1-3\nu}), \tag{84}$$

and by induction that

$$|W_0 - 1|_j = O(N^{1-3\nu+(1-\nu)j}), \quad j \geq 1. \tag{85}$$

Recalling that E_0 is given by (10) and (13), we have, by taking Fourier transforms, that

$$E_0(A^\infty) = \frac{1}{2}(A^\infty)^2 + \frac{1}{2(2\pi)^\ell}\int\left(\frac{\partial\hat{S}}{\partial\varphi}\right)^2 d\varphi = \frac{1}{2}(A^\infty)^2 + O(N^{2-4\nu}). \tag{86}$$

Next we have that

$$E_2(A^\infty) = \frac{1}{2(2\pi)^2}\int\left(\frac{\partial W_0}{\partial\varphi}\right)^2 d\varphi = O(N^{2(2-4\nu)}). \tag{87}$$

In order to obtain W_1, we regard (34) as an equation in the φ^∞-variables; the kth Fourier coefficient of $J^{-1/2}W_1$ is defined as

$$\widehat{J^{-\frac{1}{2}}W_1}(k) = \frac{-1}{2\omega^\infty \cdot k}\int e^{-ik\cdot\varphi^\infty}\left(J^{-\frac{1}{2}}\frac{\partial^2}{\partial\varphi^2}W_0\right)(\varphi(\varphi^\infty))d\varphi^\infty, \quad k \neq 0, \tag{88}$$

where

$$\omega^\infty(A^\infty) = \frac{\partial}{\partial A^\infty}E_0(A^\infty), \tag{89}$$

and where

$$\varphi^\infty = \varphi + \frac{\partial\hat{S}}{\partial A^\infty}(\varphi) \tag{90}$$

implicitly defines $\varphi = \varphi(\varphi^\infty)$. By induction one then shows that

$$\left(\frac{\partial}{\partial\varphi^\infty}\right)^j\left(J^{-\frac{1}{2}}\frac{\partial^2}{\partial\varphi^2}W_0\right)=O(N^{1-3\nu+(1-\nu)(j+2)}),\tag{91}$$

so that by integration by parts

$$\widehat{J^{-\frac{1}{2}}W_1}(k)=\frac{1}{|\omega^\infty\cdot k|}\frac{1}{\|k\|^p}O(N^{1-3\nu+(1-\nu)(p+2)}).\tag{92}$$

Consequently,

$$\left(\frac{\partial}{\partial\varphi^\infty}\right)^j\left(J^{-\frac{1}{2}}W_1\right)(\varphi(\varphi^\infty))=\frac{1}{(2\pi)^\ell}\sum(\mathrm{i}k)^j\widehat{J^{-\frac{1}{2}}W_1}(k)\mathrm{e}^{\mathrm{i}k\cdot\varphi^\infty}.\tag{93}$$

Using the previous estimate on the Fourier coefficients with $p=2\ell+j+1$, and using (4.41) of [Gallavotti] (see also (3.3) of [Thomas and Wassell])

$$\frac{1}{|\omega^\infty\cdot k|}\leq\frac{c}{N^\nu}\|k\|^\ell(\log k)^2,\tag{94}$$

we have

$$|W_1|_j=O(N^{1-4\nu+(1-\nu)(2\ell+j+3)});\tag{95}$$

here we again show by induction that φ^∞-derivatives of $J^{-1/2}W_1(\varphi(\varphi^\infty))$ are of the same order as φ-derivatives of $J^{-1/2}W_1(\varphi)$ and of $W_1(\varphi)$.

The estimates of E_2 and W_1 initiate an inductive proof of (79) and (81) of the proposition. In the induction, one notes that in the differential equation for W_n (34), $J^{-1/2}W_{n'}$, $n'=1,\ldots,n$, has 0th Fourier coefficient equal to zero, so that

$$E_{n+1}=\frac{1}{2(2\pi)^\ell}\int\frac{\partial}{\partial\varphi}J^{\frac{1}{2}}\frac{\partial}{\partial\varphi}W_{n-1}d\varphi\tag{96}$$

$$=O(N^{2-4\nu})|W_{n-1}|_1.\tag{97}$$

Also, if $D_t\equiv\frac{d}{dt}=\frac{\partial S}{\partial\varphi}\cdot\frac{\partial}{\partial\varphi}$, then one finds that

$$\left|D_t^{-1}J^{-\frac{1}{2}}\frac{\partial^2}{\partial\varphi^2}W_{n-1}\right|_j=O(N^{-\nu})|W_{n-1}|_{j+2\ell+3},\tag{98}$$

and that

$$\left|D_t^{-1}J^{-\frac{1}{2}}W_{n'}\right|_j=O(N^{-\nu})|W_{n'}|_{j+2\ell+1},\quad n'=1,\ldots,n-1.\tag{99}$$

The estimates on W_n then readily follow. □

References

Feldman, J., Knörrer, H. and Trubowitz, E. (1990): "The Perturbatively Stable Spectrum of a Periodic Schrödinger Operator," Invent. Math. Vol. 100, pp. 259–300

Gallavotti, G. (1983): "Perturbation Theory for Classical Hamiltonian Systems," in *Scaling and Self-similarity in Physics,* ed. by J. Frölich (Birkhauser, Boston) pp. 359–426

Graffi, S. and Paul, T. (1987): "The Schrödinger Equation and Canonical Perturbation Theory," Commun. Math. Phys. Vol. 108, pp. 25–40

Thomas, L. (1973): "Time-dependent Approach to Scattering from Impurities in a Crystal," Commun. Math. Phys. Vol. 33, pp. 335–343

Thomas, L. and Wassell, S.: "Stability of Hamiltonian Systems at High Energy," submitted to J. Math. Phys.

On the Magnetic Stark Resonances in Two Dimensional Case

Xue Ping WANG
Département de Mathématiques
Université de Nantes
44072 Nantes Cedex 03 FRANCE
and
Department of Mathematics
Peking University
100871 Beijing CHINA

Abstract

In this work, we study the resonances of Schrödinger operator with homogeneous magnetic and electric fields: $P = (D_x - \vec{A}(x))^2 + \vec{e}.x + V(x)$ in the case the strength of the magnetic field tends to ∞.

1 Introduction

The hydrogen atom in a strong homogeneous magnetic field is a model of "quantum chaos". See for example [HRW], [DBG1] and the references cited therein. In particular, it is said that in this problem, there exist resonances (see for example [HRW], [DBG2]). This work is an attempt to understand this phenomenom of resonances from a mathematical point of view. In fact, we shall only study a much simplified model: a two dimensional Schrödinger operator with homogeneous magnetic and electric fields

$$P = (D_x - \vec{A}(x))^2 + \vec{e}.x + V(x), x \in \mathbf{R}^2 \tag{1.1}$$

Here $\vec{A}(x) = 1/2\vec{B} \times x$, V is assumed to be a real smooth function tending to zero at the infinity, $\vec{B}$ and $\vec{e}$ are the constant magnetic and electric field, respectively. For the Schrödinger operator (1.1) in three dimensional case, there are works of some physists [CLLPT], in which they considered the case where the homogeneous magnetic and electric fields are parallel and studied the structure of diamagnetic Stark spectrum for P in the regime $|\vec{e}|/|\vec{B}| \to 0$. The structure of the spectrum of magnetic Schrödinger opertors in two dimensional case is completely different from that in three dimensional case. In particular, in two dimensional case, one can imagine that the magnetic field points to the third imaginary direction and the magnetic field is thus always perpendicular to the electric field. So in some sense, our work may be regarded as a complement to the work [CLLPT].

By an appropriate gauge transformation and change of coordinates, we can write the operator P in the form:

$$P(B,\beta) = (D_x - By)^2 + D_y^2 + \beta x + V(x,y), (x,y) \in \mathbf{R}^2, \qquad (1.2)$$

where B and β are proportionnal to the strength of the homogeneous magnetic and electric field. We are interested in the resonances of $P(B,\beta)$ in the case $\beta/B \to 0$, which we call magnetic Stark resonances. Typically, there are two situations: The first one is that the magnetic field is fixed: $B = 1$ and the strength of the electric field tends to 0, i.e., $\beta \to 0$. The second one is that the electric field is fixed and the strength of the magnetic field tends to the infinity: $\beta = 1, B \to \infty$. In the first case, the operator $P(1,\beta)$ is to be considered as a perturbation of the operator $P(1,0)$. The spectrum of $P(1,0)$ consists of the Landau levels, which are the essential spectrum of $P(1,0)$, and the discrete eigenvalues outside Landau levels. In this case, it is natural to hope that as soon as a weak homogeneous electric field switches on, the discrete eigenvalues of $P(1,0)$ will be turned into the resonances of $P(1,\beta)$. This is very similar to usual Stark resonances, on which there is now a large litterature (see for example [S1], [S2], [W1], [W2] and also references therein). We shall not study this case in details. In fact, one can show that the eigenfunctions of $P(1,0)$ associated to disrete eigenvalues outside Landau levels decay exponentially and by the method developped in the sequel of this paper, we can easily prove the existence and the stability of resonances. This paper is devoted to studying the resonances of magnetic Stark Hamiltonian in strong magnetic field: $B \to \infty, \beta = 1$. Before stating the results of this work, let us first give the definition of resonances.

Remark that having a constant double characteristic on $\{\xi = By\}$, the magnetic Schrödinger operator $(D_x - By)^2 + D_y^2$ is unstable under arbitrary analytical deformations: the domain of the distorted operator usually depends on the distortion, once the imaginary part of the deformation parameter becomes non zero. This causes difficulties in locating the essential spectrum of distorted operator. If one uses sympletic changes of coordinates ([HS2]), one can write $P(B,1)$ as

$$P(B) = B(D_y^2 + y^2) + x - \frac{1}{4B^2} \tag{1.3}$$

$$+V(x - B^{-1/2}D_y - \frac{1}{2B^2}, B^{-1}D_x + B^{-1/2}y),$$

where we denoted by $a(x, D_x), x \in \mathbf{R}^n$ the Weyl pseudodifferential operator (ΨDO, in short) with symbol $a(x,\xi)$. (Throughout this paper, we shall use Weyl quantization formula to define ΨDOs). In the form (1.3), one may try to deform x and y variables individually . But then any non linear deformation requires an analyticity of V over the whole complex space $\mathbf{C}^2$, which is unrealistic. For these reasons, we do not see for the moment how to apply the theory of Helffer-Sjöstrand [HS1] or other method of analytical distortions to study the magnetic-Stark resonances in two dimensional case. In the following, we shall just use complex dilation to define resonances of $P(B)$. Note that one can equally use complex translation in x variable to study this problem.

Assume that V extends to a holomorphic function in the region $\{z \in \mathbf{C}^2; |\Re z| \le \delta|\Im z| + R\}$ for some $\delta, R > 0$ and satisfies there:

$$|V(z)| \le C < \Re z >^{-\epsilon'}, \text{ for some } \epsilon' > 0. \tag{1.4}$$

Since we shall utilize the magnetic field to reduce the number of variables, we regard y as a dual variable of x and therefore it is more convenient to work with the operator $P(B)$ given by (1.3). In this configuration, consider the dilation $(x,y) \to (e^\theta x, e^{-\theta}y)$. The operator (1.3) becomes now

$$P(B,\theta) = B(e^{-2\theta}D_y^2 + e^{2\theta}y^2) + e^\theta x - \frac{1}{4B^2} \tag{1.5}$$

$$+V(e^\theta(x - B^{-1/2}D_y) - \frac{1}{2B^2}, e^{-\theta}(B^{-1}D_x + B^{-1/2}y))$$

which is now clearly well defined for θ in a small complex neighbourhood of 0. We still denote by $P(B,\theta)$ its closed extension from C_0^∞. The spectrum of $P(B,\theta)$ has a nice structure: For $|\theta|$ small, $\Im\theta > 0$, the essential spectrum of $P(B,\theta)$ is equal to the spectrum of $P_0(B,\theta) \equiv B(e^{-2\theta}D_y^2 + e^{2\theta}y^2) + e^\theta x$ shifted to the left by $1/(4B^2)$, which consists of an infinite number of parallel lines: $\{(2n+1)B - 1/(4B^2) + e^\theta\lambda; \lambda \in \mathbf{R}\}$, $n = 0, 1, 2, \ldots.$ Outside these lines, there may exist disrete eigenvalues created by the potential V. In this paper we define the discrete eigenvalues of $P(B,\theta)$ as resonances of $P(B)$. For the moment, we are not able to prove the absence of eigenvalues for the operator $P(B)$. So in this paper we make no distinguish between real eigenvalues and physical resonances of $P(B)$. In the following we shall study the resonances of $P(B)$ outside Landau levels $(2n+1)B, n \in \mathbf{N}$. We shall show in geometrically simple cases how the quantum effect of a strong magnetic field combines the classical contribution of total electric potential to create resonances in strong magnetic field.

Let $W(x,y) = x + V(x,y)$ be the total electric potential. For $E \neq 0$, put $\Omega_E = \{(x,y); W(x,y) = E\}$. Assume that $\Omega_E = \{(x_0,y_0)\} \cup \Gamma$, where Γ is a connected curve, $(x_0,y_0) \notin \Gamma$ and W has a local nondegenerat extremum (local maximum or minimum with definite Hessian) at (x_0,y_0):

$$W''(x_0,y_0) > 0(\text{ or } < 0). \tag{1.6}$$

and that the classical Hamiltonian $W(x,\xi)$ is non-trapping on Γ:

$$\{W(x,\xi), x.\cdot\xi\} = x\partial_x(x,\xi) - \xi\partial_\xi W(x,\xi) \neq 0 \text{ on } \Gamma. \tag{1.7}$$

Here $\{.,.\}$ denotes the Poisson bracket. The main result of this paper is the following

Theorem 1.1 *Under the assumptions (1.4), (1.6) and (1.7), for each $n \in \mathbf{N}$, define*

$$U_n = \{z \in \mathbf{C}; \Re z \in](2n+1)B+E-\frac{C}{B}, (2n+1)B+E+\frac{C}{B}[, \Im z \in]-\frac{1}{CB}, \frac{1}{CB}[\},$$

where $C > 1$ can be arbitrarily large outside a discrete set in $\mathbf{R}$. Then for B large enough, the resonances of $P(B)$ in U_n exist and are all given by complete expansions in powers of B^{-1}:

$$E_{nk}(B) \sim B(2n+1) + E$$

$$+B^{-1}/2\{\pm(2k+1)(\lambda\mu)^{1/2}+(2n+1)(\lambda+\mu)/2\}+\sum_{j\geq 2}B^{-j}e_{\pm,nkj},$$

where $e_{\pm,nkj}\in\mathbf{R},\lambda$ and μ are eigenvalues of the Hessian $W''(x_0,y_0)$ and the sign + (-) corresponds to a local minimum (maximum, resp.). In particular, the resonances of $P(B)$ in U_n are all algebraically simple and the width of resonances is of order $O(B^{-\infty})$.

REMARKS. (a). Theorem 1.1 shows that in two-dimensional magnetic resonances, there is no significant difference between a local maximum and a local minimum. It also shows that magnetic resonances have an almost ladder structure: modulo an error of order $O(B^{-1})$, one can obtain the resonances near the n-th Landau level just by translating those near the first Landau level by $2nB$.

(b). Our method in this work applies to general (non bounded) electric potential W with good geometrical properties at the infinity. We restrict ourselves to the particular structure $W=x+V$ only for the physical interest of this model. See [CLLPT].

(c). In higher even dimensional case, assume that the constant magnetic field is of maximal rank. Then we can still prove the results of Theorem 1.1. But in odd dimensional case, the situation is complete different. In particular, we do not see how to study the magnetic Stark resonances in three dimensional case.

To illustrate the results of Theorem 1.1, let us look at an example:

$$W(x,y)=x+c/(x^2+y^2+R), c\neq 0, R>0.$$

For appropriate values of $c<0$ and R (e.g., $c=-2$ and $R=1$), we can check that $\nabla W=0$ has two real solutions $(x_1,0),(x_2,0)$ with $x_2<x_1$. W has a local nondegenerate minimum at $(x_1,0)$, while $(x_2,0)$ is a saddle point for W. Put $E_j=W(x_j,0)$. Then we can verify that all the assumptions (1.5)-(1.7) are satisfied at the energy E_1. So our results show the existence of resonances near the energy $(2n+1)B+E_1$, for each n. Remark that for $E\in]E_1,E_2[$, $\Omega_E=\gamma_E\cup\Gamma_E$, where γ_E is a compact closed curve and Γ_E is unbounded. In the extreme case $E=E_2$, these two curves touch at $(x_2,0)$. We believe that near each given energy $E\in]E_1,E_2]$, there are also resonances of $P(B)$ when B is large enough.

The organization of this work is as follows: In Section 2, we reduce the resonance problem in strong magnetic field to a problem in semiclassical limit with the semiclassical parameter $h = 1/B$. This reduction is already used by Bellissard and Helffer-Sjöstrand ([B], [HS2]) in self-adjoint case with bounded total electric potential. In Section 3, we study the spectral properties of an h-pseudo-differential operator whose leading symbol is $W(e^{\theta}x, e^{-\theta}\xi)$. In particular, we shall prove the exponential decay of the eigenfunctions for this operator and establish a resolvent estimate outside the set of possible resonances. With these preparations, we shall prove the existence of resonances in Section 4 by studying a suitable Grushin problem and give the complete asymptotic expansion in Section 5 by constructing an asymptotic solution near the extreme point. In the last two Sections, many arguments are similar to semiclassical resonances of usual Schrödinger operators, so we shall sometime not give the full details of the proofs.

2 Reduction to a Semiclassical Problem

Let $E \neq 0$. To study the resonances of $P(B)$ near $(2n+1)B + E$, we shall make use of a strong field reduction by n-th eigenfunction of the harmonic oscillator. The idea is the same as in [B] and [HS2]. Since the operator in our case is neither selfadjoint nor semibounded, we need some additional estimates.

Let ϕ_n be the n-th real Hermite function:

$$(D_y^2 + y^2)\phi_n = (2n+1)\phi_n, \|\phi_n\| = 1.$$

Put $\psi_n(\theta) = e^{\theta/2}\phi_n(e^{\theta}y)$. Then $\psi_n(\theta)$ is holomorphic in θ. By deformation of the integral contour, one sees that $< \psi_n(\theta), \psi_n(\bar{\theta}) >= 1$ for all θ near 0. Let $\Pi : L^2(\mathbf{R}) \to L^2(\mathbf{R})$ be the projection defined by $\Pi u =< u, \psi_n(\bar{\theta}) >_Y \psi_n(\theta)$, here the scalar product $< .,. >_Y$ denotes the integration in y variable. We denote by the same letter the natural extension of Π on $L^2(\mathbf{R}^2)$. Put:

$$\hat{\Pi} = 1 - \Pi, \hat{P} = \hat{\Pi}P(B,\theta)\hat{\Pi} \text{ and } \hat{P}_0 = \hat{\Pi}P_0(B,\theta)\hat{\Pi}.$$

Let $\hat{R}(z)$ and $\hat{R}_0(z)$ denote the reduced resolvent of $\hat{P}$ and $\hat{P}_0$ on the range of $\hat{\Pi}$. We begin with an estimate over the free resolvent.

Lemma 2.1 *For $\theta \in \mathbf{C}$ with $|\theta|$ small and $\Im\theta > 0$, let $\Omega_1 = \{z \in \mathbf{C}; |\Re z - (2n+1)B| < 2\delta B, |\Im z| < (1-\delta)\Im\theta B/2\}, 0 < \delta < 1$. Then we have:*

$$\|\hat{R}_0(z)\| \leq C/B, \text{ uniformly in } z \in \Omega_1. \tag{2.1}$$

Let $\hat{R}_0(x; z), z \in \Omega_1$, denote the resolvent of $\hat{\Pi}(B(e^{-2\theta}D_y^2 + e^{2\theta}y^2) + e^{\theta}x)\hat{\Pi}$ on the range of $\hat{\Pi}$ in $L^2(\mathbf{R}_y)$, where x is regarded as a parameter. Then

$$\|\hat{R}_0(x; z)\|_Y \leq C/B, \quad \text{uniformly in } z \in \Omega_1 \text{ and } x \in \mathbf{R}. \tag{2.2}$$

Here $\|.\|_Y$ denotes the norm of bounded operators on $L^2(\mathbf{R}_y)$.

Proof. Making the change of the variable $x \rightarrow Bx$, $P_0(B,\theta)$ becomes $BP_0(1,\theta)$. The corresponding reduced resolvent of $P_0(1,\theta)$ is bounded on $B^{-1}\Omega_1$, which is now B-independent. (2.1) follows easily. To prove (2.2), we utilize the transformation $x \rightarrow \epsilon x + x_0$. Then the operator $P_0(B,\theta)$ is unitarily equivalent with $B(e^{-2\theta}D_y^2 + e^{2\theta}y^2) + e^{\theta}(\epsilon x + x_0)$. Let $\hat{R}_0(\epsilon, x_0; z)$ be the corresponding reduced resolvent defined as above. It follows from (2.1):

$$\|\hat{R}_0(\epsilon, x_0; z)\| \leq C/B,$$

uniformly in x_0 and $\epsilon > 0$. For each fixed B, z, x_0, $\hat{R}_0(\epsilon, x_0; z)$ converges strongly to an x-independent operator on $L^2(\mathbf{R}^2)$. The limit is clearly $\hat{R}_0(x_0; z)$. So we obtain that as operator on $L^2(\mathbf{R}^2)$, $\|\hat{R}_0(x_0, z)\| \leq C/B$, uinformly in x_0. Since $\hat{R}_0(x_0, z)$ acts only on y-variable, (2.2) follows easily. □

From (2.2), it results that $x \rightarrow \hat{R}_0(x; z)$ is a bounded symbol, each derivation on x gaining a facotr B^{-1}. In fact by rescaling x variable, one can write:

$$\hat{R}_0(x, z) = B^{-1}G(^{-1}x, B^{-1}z),$$

where $G(\zeta, \lambda), \lambda \in B^{-1}\Omega_1$, is a operator valued holomorphic function of ζ in a small θ-dependent sector in $\mathbf{C}$. It is also clear that for $x \in \mathbf{R}, |x| < C$, $\hat{R}_0(x, z)$ is well defined for $\theta = 0$ and $z \in \mathbf{R} \cap \Omega_1$, when B is large enough. In this case the reduced free resolvent is a self-adjoint operator in y variable.

Corollary 2.2 *For $B >> 1$, the reduced resolvent $\hat{R}(z)$ exists for $z \in \Omega_1$ and satisfies:*

$$\|\hat{R}(z)\| \leq C/B, \text{ uniformly in } z \in \Omega_1. \tag{2.3}$$

(2.3) follows immediately from (2.1), since the potential V is bounded. By (2.2) we can show that $\hat{R}(z)$ is an h-ΨDO with bounded symbol. See [R] for the notion of $h - \Psi DOs$. The details are omitted. Now we can proceed as in [HS2] to reduce the problem in strong magnetic field to a semiclassical one. In the following, we shall write $h = 1/B$. Consider the Grushin problem for $P(B,\theta) - z$

$$\mathcal{P}(z) = \begin{pmatrix} P(B,\theta) - z & \psi_n(\theta) \\ < ., \psi_n(\bar{\theta}) >_Y & 0 \end{pmatrix} : \mathbf{D} \oplus L^2(\mathbf{R}^2) \to L^2(\mathbf{R}^2) \oplus L^2(\mathbf{R}_x). \quad (2.4)$$

Here $\mathbf{D}$ is the domain of $P_0(B,\theta)$ and $< .,. >_Y$ denotes the scalar product in $L^2(\mathbf{R}_y)$.

Theorem 2.3 *For B large enough, the Grushin problem (2.4) is invertible for $z \in \Omega_1$ and the inverse is given by*

$$\mathcal{E}(z) = \begin{pmatrix} E(z) & E_+(z) \\ E_-(z) & E_{+-}(z) \end{pmatrix} \quad (2.5)$$

where

$$\begin{aligned} E(z) &= \hat{R}(z)a(z), E_-(z) = < a(z)., \psi_n(\bar{\theta}) >_Y, \\ E_+(z) &= -\hat{R}(z)a(z)\hat{\Pi}V_\theta\psi_n(\theta) + \psi_n(\theta)(z - (2n+1)B \\ &\quad -\frac{1}{4B^2} - e^\theta x - < V_\theta\psi_n(\theta), \psi_n(\bar{\theta}) >_Y), \\ E_{+-}(z) &= z - \{(2n+1)B - \frac{1}{4B^2} + < V_\theta\psi_n(\theta), \psi_n(\bar{\theta}) >_Y \\ &\quad + e^\theta x - < a(z)\hat{\Pi}V_\theta\psi_n(\theta)., \psi_n(\bar{\theta} >_Y\} \end{aligned}$$

Here

$$a(z) = (1 + [V_\theta, \Pi]\hat{R}(z))^{-1} \text{ and}$$

$$V_\theta = V(e^\theta(x - B^{-1/2}D_y) - 1/(2B^2), e^{-\theta}(B^{-1}D_x + B^{-1/2}y)).$$

In particular, z is an eigenvalue of $P(B,\theta)$ iff 0 is an eigenvalue of $E_{+-}(z)$.

Proof. We use an approximate inverse $\mathcal{E}_0(z)$ for $\mathcal{P}(z)$:

$$\mathcal{E}_0(z) = \begin{pmatrix} \hat{R}(z) & \psi_n(\theta) \\ <.,\psi_n(\bar{\theta}) >_Y & E^0_{+-} \end{pmatrix}$$

where

$$E^0_{+-} = z - \{(2n+1)B - 1/(4B^2) + e^\theta x + < V_\theta \psi_n(\theta), \psi_n(\bar{\theta}) >_Y\}.$$

Since as operators from $L^2(\mathbf{R}_x)$ to $L^2(\mathbf{R}^2)$,

$$V_\theta \psi_n - \Pi(V_\theta \psi_n) = [V_\theta, \Pi] = O(B^{-1/2}),$$

we can check:

$$\mathcal{P}(z)\mathcal{E}_0(z) = I + O(B^{-1/2}), \text{ and } \mathcal{E}_0(z)\mathcal{P}(z) = I + O(B^{-/2}).$$

This proves the inversibility of $\mathcal{P}(z)$. The formula for the inverse follows from a detailed computation. □

It is clear that the map $z \to \mathcal{E}(z)$ is holomorphic in $z \in \Omega_1$. For the later use, we give some more informations about $E_{+-}(z)$. Denote:

$$A(h) = e^\theta x + < V_\theta \psi_n(\theta), \psi_n(\bar{\theta} >_Y, h = \frac{1}{B}.$$

Then we can show as in [HS2] that $A(h)$ is an h-ΨDO:

$$A(h) = a(e^\theta x, he^{-\theta} D_x; h),$$

where $a(.,.;h)$ is holomorphic in x and ξ and admits a complete expansion in powers of h:

$$a(x,\xi;h) \sim \sum_{j\geq 0} h^j a_j(x,\xi), \text{ where} \tag{2.6}$$

$$a_0(x,\xi) = x + V(x,\xi), a_1(x,\xi) = (2n+1)\Delta V(x,\xi)/4.$$

Next, we remark that Π is a Weyl ΨDO and $[V_\theta, \Pi] = h^{/2} R_1(h)$, with $R_1(h)$ uniformly bounded in $L^2(\mathbf{R}^2)$. So by Corollary 2.2, we can estimate:

$$< a(z)\hat{\Pi} V_\theta \psi_n(\theta), \psi_n(\bar{\theta}) >_Y = O(h^2)$$

in $\mathcal{L}(L^2)$. From (2.3), we derive that $\hat{R}(z)$ is an h-ΨDO in x variable, whose symbol is bounded operator in y-variable. This shows that $< a(z)\hat{\Pi}V_\theta\psi_n(\theta), \psi_n(\bar{\theta}) >_Y$ is an h-ΨDO:

$$< a(z)\hat{\Pi}V_\theta\psi_n(\theta), \psi_n(\bar{\theta}) >_Y = h^2 r(e^\theta x, he^{-\theta}D; h, z),$$

where the symbol $r(.,.;h,z)$ has an asymptotic expansion of the form:

$$r(e^\theta x, e^{-\theta}\xi; h, z) \sim \sum_{j\geq 0} h^j r_j(e\theta x, e^{-\theta}\xi; he^\theta x; hz), z \in \Omega_1 \tag{2.7}$$

with r_j decaying like $O(|(x,\xi)|^{-j-1-\epsilon'})$, uniformly in h and z. Remark that it is necessary to take $\Im\theta \neq 0$ in order to make $r(.,.;h,z)$ globally defined. But it is clear that (2.3) is also valid for $x \in \mathbf{R}, |x| < C$, and $h > 0$ small enough even when $\theta = 0$. From this, one can derive that $r(x,\xi;h,z)$ is well defined for $|x| < C, \xi \in \mathbf{R}$ and is real-valued if $z \in \mathbf{R} \cap \Omega_1$. The later follows from the self-adjointness of the original operator $P(B)$.

3 Spectral Properties of A(h)

We study in this Section the leading part $A(h)$ of $E_{+-}(z)$, which should determine the leading term of the resonances of $P(B)$ in Ω. By (2.6), one sees that the essential spectrum of $A(h)$ is $\{e^\theta\lambda; \lambda \in \mathbf{R}\}$ and the discrete eigenvalues of $A(h)$ are essentially independent of θ and therefore should be considered as resonances of the ΨDO $a(x, hD; h)$. In the following, θ is fixed with $|\theta|$ small and $\Im\theta = \omega > 0$. For the simplicity of notations, we only consider the case $\Re\theta = 0$.

Let $E \neq 0$ so that (1.6) and (1.7) are satisfied. We propose to study the eigenvalus of $A(h)$ near E. Without loss, we can assume that E is a local non-degenerate minimum of $W(x,y)$. Otherwise we need only consider $-A(h)$ near $-E$. By a translation, we can also assume that $(x_0, y_0) = (0,0)$. Let $B(\delta) = \{(x,\xi) \in \mathbf{R}^2; |x| + |\xi| < \delta\}$ and $w = (e^\theta x, e^{-\theta}\xi)$. Then

$$a_0(w) - E = a_0(\Re w) - E + i\Im w \nabla a_0(\Re w) + r(w),$$

where the rest $r(w)$ is bounded by $|r(w)| < C\omega^2 < (x,\xi) >^{-\epsilon'}$. Outside $B(\delta)$, $a_0(\Re w) - E$ can vanish only when $\Re w \in \Gamma$ (see (1.6)) . By the assumption on V, Γ has an asymptote $\{x = E\}$. Near Γ, by (1.7), $\Im w.\nabla a_0(\Re w) \neq 0$. Note that for $|\xi|$ large enough, one has $|\{a_0, x.\xi\}| \geq |E|/2 > 0$ near Γ. Consequently, we obtain:

$$|a_0(w) - E| \geq c_0\{|a_0(\Re w) - E| + \omega\} \tag{3.1}$$

for $(x,\xi) \notin B(\delta)$, $\omega > 0$ small enough and for some $c_0 > 0$. By a perturbation argument in ξ variable, we can show that (3.1) remains true for $|\Im\xi| \leq \phi(x)$ and $(x, \Re\xi) \notin B(\delta)$, where ϕ is a bounded positive real function such that: (1). the symbol of $A(h)$ is holomorphic in $|\Im\xi| < \phi(x) + C$, for some $C > 0$; (2). $\phi(x) > C(\theta) > 0$ for all $x \in \mathbf{R}$, with $C(\theta)$ independent of θ for x outside a small neighbourhood of $\Gamma_1 = \{x; (x,\xi) \in \Gamma,$ for some $\xi\}$ and $C(\theta) = c'\omega$ for x near Γ_1. Making symplectic changes of coordinates, we can assume that the Hessian of a_0 at (0, 0) is diagonal. So in $B(\delta)$, one has:

$$a_0(w) = E + (\lambda e^{2\theta}x^2 + \mu e^{-2\theta}\xi^2)/2 + O((x,\xi)^3), \tag{3.2}$$

where λ and μ are the two eigenvalues of the Hessian of a_0 at (0, 0). By (3.1) and (3.2), we can construct a real smooth function f such that $f(x) > 0$ for $x \neq 0$, $f(x) = cx^2$ for x near 0 and

$$|f'(x)| \leq \phi(x) \text{ for all } x. \tag{3.3}$$

If we choose $c > 0$ small, then we can show that:

$$\begin{aligned} |a_0(e^\theta x, e^{-\theta}(\xi + if'(x))) \quad &- \quad E| \geq C\omega, \text{ for } (x,\xi) \notin B(\delta) \qquad (3.4) \\ \Re\{a_0(e^\theta x, e^{-\theta}(\xi + if'(x))) \quad &- \quad E\} \geq c'(x^2+\xi^2), \text{ for } (x,\xi) \in B(\delta). \end{aligned}$$

In the above analysis, x and ξ play the same role. In particular, we can also derive from (3.1) that there exists a positive function $g(\xi)$ with $g(\xi) = c\xi^2$, for ξ near 0 so that (3.4) holds for $\xi \in \mathbf{R}$ and $\Im x = g'(\xi)$.

Theorem 3.1 *Let $\lambda(h) \in \{z \in \mathbf{C}; |z - E| < Ch\}, (C > 0)$, be an eigenvalue of $A(h)$. Let $u(h)$ be an associated eigenfunction: $A(h)u(h) = \lambda(h)u(h)$, $\|u(h)\| = 1$. Then there exists $C > 0$,*

$$\|e^{f/h}u(h)\| \leq C, \text{ uniformly in } h > 0 \tag{3.5}$$

Here f is the function constructed above.

Proof. Since f is bounded, $u_f \equiv e^{f/h}u(h)$ is in L^2. We want to estimate the h-dependence of its norm. Put: $A_f = e^{f/h}A(h)e^{-f/h}$. Then

$$(A_f - \lambda(h))u(h) = 0. \tag{3.6}$$

Since the symbol $a(.,.;\theta,h)$ of $A(h)$ is holomorphic in a neighbourhood of $\{|\Im\xi| \leq \phi(x)\}$, by a change of integral contour in ξ, we obtain

$$\begin{aligned} A_f g(x) &= (2\pi h)^{-1} \int\int_{\mathbf{R}^2} e^{(i(x-y).\xi + f(x)-f(y))/h} a((x+y)/2,\xi;\theta,h)g(y)dyd\xi \\ &= (2\pi h)^{-1} \int\int_{\mathbf{R}^2} e^{i(x-y).(\xi - if'(x,y))/h} a((x+y)/2,\xi;\theta,h)g(y)dyd\xi \\ &= (2\pi h)^{-1} \int\int_{\mathbf{R}^2} e^{i(x-y).\xi/h} a((x+y)/2,\xi + if'(x,y);\theta,h)g(y)dyd\xi \end{aligned}$$

for $g \in C_0^\infty$. Here $f'(x,y)$ is defined by $f(x)-f(y) = (x-y)f'(x,y)$. we have used the analyticity and decay assumptuons on V to show that the integral in ξ over $\mathbf{R}$ and over $\Im\xi = f'(x,y)$ are equal. Now it is clear that A_f is an h-ΨDO , whose h- principal symbol is given by $a_0(e^\theta x, e^{-\theta}(\xi + if'(x)))$.

Let $0 \leq \chi_j \leq 1, j = 1,2$ be smooth functions such that supp $\chi_1 \subset B(\delta), \chi_1 = 1$ near (0, 0) and $\chi_1^2 + \chi_2^2 \equiv 1$. By the analysis made above, the symbol of $A_f - \lambda(h)$ is elliptic on supp χ_2. By constructing an approximate inverse for $A_f - \lambda(h)$, we derive from (3.6) that

$$\{\chi_2(x,hD) + O(h)\}u(h) = 0. \tag{3.7}$$

On supp χ_1, we use the second estimate in (3.4):

$$\Re < (A_f - \lambda(h))u^1, u^1 > \geq c < (h^2D^2 + x^2 - Ch)u^1, u^1 >, \tag{3.8}$$

for some $C, c > 0$. Here $u^1 = \chi_1(x,hD)u_f$. We decompose the last scalar product in (3.8) into twe parts according to: $|x| > Rh^{1/2}$ and $|x| \leq Rh^{1/2}$, $R >> 1$. Then one has:

$$\Re < (A_f - \lambda(h))u^1, u^1 > \geq (R-C)h < u^1, u^1 > -(R+C)h \int_{\{|x|<Rh^{1/2}\}} |u^1(x)|^2 dx.$$

Since $u(h)$ is normalized and $f(x)/h$ is bounded when $|x| < Rh^{1/2}$, we can show that the last integral in the above inequality is bounded by C_1 uniformly in h. Since $(A_f - \lambda(h))u^1 = O(h)u_f$, we derive from (3.6) and (3.8) that

$$(R - C)\|u^1\|^2 \leq C_R + c\|u_f\|\|u^1\|. \tag{3.9}$$

From here we derive: $\|u^1\|^2 \leq C'_R + c'/R\|u_f\|^2$. Since $\chi_1(x,hD)^2 + \chi_2(x,hD)^2 \geq 1/2$ in the sense of selfadjoint operators for $h > 0$ small enough, it results from (3.7) and (3.9):

$$\|u_f\|^2 \leq C''_R + c''/R\|u_f\|^2.$$

(3.5) follows when we choose R large enough. □

REMARK. Since the Fourier transform exchanges the role of x and ξ, making use of the remark preceeding Theorem 3.1, we can show that (3.5) also holds in ξ variable:

$$\|e^{g(hD)/h}u(h)\| \leq C$$

uniformly in h, where g is given before Theorem 3.1. This shows that the energy of eigenfunctions are microlocally exponentially concentrated near (0,0).

Theorem 3. 1 is important, because it allows us to determine the eigenvalues of $A(h)$ just by computing the microlocalized version of $A(h)$ near (0,0). From now on, we denote: $D = \{z \in \mathbf{C}; |z - E| \leq C_0 h\}$, where C_0 is a large fixed constant. Near (0,0), $a_0(x,\xi) = E + (\lambda x^2 + \mu\xi^2)/2 + O((x,\xi)^3)$. Put

$$A_0(h) = (\lambda e^{2\theta}x^2 + \mu h^2 e^{-2\theta}D^2)/2 + h(2n+1)\Delta V(0,0)/4.$$

The eigenvalues of $A_0(h)$ are independent of θ and are given by $he_k, k = 0,1,2...,$ where

$$e_k = (2k+1)(\lambda\mu)^{1/2}/2 + (2n+1)(\lambda+\mu)/4. \tag{3.10}$$

Define $D_0 = \{z \in D; |z - he_k| > \epsilon h, \text{ for any } k\}$. The following result shows that there is no spectrum of $A(h)$ in D_0.

Proposition 3.2 *For $h > 0$ small enough, D_0 is contained in the resolvent set of $A(h)$ and $\|(A(h) - z)^{-1}\| \leq ch^{-1}, z \in D_0$.*

Proof. Let $\chi_j, j = 1,2$, be a partition of unity on $\mathbf{R}^2$ with $\chi_1 = 1$ near 0 and 0 outside an ϵ- neighbourhood of 0. Let B_1 the h-ΨDO with symbol $\chi_2(a_0(\theta) - z)^{-1}$, where $a_0(\theta)$ is the leading symbol of $A(h)$ and $z \in D$. Then

$$B_1(A(h) - z) = \chi_2(x,hD) + O(h), \text{ for any } z \in D. \tag{3.11}$$

Let $R_0 = (A_0(h) + E - z)^{-1}$. Then a scaling argument shows

$$\|x^j (hD)^k R_0\| \leq C h^{(k+j)/2-1} \tag{3.12}$$

for $j + k \leq 2$. On supp χ_1, we compare $A(h)$ with $A_0(h)$:

$$\chi_1(x, hD) R_0(A(h) - z) = \chi_1(x, hD)\{1 + R_0(A(h) - A_0(h) - E)\} \tag{3.13}$$

Let χ_3 be equal to 1 near support of χ_1 and 0 outside a 2ϵ-neighbourhood of (0, 0). Making use of (3.12) and an inductive argument, we can show that $\chi_1(x, hD)R_0(1 - \chi_3(x, hD)) = O(h^\infty)$ On supp χ_3, we expand the symbol of $A(h)$ near (0, 0). Again using (3.12) and the fact that $|x| + |\xi| < 2\epsilon$ on supp χ_3, we have : $\|R_0(A(h) - A_0(h) - E)\chi_3(x, hD)\| \leq C\epsilon + C_\epsilon h$ where $C > 0$ is independent of ϵ and h. From (3.11) and (3.13) it results

$$Ch^{-1}\|A(h) - z)u\| \geq (1 - C\epsilon - C_\epsilon h)\|u\| \geq \|u\|/2,$$

for $z \in D_0$. Taking $\epsilon > 0$ and $h > 0$ small enough, this proves that $A(h) - z$ is inversible , because the spectrum of $A(h)$ in D is discrete. The resolvent estimate follows easily. □

Theorem 3.3 *Let D be defined as above with $C_0 \neq e_k, \forall k \in \mathbf{N}$. Then for $h > 0$ small enough, there is a one-to-one correspondence between the eigenvalues of $A(h)$ in D and the e_k's such that $e_k < C_0$. After an appropriate arrangement, the j-th eigenvalue $E_j(h)$ of $A(h)$ is given by:*

$$E_j(h) = E + he_j + O(h^{3/2}). \tag{3.14}$$

Proof. With the notations of Proposition 3.2, one has:

$$(A(h) - z)^{-1} = B_1(z) + \chi_1(x, hD)R_0$$

$$+\{O(h) - \chi_1(x, hD)R_0(A(h) - A_0 - E)\chi_3(x, hD)\}(A(h) - z)^{-1}$$

for $z \in D_0$. Now fix an e_j with $e_j < C_0$. Denote by γ the contour $\{|z - E - he_j| = \epsilon h\}$ with $\epsilon > 0$ small enough. Define $\Pi(h)$ by

$$\Pi(h) = \frac{1}{2\pi i} \oint_\gamma (z - A(h))^{-1} dz.$$

Since $B_1(z)$ is holomorphic inside γ, it results from the above resolvent expression and Proposition 3.2 that $\Pi(h) = \chi_1(x, hD)\Pi_0(h)+o(1)$, where $\Pi_0(h)$ is the spectral projector of A_0 associated to e_j and $o(1)$ is a bounded operator whose norm can be made arbitrarily small by the argument used in the proof of Proposition 3.2. Since $\Pi(h)$ is a projector, we deduce that $\Pi(h)$ is of the same rank as $\chi_1(x, hD)\Pi_0(h)$: Rank $\Pi(h) = 1$. This shows that $A(h)$ has just a simple eigenvalue inside γ. Using the exponential decay of the eigenfunctions, we can show that this eigenvalue is given by (3.14). The details are omitted. □

By constructing an asymptotic solution for each eigenvalue of $A(h)$ in D, we can show that the eugenvalues of $A(h)$ in D have a complete asymptotic expansion in powers of h. The details are now standard and therefore omitted here. See also the construction of asymptotic solutions given in Section 5.

4 Existence of Resonances

With the preparations of Section 3, we can easily prove the existence of resonances of $P(B)$ near the eigenvalues of $A(h)$. To show this, we work in a neighbourhood of an eigenvalue $E_j(h)$ of $A(h)$ given by (3.14). Let Ω be defined by: $\Omega = \{z; |z - E_j(h)| < \epsilon h\}$ where $\epsilon > 0$ is small enough. Let u be an eigenfunction of $A(h)$: $A(h)u(h) = E_j(h)u$, $\|u\| = 1$. Let $A(h)^*$ be the adjoint of $A(h)$. Take u^* an eigenfunction of $A(h)^*$ such that

$$A(h)^*u^* = \bar{E}_j(h)u^*, < u, u^* >= 1.$$

Let Π' be the spectral projector of $A(h)$ associated to the eigenvalue $E_j(h)$ defined by: $\Pi' f =< f, u^* > u$. Put $\hat{\Pi}' = 1 - \Pi'$ and $A' = \hat{\Pi}' A(h)\hat{\Pi}'$. Then by Theorem 3.3 one can show that the reduced resolvent $R'(z) \equiv (A' - z)^{-1}\hat{\Pi}'$ exists and is holomorphic for $z \in \Omega$.

Lemma 4.1 *For $h > 0$ small enough, we have: $\|R'(z)\| \leq Ch^{-1}$, uniformly in $z \in \Omega$.*

Lemma 4.1 follows from the similar estimate on the reduced resolvent in Ω of $A_0(h)$ introduced in Section 3. The details are the same as the proof of

Proposition 3.2 and are omitted here. Let $E_{+-}(z)$ be constructed in Theorem 2.3. To find the values of $z \in \Omega$, for which 0 is an eigenvalue of $E_{+-}(z)$, let us consider the Grushin problem:

$$\mathcal{P}^1(z) = \begin{pmatrix} E_{+-}(z) + (2n+1)B & u \\ < ., u^* > & 0 \end{pmatrix} : D \oplus \mathbf{C} \to L^2(\mathbf{R}) \oplus \mathbf{C}, \qquad (4.1)$$

where D is the domain of $E_{+-}(z)$ which is equal to the domain of the multiplication by x. Since $E_{+-}(z) = z - ((2n+1)B + A(h) + O(h^2))$ uniformly in $z \in \Omega$, we can easily construct an inverse for (4.1). Consider the operator:

$$\mathcal{E}_0(z) = \begin{pmatrix} -R'(z) & u \\ < ., u^* > & E_j(h) - z \end{pmatrix} : L^2(\mathbf{R}) \oplus \mathbf{C} \to D \oplus \mathbf{C}.$$

By Lemma 3.1 it is easy to check that

$$\mathcal{E}_0(z)\mathcal{P}^1(z) = 1 + O(h) \text{ and } \mathcal{P}^1(z)\mathcal{E}_0(z) = 1 + O(h), z \in \Omega.$$

So the inverse of $\mathcal{P}^1(z)$ exists and $\mathcal{P}^1(z)^{-1} = \mathcal{E}_0(z) + O(h)$.

Theorem 4.2 *For $h > 0$ small enough, there exists only one resonance $z_j(h)$ of $P(B)$ in $\{z \in \mathbf{C}; |z - (2n+1)B - E_j(h)| < \epsilon h\}$. In addition $z_j(h)$ is algebraically simple and is given by*

$$z_j(h) = (2n+1)B + E + he_j + O(h^2). \qquad (4.2)$$

Proof. To give the existence and the location of resonances inside Ω, we need only compute the zeros of the element $F(z)$ in the right lower corner of $\mathcal{P}^1(z)^{-1}$, which is now a holomorphic function of $z \in \Omega$. See [HS1]. A direct computation gives:

$$\mathcal{P}^1(z)^{-1} = \mathcal{E}_0(z)(1 + K(z))^{-1},$$

where $K(z)$ is given by

$$K(z) = \begin{pmatrix} r(z,h)R'(z) & r(z,h)u \\ 0 & 0 \end{pmatrix}$$

with $r(z,h) = -\{E_{+-}(z) + (2n+1)B - z + A(h)\}$. Since $r(z,h)u = O(h^2)$ as operator from $\mathbf{C}$ to L^2, we obtain $F(z) = E_j(h) - z + < r(z,h)u, u^* > + O(h^2) = E_j(h) - z + O(h^2)$. Consequently, we derive that $F(z)$ has just a simple zero near $E_j(h)$ which is given by $E_j(h) + O(h^2)$ (see also [HS1]). The desired result follows from Theorem 3.3. □

From the proof of Theorem 4.1, it is clear that the resonances of $P(B)$ in $(2n+1)B + D$, where D is defined in Section 3, are all given by the formula (4.2) with $e_j < C_0$.

5 Construction of Asymptotic Solutions

By the results of Section 2, we can write

$$E_{+-}(z) = z - (2n+1)B - A(\theta,h) - h^2 r(e^\theta x, he^{-\theta}D; h, z),$$

where $r(z,\theta,h) \equiv r(e^\theta x, he^{-\theta}D; h,z)$ is an h-ΨDO. Different from (2.7), the insignificant term $-1/(4B)^2$ is now included in $r(z,\theta,h)$. Let Ω be the neighbourhood of $E_j(h)$ defined in Section 4. We want to construct an asymptotic solution for the non-linear eigenvalue problem:

$$A(\theta,h) + h^2 r(z,\theta,h))u = zu, z \in \Omega. \tag{5.1}$$

The same problem was studied in [M], where the leading part of $A(h)$ is a semiclassical Schrödinger operator. So we shall only give details on the construction of the first few terms for an asymptotic solution of (5.1). The arguments of [M] can be carried over for the remaining parts.

Lemma 5.1 *For any $R > 0$, there exist $h_0 > 0$ and an h-ΨDO $r'(z,\theta,h)$ depending holomorphically on θ near $0 \in \mathbf{C}$ such that for any $\chi \in C_0^\infty$ with supp $\chi \subset]-R, R[$, one has:*

$$\|\chi(x)(r(z,\theta,h) - r'(z,\theta,h))\| = O(h^\infty), \tag{5.2}$$

for $\Im\theta > 0$. In addition, $r'(z,\theta,h)$ is selfadjoint if z and θ are real.

Proof. Remark that $r(z,\theta,h)$ is only defined when $\Im\theta > 0$. But by the remark following (2.7), its symbol $r(x,\xi;z,\theta,h)$ is well defined for $|x| < R, z \in \Omega$ and θ in a small neighbourhood of 0, if $h > 0$ is small enough. So we only need to take $r'(z,\theta,h)$ an h-ΨDO with symbol $\chi'(x)r(x,\xi;z,\theta,h)$, where χ' is a cut-off function for $[-R,R]$. □

By Lemma 5.1, we are led to consider the problem:

$$(A(0,h) + h^2 r'(z,0,h))u = zu, z \in \Omega, \tag{5.3}$$

where $A(0,h)$ and $r'(z,0,h)$ are now selfadjoint, if z is real. We want to construct an asymptotic solution of (5.3) in the form

$$z \sim \sum h^j \lambda_j, u \sim e^{-d(x)/h} \sum h^j u_j,$$

with $\lambda_j \in \mathbf{R}$. To begin with, we determine the phase function $d(.)$. Write the total symbol of $A(0,h) + h^2 r'(z,0,h)$ as:

$$a_0(x,\xi) + ha_1(x,\xi) + \sum_{j\geq 2} h^j a_j(x,\xi;z),$$

where a_0, a_1 are given by (2.6). Near (0,0), a_0 can be written in the form:

$$a_0(x,\xi) = E + \frac{1}{2}(\lambda x^2 + \mu\xi^2) + O((x,\xi)^3).$$

Here $\lambda, \mu > 0$, since we have assumed that a_0 has a local non-degenerate minimum at (0,0). On the set $a_0(x, i\xi) = E$ for (x,ξ) near (0, 0), one has:

$$\xi = \pm(\lambda/\mu)^{1/2} x + o(x). \tag{5.4}$$

Note that on the above set, $\xi \neq 0$ if $x \neq 0$, because $(x,0)$ is not in the energy surface Ω_E for x near 0. On the set $a_0(x,i\xi) = E$, one has $\partial_\xi a_0(x,i\xi) \neq 0$ for $x \neq 0$ and x near 0. This means that in a small neighbourhood of 0, there exists a unique function $\xi(x)$ defined for $x \neq 0$ such that $x\Re\xi(x) > 0$ for x real and

$$a(x, i\xi(x)) = E \text{ for } x \neq 0 \text{ small.} \tag{5.5}$$

From (5.4), it results that $\xi(x) = (\lambda/\mu)^{1/2}x + o(x)$. So $\xi(x)$ can be defined by continuity at $x = 0 : \xi(0) = 0$. Making use of the analyticity of V, we can show that $\xi(x)$ thus defined extends to a holomorphic function of x in a small complex neighbourhood of 0. Now choose $d(.)$ so that

$$d'(x) = \xi(x), d(0) = 0$$

Then $d(.)$ is holomorphic in x near 0 and $\Re d(x) > 0$ for x real with $x \neq 0$. We extend $d(.)$ smoothly so that it equals a small positive constant outside a neighbourhood of 0. Now as shown in Section 3, $e^{d/h}A(0,h)e^{-d/h}$ is an h-ΨDO and we can formally expand it in the form:

$$e^{d/h}A(0,h)e^{-d/h} \sim \sum_j h^j b_j(x,D),$$

where $b_j(x,D)$ is a differential operator of order j. In particular, we have:

$$b_0(x,D) = a_0(x, id'(x)) \tag{5.6}$$

$$b_1(x,D) = -i(\partial_\xi a_0(x,id'(x))\partial_x + 1/2d''(x)\partial_\xi^2(x,id'(x)) + a_1(x,id'(x))$$

Note that $a_0(x,id'(x)) = E$ for x near 0. Now to find a formal solution for $A(0,h)u = zu$ in the form

$$z \sim \sum h^j \lambda_j, u \sim e^{-d/h}\sum h^j u_j, \text{ with } \lambda_0 = E,$$

we need only to solve the transport equations

$$(b_1(x,D) - \lambda_1)u_{l-1} = -\sum_{k+j=l,k>1}(b_k(x,D) - \lambda_k)u_j, l = 1,2,3,... \tag{5.7}$$

For x near 0, these transport equations can be easily solved by an induction in l. Note that $b_1(x,D)$ can be written as:

$$b_1(x,D) = xc_0(x)\partial_x + c_1(x),$$

where $c_j(0) \neq 0$, for $j = 0,1$. If we look for solution to (5.7) with $l = 1$ in the form $u_0 = x^m\phi_0(x)$ with $\phi_0(0) \neq 0$, then we must take

$$\lambda_1 = mc_0(0) + c_1(0)$$

which is nothing else but e_m given in (3.10). In this case $\phi_0(x)$ is uniquely determined by

$$(xc_0(x)\partial_x + mc_0(x) + c_1(x) - \lambda_1)\phi_0 = 0, \phi_0(0) = 1.$$

Clearly ϕ_0 exists and is holomorphic in x near 0. In the case $m = 0$, one can directly apply the method of [H] to determine u_l for $l \geq 1$. In general case, we look for solutions to (5.7) with $l \geq 2$ in the form $u_{l-1} = p_{l-1}(x) + x^m \phi_{l-1}(x)$, where p_{l-1} is a polynomial of degree $m-1$. Assume that we have determined λ_j for all $j < l$ and u_j for all $j < l-1$. To determine ϕ_{l-1} and the coefficients of p_{l-1}, we expand the functions in (5.7) into the form

$$a_0 + a_1 x + \ldots + a_{m-1} x^{m-1} + x^m f_m(x),$$

where a_j are constants. Then (5.7) becomes

$$(xc_0(x)\partial_x + c_1(x) - \lambda_1)(p_{l-1} + x^m \phi_{l-1}) = q_l(x) + x^m(g_l(x) + \lambda_l \phi_0(x)) \quad (5.8)$$

where q_l is a polynomial of degree $m-1$. Since $kc_0(0) + c_1(0) - \lambda_1 \neq 0$ for $k \leq m-1$, (5.8) uniquely determines $a_j, j = 0, 1, \ldots, m-1$. In addition, comparing the coefficients before x^m of the both sides of (5.8)(after having expanded each term in powers of x), we see that λ_l is given by:

$$\lambda_l = -g_l(0) + \sum_{j=1}^{m-1} (m-j) c_0^{(j)}(0) a_{m-j}/j! + \partial_x^m (c_1 p_{l-1})(0)/m!$$

and ϕ_{l-1} should verify

$$(xc_0(x)\partial_x + mc_0(x) + c_1(x) - \lambda_1)\phi_{l-1} = r_{ml} + g_l + \lambda_l \phi_0 \quad (5.9)$$

with $r_{lm}(0) = -\lambda_l - g_l(0)$. We can determine ϕ_{l-1} from (5.9) by adding the condition $\phi_{l-1}(0) = 0$. This shows that for each m, we can construct an asymptotic solution near 0 to $A(0,h)u = zu$ in the form

$$z^m \sim \sum h^j \lambda_j^m, u^m \sim e^{-d/h} \sum h^j u_j^m$$

with $\lambda_0^m = E$ and $\lambda_1^m = e_m$. All λ_j^m are real, since $A(0,h)$ is self-adjoint. Now put $z_0 = z^m, v_0 = u^m$. We can now construct by an induction in k an asymptotic solution of the linear eigenvalue problem in $z_k \in \Omega$:

$$(A(0,h) - z_k + h^2 r'(z_{k-1}, 0, h))v_k \sim O(e^{-d/h} h^\infty), \quad (5.10)$$

for $k = 1, 2\ldots$, in the form: $z_k \sim \sum h^j \lambda_{jk}, v_k \sim e^{-d/h} \sum h^j v_{jk}$ with $\lambda_{k0} = E$ and $\lambda_{k1} = e_m$. By the argument of [M], one can show that we do not need to modify the k (resp. $k+1$) first terms in the semiclassical expansions of v_{k-1} (resp. z_k) in order to obtain an asymptotic solution (v_k, z_k) to (5.10). In addition, all λ_{kj} are real and all v_{kj} are holomorphic near 0. Define $z'(h)$ and $u'(h)$ such that:

$$z'(h) \sim \sum_{j=0}^{\infty} h^j \lambda_{jj}, u'(h) \sim h^{-m/4} e^{-d(x)/h} \sum_{j=0}^{\infty} h^j v_{jj}(x).$$

This gives us an asymptotic solution to (5.3). To obtain an asymptotic solution to (5.1), we put:

$$u(x,h,\theta) = \chi(x) u'(e^{\theta} x, h), \Im\theta > 0, \tag{5.11}$$

where χ is a cut-off function which equals 1 near 0. Applying Lemma 5.1, we have proven the following

Theorem 5.2 *Let Ω be a small neighbourhood of $E_j(h)$ defined in Section 4. There exist $z'(h) \in \Omega$ in the form $z'(h) \sim \sum h^j \mu_j, \mu_j \in \mathbf{R}$, and $u(h,\theta)$ with $\|u(h,\theta)\| \geq c > 0$, such that:*

$$E_{+-}((2n+1)B + z'(h)) u(h,\theta) = O(h^{\infty}).$$

Now to finish the proof of Theorem 1.1, we just need to study a Grushin problem for $E_{+-}((2n+1)B + z), z \in \Omega$, similar to (4.1), but with $u(h,\theta)$ given by (5.11) and $u(h,\theta)^* = u(h,\bar{\theta}) / < u(h,\theta), u(h,\bar{\theta} >$. The details are omitted.

Finally we remark that several questions related to the present work remain open. In particular, it seems that the absence of eigenvalues for the class of Hamiltonians considered in this paper is not yet established. It is also interesting to give precise estimates on the width of resonances which we believe is exponentially small. Finally we indicate that the imaginary part of resonances might be positive or negative, depending on whether $E > 0$ or $E < 0$. This situation should be compared with the resonances of Dirac operator, where the choice is often made so that the product of the energy and the width of resonances is positive.

Acknowledgements. I would thank Professor B. Helffer for indicating me the references [DBG1] and [DBG2] and Professor E. Balslev for his hospitality during my visit in Aarhus.

References

[AC] J. Aguilar and J.M. Combes, A class of analytic perturbations for one-body Schrödinger Hamiltonians, Comm. Math. Phys. 22(1971), 269-279

[BC] E. Balslev and J.M. Combes, Spectral properties of many-body Schrödinger operators with dilation analytical interactions, Comm. Math. Phys. 22(1971), 180-194

[B] J. Bellissard, C^*-algebras in solid state physics -$2D$ electrons in a uniform magnetic field, in Warwick Conference on Operator Algebras, July 1987

[CLLPT] P. Cacciani, S. Liberman, E. Luc-Koenig, J. Pinard and C. Thomas, J. of Phys. B19(1986), L519; J. of Phys. B21(1988), 3473

[DBG1] D. Delande, A. Bommier,J.C. Gay, L'atome d'hydrogène en champ magnétique intense: un modèle de chaos quantique, in Physique Quantique Fondamentale, pp 49-54, 1988

[DBG2] D. Delande A. Bommier and J.C. Gay, Positive energy spectrum of the hydrogen atom in a magnetic field, preprint 1990

[HRW] H. Hasegawa, M. Robnik and G. Wunner, Classical and quantal chaos in the diamagnetic Kepler problem, Prog. of Theoretical Phys., Suppl. No. 98, 1989, 198-287

[H] B. Helffer, Semiclassical Ananlysis for the Schrödinger Operator and Applications, Lecture Notes in Math., No.1336, Springer-Verlag, 1988

[HS1] B. Helffer and J. Sjöstrand, Résonances en limite semiclassique, Bull. Soc. Math. France, Suppl. No. 24/25, 1986

[HS2] B. Helffer and J. Sjöstrand, Equation de Schrödinger avec champ magnétique et équation de Harper, pp 118-197 in Lecture Notes in Phys. No. 345, 1989

[M] A. Martinez, Résonances dans l'approximation de Born-Oppenheimer II -Largeur des résonances, Comm. Math. Phys., 1991

[R] D. Robert, Autour de l'Approximation Semiclassique, Progress in Math. No.68, Birkhaüser 1986

[S1] I.M. Sigal, Sharp exponential bounds on resonances states and width of resonances, Adv. in Appl. Math., 1988

[S2] I.M.Sigal, Geometric theory of Stark resonances in multi-electrin systems, Comm. in Math. Phys., 19(1988), 287-314

[W1] X.P. Wang, Bounds on width of resonances for Stark Hamiltonians, Acta Math. Sinica, New Series 6(1990), 100-119

[W2] X.P. Wang, Resonances of N-body Schrödinger operators with Stark effect, in Annual Report of Inst. Math., Academia Sinica, 1987, pp. 108-126, revised version in: Ann. Inst. H. Poincaré, 52(1990), 1-30; On resonances of generalized N- body Stark Hamiltonians, to appear in J. of Operator Theory

RADIATION CONDITIONS AND SCATTERING THEORY FOR N-PARTICLE SCHRÖDINGER OPERATORS

D. Yafaev *
Université de Nantes

The correct form of radiation conditions is found in scattering problem for N-particle quantum systems. The estimates obtained allow us to give an elementary proof of asymptotic completeness for such systems in the framework of the theory of smooth perturbations.

1. One of the main problems of scattering theory is a description of asymptotic behaviour of N interacting quantum particles for large times. The complete classification of all possible asymptotics (channels of scattering) is called asymptotic completeness. The final result can easily be formulated in physics terms. Two particles can either form a bound state or are asymptotically free. In case $N \geq 3$ a system of N particles can also be decomposed asymptotically into its subsystems (clusters). Particles of the same cluster form a bound state and different clusters do not interact with each other.

There are two essentially different approaches to a proof of asymptotic completeness for multiparticle ($N \geq 3$) quantum systems. The first of them, started by L. D. Faddeev [1], relies on the detailed study of a set of equations derived by him for the resolvent of the corresponding Hamiltonian. This approach was developped in [1] for the case of three particles and was further elaborated in [2, 3]. The attempts [4, 5] towards a straightforward generalization of Faddeev's method to an arbitrary number of particles meet with numerous difficulties. However, the results of [6] for weak interactions are quite elementary.

Another approach relies on the commutator method [7] of T. Kato. In the theory of N-particle scattering it was introduced by R. Lavine [8, 9] for repulsive potentials. A proof of asymptotic completeness in the general case is much more complicated and is due to I. Sigal and A. Soffer [10] (see also the article [11] by J. Derezinski for the proof of intermediary analytical results). In the recent paper [12] G. M. Graf gave an accurate proof of asymptotic

*Permanent address: Math. Inst., Fontanka 27, St. Petersburg, 191011 USSR

completeness in the time-dependent framework. The distinguishing feature of [12] is that all intermediary results are also purely time-dependent and most of them have a direct classical interpretation. Papers [10, 12] were to a large extent inspired by V. Enss (see e.g. [13]) who was the first to apply a time-dependent technique for the proof of asymptotic completeness.

The aim of this article is to outline an elementary proof of asymptotic completeness (Theorem 7) for N-particle Hamiltonians with short-range potentials which fits into the theory of smooth perturbations [7, 14]. Our approach hinges on new estimates which establish some kind of radiation conditions (for the precise statement, see Theorem 8) for N-particle systems. We omit some details. However, basic intermediary results are formulated and their proofs are sketched. The present exposition is close to the talk given by the author in Ecole Polytechnique.

2. Let us briefly recall some basic definitions of the scattering theory. For a self-adjoint operator H in a Hilbert space $\mathcal{H}$ we introduce the following standard notation: $\mathcal{D}(H)$ is its domain; $\sigma(H)$ is its spectrum; $E(\Omega; H)$ is the spectral projection of H corresponding to a Borel set $\Omega \subset \mathbf{R}$; $\mathcal{H}^{(ac)}(H)$ is the absolutely continuous subspace of H; $P^{(ac)}(H)$ is the orthogonal projection on $\mathcal{H}^{(ac)}(H)$; $\mathcal{H}^{(p)}(H)$ is the subspace spanned by all eigenvectors of the operator H; $\sigma^{(p)}(H)$ is the spectrum of the restriction of H on $\mathcal{H}^{(p)}(H)$, i.e. $\sigma^{(p)}(H)$ is the closure of the set of all eigenvalues of H. Norms of vectors and operators in different spaces are denoted by the same symbol $\|\cdot\|$; I is always the identity operator; C and c are positive constants whose precise values are of no importance.

Let K be H-bounded operator. It is called H-smooth (in the sense of T. Kato) on a Borel set $\Omega \subset \mathbf{R}$ if for every $f = E(\Omega; H)f \in \mathcal{D}(H)$

$$\int_{-\infty}^{\infty} \|K \exp(-iHt)f\|^2 \, dt \le C\|f\|^2.$$

Let now H_j, $j = 1, 2$, be a couple of self-adjoint operators and let J be a bounded operator in a Hilbert space $\mathcal{H}$. The wave operator for the pair H_1, H_2 and the "identification" J is defined by the relation

$$W^{\pm}(H_2, H_1; J) = s - \lim_{t \to \pm\infty} \exp(iH_2 t) J \exp(-iH_1 t) P^{(ac)}(H_1) \tag{0.1}$$

under the assumption that this strong limit exists. We emphasize that all definitions and considerations for " + " and " − " are independent of each other. Clearly, for every $f_2 = W^{\pm}(H_2, H_1; J) f_1$

$$\exp(-iH_2 t) f_2 \sim J \exp(-iH_1 t) f_1, \quad t \to \pm\infty,$$

where " $\sim$ " means that the difference between left and right sides tends to zero. If the wave operator (1) exists, then the intertwining property

$$E_2(\Omega) W^{\pm}(H_2, H_1; J) = W^{\pm}(H_2, H_1; J) E_1(\Omega)$$

($\Omega \subset \mathbf{R}$ is any Borel set and $E_i(\Omega) = E(\Omega; H_i)$) holds. It follows that the range $R(W^\pm(H_2, H_1; J))$ of the operator (1) is contained in $\mathcal{H}^{(ac)}(H_2)$ and its closure is an unvariant subspace of H_2. Moreover, if the wave operator is isometric on some subspace $\mathcal{H}_1$, then the restrictions of H_1 and H_2 on the subspaces $\mathcal{H}_1$ and $\mathcal{H}_2 = W^\pm(H_2, H_1; J)\mathcal{H}_1$ respectively are unitarily equivalent. This equivalence is realized by the wave operator.

We need the following sufficient condition of existence of wave operators (see e.g. [15]).

Proposition 0.1 *Let an operator $\mathcal{J}$ be H_1-bounded and let its adjoint $\mathcal{J}^*$ be H_2-bounded. Suppose that*

$$H_2\mathcal{J} - \mathcal{J}H_1 = \sum_n K_{2,n}^* K_{1,n}, \quad n = 1, \ldots, N < \infty,$$

where the operators $K_{j,n}$ are H_j-bounded and are H_j-smooth on some bounded interval Λ. Then the wave operators

$$W^\pm(H_2, H_1; \mathcal{J}E_1(\Lambda)), \quad W^\pm(H_1, H_2; \mathcal{J}^*E_2(\Lambda))$$

exist.

3. We consider the Schrödinger operator $H = T + V$ where $T = -\Delta$ and V is multiplication by a real function $V(x)$ defined as follows. Suppose that some finite number of subspaces X^α of $X := \mathbf{R}^d$ is given and let x^α be the orthogonal projection of $x \in X$ on X^α. Set

$$V(x) = \sum_\alpha V^\alpha(x^\alpha), \tag{0.2}$$

where V^α are decreasing functions of variables x^α. Note that in the two-particle case the sum (2) consists of a single term V^α with $X^\alpha = X$. Clearly, $V^\alpha(x^\alpha)$ tends to zero as $|x| \to \infty$ outside of any neighbourhood (neighbourhoods of all subspaces are supposed to be conical) of $X_\alpha = X \ominus X^\alpha$ and $V^\alpha(x^\alpha)$ is constant on planes parallel to X_α. Due to this property the structure of the spectrum of H is much more complicated than in the two-particle case. Operators H studied here were introduced in [16] and are natural generalizations of N-particle Hamiltonians. Consideration of more general class of operators allows to unravel better the geometry of the problem.

We assume that each V^α is a sum of short-range V_s^α and long-range V_l^α terms:

$$V^\alpha = V_s^\alpha + V_l^\alpha. \tag{0.3}$$

We say that a potential V^α is short-range if $V_l^\alpha = 0$. It is convenient to split all conditions on V^α into two parts (the first has a preliminary nature). To formulate them we need to introduce the operator $T^\alpha = -\Delta_{x^\alpha}$ in the space $\mathcal{H}^\alpha = L_2(X^\alpha)$.

Assumption 0.2 *Operators*

$$V^\alpha(T^\alpha+I)^{-1}, \quad (|x^\alpha|+1)V_s^\alpha(T^\alpha+I)^{-1}, \quad (|x^\alpha|+1)|\nabla V_l^\alpha|(T^\alpha+I)^{-1}$$

are compact in the space $\mathcal{H}^\alpha$.

Assumption 0.3 *For some* $\rho > 1$ *operators*

$$(|x^\alpha|+1)^\rho V_s^\alpha(T^\alpha+I)^{-1}, \quad (|x^\alpha|+1)^\rho|\nabla V_l^\alpha|(T^\alpha+I)^{-1}$$

are bounded in the space $\mathcal{H}^\alpha$.

Compactness of $V^\alpha(T^\alpha+I)^{-1}$ ensures that the operator H is self-adjoint on the domain $\mathcal{D}(H)=\mathcal{D}(T)=:\mathcal{D}$ in the Hilbert space $\mathcal{H}=L_2(\mathbf{R}^d)$ and H is semi-bounded from below. Let us set

$$E(\cdot)=E(\cdot;H), \quad U(t)=\exp(-iHt).$$

We prove asymptotic completeness under the assumption that V^α are short-range functions of x^α but many intermediary results (in particular, radiation conditions-estimates) are as well true for long-range potentials.

The spectral theory of the operator H starts with the following geometrical construction. Let us introduce the set $\mathcal{X}$ of linear sums

$$X^a = X^{\alpha_1} \vee X^{\alpha_2} \vee \ldots \vee X^{\alpha_k}$$

of subspaces X^{α_j}. The zero subspace $X^0=\{0\}$ is included in the set $\mathcal{X}$ and X itself is excluded. However, without loss of generality, we may always assume that X is a linear sum of subspaces X^α. The index a (or b) labels all subspaces $X^a \in \mathcal{X}$ and can be interpreted as the collection of all those α_j for which $X^{\alpha_j} \subset X^a$. Let x^a and x_a be the orthogonal projections of $x \in X$ on the subspaces X^a and

$$X_a := X \ominus X^a = X_{\alpha_1} \cap X_{\alpha_2} \cap \ldots \cap X_{\alpha_k}$$

respectively. Since $X = X_a \oplus X^a, \mathcal{H}$ splits into a tensor product

$$L_2(X) = L_2(X_a) \otimes L_2(X^a). \tag{0.4}$$

Let us introduce for each a an auxiliary operator

$$H_a = T + V^a, \quad V^a = \sum_{X^\alpha \subset X^a} V^\alpha, \tag{0.5}$$

with a potential V^a which does not depend on x_a. In the representation (4)

$$H_a = T_a \otimes I + I \otimes H^a, \tag{0.6}$$

where $T_a = -\Delta_{x_a}$ acts in the space $\mathcal{H}_a = L_2(X_a)$ and

$$H^a = T^a + V^a, \quad T^a = -\Delta_{x^a}$$

are the operators in the space $\mathcal{H}^a = L_2(X^a)$. Set $\mathcal{H}^0 = \mathbb{C}, V^0 = 0, H^0 = 0$. In the multiparticle terminology, index a parametrizes decompositions of an N-particle system into noninteracting clusters. The operator H^a corresponds to the Hamiltonian of clusters with their centers-of-mass fixed at the origin; T_a is the kinetic energy of the center-of-mass motion of these clusters. Thus the operator H_a describes an N-particle system with interactions between different clusters neglected.

Scattering theory for the Hamiltonian H is formulated in terms of eigenvalues λ_n^a and eigenfunctions $\psi_n^a(x^a)$ of the operators H^a. Denote by P^a the orthogonal projection in $\mathcal{H}^a$ on the subspace $\mathcal{H}^{(p)}(H^a)$ and let $P_a = I \otimes P^a$. Clearly, the orthogonal projection P_a commutes with H_a and its functions. Set also $H_0 = T, P_0 = I$. The union over all a

$$\Upsilon_0 = \bigcup \sigma^{(p)}(H^a)$$

is called the set of thresholds for the operator H.

We need the following basic result (see [17, 18, 19]) of spectral theory of multiparticle Hamiltonians. It is formulated in terms of the auxiliary operator

$$A = \sum_j (x_j D_j + D_j x_j), \quad D_j = -i\partial_j, \quad \partial_j = \partial/\partial x_j, \quad j = 1, \ldots, d.$$

Proposition 0.4 *Let Assumption 2 hold. Then the set Υ_0 is closed and countable and eigenvalues of H may accumulate only at Υ_0 so that the "exceptional" set $\Upsilon = \Upsilon_0 \cup \sigma^{(p)}(H)$ is also closed and countable. Furthermore, for every $\lambda \in \mathbf{R} \setminus \Upsilon$ there exists a small interval $\Lambda_\lambda \ni \lambda$ such that the estimate (the Mourre estimate) for the commutator holds:*

$$i([H, A]u, u) \geq c\|u\|^2, \quad c = c_\lambda > 0, \quad u \in E(\Lambda_\lambda)\mathcal{H}. \tag{0.7}$$

Let Q be multiplication by $(x^2+1)^{1/2}$. *Below Λ is always an arbitrary bounded interval such that $\bar{\Lambda} \cap \Upsilon = \emptyset$*, where $\bar{\Lambda}$ is the closure of Λ. One of the main consequences of (7) is the following

Proposition 0.5 *Let Assumptions 2 and 3 hold. Then for any $r > 1/2$ the operator $Q^{-r}(T+I)$ is H-smooth on Λ.*

This assertion is usually called the limiting absorption principle. Its proof under Assumptions 2 and 3 can be found in [20].

Corollary 0.6 *The operator H is absolutely continuous on $E(\Lambda)\mathcal{H}$. In particular, it does not have any singular continuous spectrum, i.e.*

$$\mathcal{H} = \mathcal{H}^{(p)}(H) \oplus \mathcal{H}^{(ac)}(H).$$

The basic result of the scattering theory for N-particle Hamiltonians is the following

Theorem 0.7 *Suppose that functions V^α satisfy Assumptions 2 and 3 and are short-range, i.e. $V^\alpha = V_s^\alpha$. Then the wave operators*

$$W_a^\pm = W^\pm(H, H_a; P_a) \tag{0.8}$$

exist and are isometric on the ranges $R(P_a)$ of P_a. The ranges $R(W_a^\pm)$ of $W_a^\pm$ are mutually orthogonal and the asymptotic completeness holds:

$$\sum_a \oplus R(W_a^\pm) = \mathcal{H}^{(ac)}(H). \tag{0.9}$$

Our assumptions on V^α are somewhat larger than those of I. M. Sigal and A. Soffer [10] or G .M. Graf [12] since we do not require anything about derivatives of V^α.

Theorem 7 gives the complete spectral analysis of the operator H. Actually, by the relation (9), H is the orthogonal sum of its restrictions on different subspaces $R(W_a^\pm)$. In virtue of the intertwining property $HW_a^\pm = W_a^\pm H_a$, each of these restrictions is unitarily equivalent to the operator H_a considered in the space $P_a\mathcal{H}$. Actually, if $f \in \mathcal{H}^{(ac)}(H)$ and $f_a^\pm = (W_a^\pm)^* f \in R(P_a)$, then

$$f = \sum_a W_a^\pm f_a^\pm \quad \text{and} \quad Hf = \sum_a W_a^\pm H_a f_a^\pm. \tag{0.10}$$

Furthermore, according to (6)

$$H_a f_a^\pm = \sum_n (T_a + \lambda_n^a) f_{a,n}^\pm \otimes \psi_n^a, \quad \text{if} \quad f_a^\pm = \sum_n f_{a,n}^\pm \otimes \psi_n^a. \tag{0.11}$$

Thus the absolutely continuous part of H is unitarily equivalent to the orthogonal sum of the "free" operators T_a shifted by the eigenvalues of the operators H^a.

Theorem 7 describes also the asymptotics as $t \to \pm\infty$ of the evolution $U(t)f$ governed by the Hamiltonian H. Indeed, the first equality (10) ensures that

$$U(t)f \sim \sum_a U_a(t) f_a^\pm, \quad t \to \pm\infty, \quad U_a(t) = \exp(-iH_a t), \tag{0.12}$$

which is basically equivalent to the asymptotic completeness. In virtue of (11) functions $U_a(t)f_a^\pm$ admit an explicit representation:

$$U_a(t) f_a^\pm = \sum_n \exp(-i(T_a + \lambda_n^a)t) f_{a,n}^\pm \otimes \psi_n^a. \tag{0.13}$$

4. The main analytical result of our approach is formulated as certain estimates which we call radiation conditions-estimates. Compared to the limiting absorption principle they give an additional information on the asymptotic behaviour of a quantum system for large distances and large times. To give the precise formulation let us introduce the gradient ∇_{x_a} in the variable x_a (in particular, $\nabla_{x_0} = \nabla$ is the gradient in the variable $x_0 = x$) and its angular part $\nabla_{x_a}^{(s)}$:

$$(\nabla_{x_a}^{(s)}u)(x) = (\nabla_{x_a}u)(x) - |x_a|^{-2}\langle(\nabla_{x_a}u)(x), x_a\rangle x_a. \qquad (0.14)$$

Theorem 0.8 *Let Γ_a be a closed cone in $\mathbf{R}^d$ such that $\Gamma_a \cap X_b = \{0\}$ if $X_a \not\subset X_b$. Denote by χ_a the characteristic function of Γ_a. Suppose that V^α are defined by (3) where V_s^α and V_l^α satisfy Assumptions 2 and 3. Then for all a the operators*

$$G_a = \chi_a Q^{-1/2}\nabla_{x_a}^{(s)}$$

are H-smooth on Λ.

Remark. It is easy to see that

$$|\nabla_{x_b}^{(s)}u| \le |\nabla_{x_a}^{(s)}u|, \quad \text{if} \quad X_b \subset X_a.$$

Therefore Theorem 8 gives us more information about $U(t)f$ in the cone Γ_a than in Γ_b. In particular, the most complete information is obtained in the cone Γ_0 which does not intersect any $X_a \neq X$. On the contrary, the result of Theorem 8 is trivial for a such that $\dim X_a = 1$.

Remark. The notion of H-smoothness can be equivalently reformulated in terms of the resolvent of H. Thus radiation conditions-estimates given by Theorem 8 also admit a stationary formulation.

Remark. In the two-particle case the result of Theorem 8 reduces to H-smoothness of the operator $Q^{-1/2}\nabla^{(s)}$ on any bounded positive interval separated from the point 0. This corresponds to the angular part of the usual form of the radiation conditions-estimates (see e.g. [21]). Note also that the result of Proposition 5 (limiting absorption principle) definitely fails if $r = 1/2$ (even in the free case $H = H_0$). Thus the differential operator $\nabla^{(s)}$ improves the fall-off of $(U(t)f)(x)$ at infinity.

The proof of Theorem 8 relies on consideration of the commutator of H with a first-order differential operator

$$M = \sum_j (m_j D_j + D_j m_j), \quad m_j = \partial m / \partial x_j,$$

with *bounded* coefficients m_j. Actually, the function m is chosen as C^∞-homogeneous function of degree 1 (all properties of $m(x)$ are formulated for

$|x| \geq 1$; in some neighbourhood of $x = 0$ we require that $m(x) = 0$). In the two-particle case we can set $m(x) = \mu_0|x|$. In the N-particle case it should be modified in such a way that $m(x) = m(x_a)$ in some neighbourhood of each X_a. If, furthermore, x is separated from all X_b such that $X_a \not\subset X_b$, then $m(x) = \mu_a|x_a|$. Note also that, by the construction of $m(x)$, a cone $\tilde{\Gamma}_a$, where $m(x) = \mu_a|x_a|$, can be made arbitrary close to all $X_b \subset X_a$. Finally, $m(x)$ should be a convex and positive function (for $|x| \geq 1$). Such a function $m(x)$ can be constructed averaging over all ε_a of the family of functions

$$\max_a\{(1+\varepsilon_a)|x_a|\}, \tag{0.15}$$

where ε_a are suitably chosen small positive numbers. These functions satisfy all properties listed above except smoothness which is recovered by integration of (15) with some smooth functions of variables ε_a. We emphasize that only properties of $m(x)$ for large $|x|$ are essential. In a bounded domain $m(x)$ can be arbitrary.

The commutator $[V, M]$ is small in the following sense.

Proposition 0.9 *Suppose that V^α is defined by (3) where V_s^α and V_l^α satisfy Assumptions 2 and 3. Let m be C^∞-homogeneous function of degree 1 (for $|x| \geq 1$) and let $m(x) = m(x_\alpha)$ in some neighbourhood of X_α (for $|x| \geq 1$). Then*

$$|([V^\alpha, M]u, u)| \leq C\|Q^{-r}(T+I)u\|^2, \quad u \in \mathcal{D}, \quad 2r = \rho.$$

The proof of this assertion is based on the following observation. The potential V^α depends on x^α and is concentrated in a neighbourhood of X_α. In this region m depends on x_α only so that V^α and M essentially commute. This is similar to the idea of G. M. Graf applied in [12] in the time-dependent context.

The commutator of T with M equals

$$i[T, M] = 4\sum_{j,k} D_j m_{jk} D_k - (\Delta^2 m), \quad m_{jk} = \partial^2 m/\partial x_j \partial x_k. \tag{0.16}$$

Since the matrix $\{m_{jk}(x)\}$ is nonnegative for $|x| \geq 1$ and

$$\Delta^2 m(x) = O(|x|^{-3}), \quad |x| \to \infty, \tag{0.17}$$

it follows that

$$i([T, M]u, u) \geq \sum_{j,k} \int_\Gamma m_{jk} u_j \bar{u}_k dx - c\int_{|x|<1} |\nabla u|^2 dx - c\|Q^{-3/2}u\|^2, \tag{0.18}$$

where Γ is any region lying outside of the unit ball. Let $\tilde{\chi}_a$ be the characteristic function of the region $\tilde{\Gamma}_a$ where $m(x) = \mu_a|x_a|$. Simple calculations show that

$$\sum_{j,k} m_{jk} u_j \bar{u}_k = \mu_a|x_a|^{-1}|\nabla_{x_a}^{(s)} u|^2, \quad x \in \tilde{\Gamma}_a. \tag{0.19}$$

So combining Proposition 9 with (18) and (19) we arrive at the estimate (we assume here that $\rho \leq 3$)

$$i([H,M]u,u) \geq 4\mu_a \|\tilde{\chi}_a Q^{-1/2} \nabla^{(s)}_{x_a} u\|^2 - c\|Q^{-r}(T+I)f_t\|^2, \quad 2r = \rho. \quad (0.20)$$

The H-smoothness of the operator $\tilde{\chi}_a Q^{-1/2} \nabla^{(s)}_{x_a}$ can be standardly deduced from (20). Indeed, applying (20) to $u = f_t = U(t)f$ and integrating the identity

$$d(MU(t)f, U(t)f)/dt = i([H,M]f_t, f_t),$$

we obtain that

$$\int_{t_1}^{t_2} \|\tilde{\chi}_a Q^{-1/2} \nabla^{(s)}_{x_a} f_t\|^2 dt \leq C(|(Mf_t, f_t)\Big|_{t_1}^{t_2}| + \int_{t_1}^{t_2} \|Q^{-r}(T+I)f_t\|^2 dt).$$

If $f = E(\Lambda)f$, then the first term in the right side is bounded by $C\|f\|^2$ because the operator $ME(\Lambda)$ is bounded. The second term admits the same estimate by Proposition 5. Finally, H-smoothness of the operator $\tilde{\chi}_a Q^{-1/2} \nabla^{(s)}_{x_a}$ ensures H-smoothness of $\chi_a Q^{-1/2} \nabla^{(s)}_{x_a}$ since the region $\tilde{\Gamma}_a$ can be chosen arbitrary close to the subspaces $X_b \subset X_a$.

5. Our proof of asymptotic completeness demands preliminary consideration of auxiliary wave operators with identifications which are first-order differential operators

$$M^{(a)} = \sum_j (m_j^{(a)} D_j + D_j m_j^{(a)}), \quad m_j^{(a)} = \partial m^{(a)}/\partial x_j.$$

We set $m^{(a)}(x) = \eta^{(a)}(x) m(x)$ where the function m was constructed in the previous part and

$$\sum_a \eta^{(a)}(x) = 1. \quad (0.21)$$

We require that each $\eta^{(a)} \in C^\infty(\mathbf{R}^d \setminus \{0\})$, $\eta^{(a)}$ is homogeneous function of degree 0 and $\eta^{(a)}(x) = 0$ in some neighbourhoods of all X_b such that $X_a \not\subset X_b$. If $X_a \subset X_b$, then $\eta^{(a)}$ should not depend on x^b in some neighbourhod of X_b. The partition of unity with such properties can be constructed by the same procedure as the function m. Actually, we define first a non-smooth function

$$\theta((1+\varepsilon_a)|x_a| - \max_{b, b\neq a}\{(1+\varepsilon_b)|x_b|\}), \quad (0.22)$$

where $\theta(s) = 1$ for $s \geq 0$ and $\theta(s) = 0$ for $s < 0$, and then obtain $\eta^{(a)}(x)$ by means of averaging of (22) over all ε_b and ε_a.

Theorem 0.10 *Suppose that functions V^α satisfy the assumptions of Theorem 7 and $E_a(\Lambda) = E_a(\Lambda; H_a)$. Then for all a the wave operators*

$$W^\pm(H, H_a; M^{(a)} E_a(\Lambda)), \quad W^\pm(H_a, H; M^{(a)} E(\Lambda)) \quad (0.23)$$

exist.

To prove this theorem we verify that the triple $H^a, H, M^{(a)}$ satisfies on Λ the conditions of Proposition 1. Actually, according to (5)

$$HM^{(a)} - M^{(a)}H_a = [T, M^{(a)}] + \sum_{X_a \subset X_\alpha} [V^\alpha, M^{(a)}] + \sum_{X_a \not\subset X_\alpha} V^\alpha M^{(a)}.$$

The second and third terms can be estimated with the help of limiting absorption principle. Indeed, by the construction of m and $\eta^{(a)}$, the function $m^{(a)}$ depends only on x_α in some neighbourhood of X_α. Therefore by Proposition 9

$$[V^\alpha, M^{(a)}] = (T+I)Q^{-r}BQ^{-r}(T+I), \quad 2r = \rho,$$

with a bounded operator B. The operator $Q^{-r}(T+I)$ is H- and H_a-smooth according to Proposition 5. A similar representation for $V^\alpha M^{(a)}$, $X_a \not\subset X_\alpha$, can be obtained due to the fact that $\eta^{(a)}(x) = 0$ in some neighbourhood of X_α where V^α is concentrated. We emphasize that the short-range assumption is used for the estimate of this term only.

The commutator $i[T, M^{(a)}]$ is again defined by (16) with m replaced by $m^{(a)}$. The function $\Delta^2 m^{(a)}(x)$ satisfies (17) and hence can be taken into account by Proposition 5. In order to estimate the operator

$$L^{(a)} = \sum_{j,k} D_j m_{jk}^{(a)} D_k$$

we use Theorem 8. Let $\lambda_n^{(a)}(x)$ and $p_n^{(a)}(x)$ be eigenvalues and normalized eigenvectors of the symmetric matrix $\mathbf{M}^{(a)}(x) = \{m_{jk}^{(a)}(x)\}$. Clearly, $\lambda_n^{(a)}(x)$ are homogeneous (for $|x| \geq 1$) functions of order -1 and $p_n^{(a)}(x)$ – of order 0. Diagonalizing $\mathbf{M}^{(a)}(x)$ we find that $L^{(a)} = (K_2^{(a)})^* K_1^{(a)}$, where

$$(K_j^{(a)}u)(x) = \sum_n \nu_{n,j}^{(a)}(x)\langle \nabla u(x), p_n^{(a)}(x)\rangle p_n^{(a)}(x), \quad j = 1, 2,$$

and $\nu_{n,1}^{(a)} = |\lambda_n^{(a)}|^{1/2}$, $\nu_{n,1}^{(a)}\nu_{n,2}^{(a)} = \lambda_n^{(a)}$. To prove smoothness of the operators $K_j^{(a)}$ we need the following elementary observation.

Lemma 0.11 *Suppose that $m(x) = m(x_b)$ is a smooth homogeneous (for $|x| \geq 1$) function of degree 1 in some cone Γ. Let $\lambda_n(x)$ and $p_n(x)$ be eigenvalues and eigenvectors of the symmetric matrix $\mathbf{M}(x) = \{m_{jk}(x)\}$. Then vectors $p_n(x)$, $x \in \Gamma$, $|x| \geq 1$, corresponding to $\lambda_n(x) \neq 0$, belong to X_b and are orthogonal to x_b.*

By our construction of m and $\eta^{(a)}$ we can find cones Γ_b such that $\bigcup_b \Gamma_b = X$, Γ_b satisfy the condition of Theorem 8 and $m^{(a)}(x) = m^{(a)}(x_b)$ if $x \in \Gamma_b$. According to Lemma 11 and the definition (14)

$$\langle (\nabla u)(x), p_n^{(a)}(x)\rangle = \langle (\nabla_{x_b}^{(s)} u)(x), p_n^{(a)}(x)\rangle, \quad x \in \Gamma_b \quad |x| \geq 1, \quad \lambda_n^{(a)}(x) \neq 0.$$

It follows that

$$|(K_j^{(a)}u)(x)| \leq C|x|^{-1/2}|(\nabla_{x_b}^{(s)}u)(x)|, \quad x \in \Gamma_b, \quad |x| \geq 1.$$

Thus H- and H_a-smoothness of the operators $K_j^{(a)}$ is ensured by Theorem 8. Putting all things together we arrive at Theorem 10.

6. Our goal now is to deduce Theorem 8 from Theorem 9. To that end we introduce the observable

$$M^{\pm}(\Lambda) := W^{\pm}(H, H; ME(\Lambda)). \tag{0.24}$$

Its basic properties are formulated in the following

Theorem 0.12 *Let M be the same operator as in part 4. Suppose that functions (3) satisfy Assumptions 2 and 3. Then the wave operators (24) exist, are self-adjoint and commute with H. Furthermore, their ranges*

$$R(M^{\pm}(\Lambda)) = E(\Lambda)\mathcal{H}. \tag{0.25}$$

Existence of the wave operators (24) can be verified quite similarly to Theorem 10. The intertwining property of wave operators implies that $M^{\pm}(\Lambda)$ commutes with H. We shall show that $\pm M^{\pm}(\Lambda)$ is positively definite on the subspace $E(\Lambda)\mathcal{H}$. Note the identity

$$d(mf_t, h_t)/dt = i([H, m]f_t, h_t) = i([T, m]f_t, h_t) = (Mf_t, h_t), \tag{0.26}$$

where $f_t = U(t)f$ and $h_t = U(t)h$. Element $h \in \mathcal{H}$ is arbitrary and f belongs to some dense in $\mathcal{H}$ set so that mf_t are well-defined. Integrating (26) and taking into account existence of the wave operators (24) we find that

$$U^*(t)mU(t)f = t\, M^{\pm}(\Lambda)f + o(|t|), \quad t \to \pm\infty, \tag{0.27}$$

and, in particular,

$$\|mf_t\| = \|M^{\pm}(\Lambda)f\|\, |t| + o(|t|). \tag{0.28}$$

Since $m(x) \geq 0$, (27) ensures that $\pm M^{\pm}(\Lambda) \geq 0$. To prove that $\pm M^{\pm}(\Lambda)$ is positively definite we use Proposition 4. In virtue of the identity $i[H, Q^2] = 2A$, it follows from (7) that

$$2^{-1}d^2(Q^2 f_t, f_t)/dt^2 = d(Af_t, f_t)/dt = (i[H, A]f_t, f_t) \geq c\,\|f\|^2,$$

if f belongs to the subspace $E(\Lambda)\mathcal{H}$. Integrating twice this inequality we find that for sufficiently large $|t|$

$$\|Qf_t\| \geq c\,|t|\,\|f\|. \tag{0.29}$$

Comparing (28) with (29) and considering that $m(x) \geq m_0|x|$, $m_0 > 0$, for $|x| \geq 1$, we obtain the inequality

$$\|M^{\pm}(\Lambda)f\| \geq c\,\|f\|, \quad c = c(\Lambda) > 0.$$

Thus $\pm M^{\pm}(\Lambda)$ is positively definite on $E(\Lambda)\mathcal{H}$. In particular, (24) holds.

7. The difficult part of Theorem 7 is, of course, asymptotic completeness. We start with its proof in the form (12).

Theorem 0.13 *Under the assumptions of Theorem 7 for every $f = E(\Lambda)f$ there exist elements $f_a^{\pm}$ such that the relation (12) is fulfilled.*

Proof can be easily deduced from the results obtained. Let M and $M^{(a)}$ be the operators introduced in parts 4 and 5 respectively. According to (21)

$$\sum_a M^{(a)} = M. \tag{0.30}$$

By Theorem 12 every $f \in E(\Lambda)\mathcal{H}$ admits the representation $f = M^{\pm}(\Lambda)f^{\pm}$, $f^{\pm} \in E(\Lambda)\mathcal{H}$, so that the asymptotic relation

$$U(t)f \sim MU(t)f^{\pm}, \quad t \to \pm\infty, \tag{0.31}$$

holds. On the other hand, Theorem 10 ensures that for every a

$$M^{(a)}U(t)f^{\pm} \sim U_a(t)f_a^{\pm}, \quad t \to \pm\infty, \tag{0.32}$$

where

$$f_a^{\pm} = W^{\pm}(H_a, H; M^{(a)}E(\Lambda))f^{\pm}.$$

Summing up the relations (32) and taking into account (30) we find that

$$MU(t)f^{\pm} \sim \sum_a U_a(t)f_a^{\pm}, \quad t \to \pm\infty.$$

Comparing it with (31) we arrive at (12).

It remains to derive Theorem 7 from Theorem 13. Note that in the proof of Theorem 13 we have used only existence of the second set of the wave operators (23). Using existence of the first set of these operators and the equality (13) we can prove that the wave operators $W^{\pm}(H, H_a; ME_a(\Lambda))$ also exist. This implies existence of $W^{\pm}(H, H_a; E_a(\Lambda))$ because, according to Theorem 12, for every $g \in E_a(\Lambda)\mathcal{H}$ we can find elements $g_a^{\pm}$ such that

$$U_a(t)g \sim MU_a(t)g_a^{\pm}, \quad t \to \pm\infty.$$

Elements $g = E_a(\Lambda)g$ for different Λ, $\bar{\Lambda} \cap \Upsilon = \emptyset$, are dense in the space $\mathcal{H} = \mathcal{H}^{(ac)}(H_a)$ so that operators $W^{\pm}(H, H_a; I)$ exist. In virtue of (13) orthogonality of ranges of the wave operators (8) is an automatic consequence (see e.g. [15]) of their existence.

Admitting that Theorem 7 is already proven for all operators H^a and taking into account (6) we can rewrite (12) as

$$U(t)f \sim \sum_a U_a(t)\tilde{f}_a^{\pm}, \quad t \to \pm\infty, \quad \text{where} \quad \tilde{f}_a^{\pm} \in P_a\mathcal{H}.$$

Multiplying both sides by $U^*(t)$ we obtain that every $f \in E(\Lambda)\mathcal{H}$ belongs to the left side of (9). Since elements $f = E(\Lambda)f$ are dense in the subspace $\mathcal{H}^{(ac)}(H)$, this concludes our proof of the asymptotic completeness.

Acknowledgements. This work was partially done during author's stays in California Institute of Technology and University of Aarhus. It is a pleasure to thank B. Simon and E. Balslev for hospitality and G. M. Graf and participants of the workshop on Schrödinger operators in Aarhus for many stimulating discussions.

REFERENCES

[1] L. D. Faddeev, *Mathematical Aspects of the Three Body Problem in Quantum Scattering Theory*, Trudy MIAN **69**, 1963. (Russian)

[2] J. Ginibre and M. Moulin, Hilbert space approach to the quantum mechanical three body problem, Ann. Inst. H.Poincaré, A **21**(1974), 97-145.

[3] L. E. Thomas, Asymptotic completeness in two- and three-particle quantum mechanical scattering, Ann. Phys. **90** (1975), 127-165.

[4] K. Hepp, On the quantum-mechanical N-body problem, Helv. Phys. Acta **42**(1969), 425-458.

[5] I. M. Sigal, *Scattering Theory for Many-Body Quantum Mechanical Systems*, Springer Lecture Notes in Math. **1011**, 1983.

[6] R. J. Iorio and M. O'Carrol, Asymptotic completeness for multi-particle Schrödinger Hamiltonians with weak potentials, Comm. Math. Phys. **27**(1972), 137-145.

[7] T. Kato, Smooth operators and commutators, Studia Math. **31**(1968), 535-546.

[8] R. Lavine, Commutators and scattering theory I: Repulsive interactions, Comm. Math. Phys. **20**(1971), 301-323.

[9] R. Lavine, Completeness of the wave operators in the repulsive N-body problem, J. Math. Phys. **14** (1973), 376-379.

[10] I. M. Sigal and A. Soffer, The N-particle scattering problem: Asymptotic completeness for short-range systems, Ann. Math. **126**(1987), 35-108.

[11] J. Derezinski, A new proof of the propagation theorem for N-body quantum systems, Comm. Math. Phys. **122** (1989), 203-231.

[12] G. M. Graf, Asymptotic completeness for N-body short-range quantum systems: A new proof, Comm. Math. Phys. **132** (1990), 73-101.

[13] V. Enss, Completeness of three-body quantum scattering, in: *Dynamics and processes*, P. Blanchard and L. Streit, eds., Springer Lecture Notes in Math. **103** (1983), 62-88.

[14] T. Kato, Wave operators and similarity for some non-self-adjoint operators, Math. Ann. **162** (1966), 258-279.

[15] M. Reed and B. Simon, *Methods of Modern Mathematical Physics* III, Academic Press, 1979.

[16] S. Agmon, *Lectures on Exponential Decay of Solutions of Second-Order Elliptic Equations*, Math. Notes, Princeton Univ. Press, 1982.

[17] E. Mourre, Absence of singular spectrum for certain self-adjoint operators, Comm. Math. Phys. **78** (1981), 391-400.

[18] P. Perry, I. M. Sigal and B. Simon, Spectral analysis of N-body Schrödinger operators, Ann. Math. **144** (1981), 519-567.

[19] R. Froese, I. Herbst, A new proof of the Mourre estimate, Duke Math. J. **49** (1982), 1075-1085.

[20] D. R. Yafaev, Remarks on the spectral theory for the multiparticle type Schrödinger operator, J. Soviet Math. **31** (1985), 3445-3459 (translated from Zap. Nauchn. Sem. LOMI **133** (1984), 277-298).

[21] Y. Saito, *Spectral Representation for Schrödinger Operators with Long-Range Potentials*, Springer Lecture Notes in Math. **727**, 1979.

GEVREY FREQUENCY SET AND SEMI–CLASSICAL BEHAVIOUR OF WAVE PACKETS

KENJI YAJIMA

Department of Mathematics, University of Tokyo
3-8-1 Komaba, Meguroku, Tokyo 153, Japan

§1. Introduction, Theorems.

We begin with a brief review of the method of WKB-approximation for the Cauchy problem for time dependent Schrödinger equations

$$ih\partial_t u - H^h(t)u = 0, \quad H^h(t) = (1/2)(-ih\partial_x - A(t,x))^2 + V(t,x). \tag{1.1}$$

It is a method to construct an approximate solution $u_W(t,x)$, called WKB-solution, of (1.1) with coherent initial condition $u(0,x) = \exp(iS(x)/h)f(x)$ in the form of asymptotic series in h:

$$u_W(t,x) \sim e^{iS(t,x)/h}\sum_{j=0}^{\infty} h^j u_j(t,x), \tag{1.2}$$

and its merit is that $S(t,x)$ and $u_j(t,x)$, $j = 0,1,...$, in the series (1.2) are given in terms of classical mechanical quantities in a rather simple manner. We briefly describe the method, for sufficiently small $t > 0$, assuming that *supp f* is contained in a fixed open set K with compact closure.

We denote the classical Hamiltonian associated with (1.1) by

$$H(t,x,p) = (1/2)(p - A(t,x))^2 + V(t,x) \tag{1.3}$$

and by $\Gamma(t,t_0)$ the associated Hamiltonian flow:

$$\Gamma(t,t_0) : (y,\eta) \to (x(t,t_0,y,\eta), p(t,t_0,y,\eta))$$

where $(x(t,t_0,y,\eta), p(t,t_0,y,\eta))$ is the solution of the Hamiltonian equations

$$\dot{x}(t) = (\partial_p H)(t,x(t),p(t)), \quad \dot{p}(t) = -(\partial_x H)(t,x(t),p(t)) \tag{1.4}$$

with initial condition $(x(t_0),p(t_0)) = (y,\eta)$. When $t_0 = 0$, we shall often omit the variable t_0 and write $\Gamma(t) = \Gamma(t,0)$, $x(t,y,\eta) = x(t,0,y,\eta)$ and etc. For a smooth

Typeset by $\mathcal{A}_{\mathcal{M}}\mathcal{S}$-TEX

real function $S(x)$ and $D \subset \mathbf{R}^n$, we denote by $\Lambda(S, D) = \{(x, \nabla S(x)) : x \in D\}$ the Lagrangian manifold with the generating function $S(x)$ over D. $\Lambda(S) = \Lambda(S, \mathbf{R}^n)$ and

$$\Omega_t(S, D) = \{x = x(t, y, \nabla S(y)) : y \in D\}. \tag{1.5}$$

It is well known that for sufficiently small $T > 0$, the map

$$K \ni y \to x = x(t, y, \nabla S(y)) \in \Omega_t(S, K)$$

is diffeomorphic for all $0 \leq t \leq T$. The function $S(t, x)$ in (1.2) is given by the generating function of $\Gamma(t)\Lambda(S, supp\, f)$ defined, for $x = x(t, y, \nabla S(y)) \in \Omega_t(S, K)$, by

$$S(t, x) = S(y) + \int_0^t L(s, q(s), v(s))ds, \tag{1.6}$$

where $q(s) = x(s, y, \nabla S(y)), v(s) = p(s, y, \nabla S(y)) - A(s, x(s, y, \nabla S(y)))$ and $L(t, q, v) = (1/2)v^2 - V(t, q) + v \cdot A(t, q)$ is the Lagrangian function associated with (1.3). Thus the integral in (1.6) is the action integral along the trajectory $\{\Gamma(s)(y, \nabla S(y)), 0 \leq s \leq t\}$; $u_j(t, x)$, $j = 0, 1, ...,$ are the solutions of the transport equations associated with $S(t, x)$:

$$\partial_t u_j + (\nabla_x S - A) \cdot \nabla_x u_j + (1/2)(\Delta_x S - div_x A)u_j = (i/2)\Delta_x u_{j-1} \tag{1.7}$$

with initial conditions $u_0(0, x) = f(x)$ and $u_j(0, x) = 0$, $j = 1, ...,$ respectively. Here we set $u_{-1}(t, x) \equiv 0$ and $S = S(t, x), A = A(t, x)$. (1.7) can be solved for $u_j(t, x)$ to yield

$$u_0(t, x) = \{\det(\partial x/\partial y)(t)\}^{-1/2} f(y), \quad x = x(t) = x(t, y, \nabla S(y)) \tag{1.8a}$$

and, inductively for $j \geq 1$,

$$u_j(t, x) = (i/2)\{\det(\partial x/\partial y)(t)\}^{-1/2} \int_0^t \{\det(\partial x/\partial y)(s)\}^{1/2} \Delta u_{j-1}(s, x(s))ds. \tag{1.8b}$$

It is well known (cf. Lemma 2.4, (2) below) under fairly general conditions that $u_W(t, x)$ thus constructed is indeed an approximate solution as $h \to 0$ in the sense that each first N-terms of the series (1.2) satisfy

$$\|U^h(t, 0)(e^{iS/h} f)(x) - e^{iS(t,x)/h} \sum_{j=0}^{N} h^j u_j(t, x)\|_{L^2(\mathbf{R}^n)} \leq C_N h^{N+1},$$

where $U^h(t, s)$ is the propagator for (1.1). This has been considered to be a confirmation of the standard semi-classical convention (cf. Maslov [Mas], Yajima [Y]): The coherent state $\exp(iS(x)/h)f(x)$ represents in the limit $h \to 0$ the ensemble of classical particles sitting on the Lagrangian manifold $\Lambda(S, supp\, f)$ with density $|f(x)|^2 dx$ and the state after time t, $U^h(t, 0)(e^{iS(x)/h} f)$, is again a coherent state representing those sitting on its image $\Gamma(t)\Lambda(S, supp\, f)$ under the classical flow $\Gamma(t)$, corrections being given by higher oder terms.

Note that all $u_j(t,x)$ are supported by $\Omega_t(S, supp\, f)$, hence, as $h \to 0$, $U^h(t,0)(e^{iS/h}f)$ decays faster than any power of h in the *classically forbidden region* $\Omega_t(S, supp\, f)^c$. If $h > 0$, however, $U^h(t,0)(e^{iS/h}f)$ can not vanish on $\Omega_t(S, supp\, f)^c$ identically and some of its part tunnels out of $\Omega_t(S, supp\, f)$. In this note we are interested in estimating how much of it indeed tunnels out of this classical region, assuming that $S(x)$ and the coefficients of (1.1) are analytic in the spatial variable x.

It is clear that how much tunnels out is related to how sharply the the initial state $e^{iS(x)/h}f(x)$ is localized on $\Lambda(S, supp\, f)$ as $h \to 0$, which, of course, is determined by the smoothness of $f(x)$, more specifically, the decay property of the Fourier transform $\hat{f}(\xi)$ as $|\xi| \to \infty$. As was described above, the C^∞ property of S and f leads to

$$\|U^h(t,0)(e^{iS(x)/h}f(x))\|_{L^2(\Omega_t(S,supp\, f)^c)} = O(h^\infty),$$

and we believe that this is the best possible estimate under the condition. Thus it is natural to ask whether or not the more smoothness of the initial wave functions $e^{iS(x)/h}f(x)$ leads to the faster decay of $\|U^h(t,0)(e^{iS/h}f)\|_{L^2(\Omega_t(S,supp\, f)^c)}$ as $h \to 0$. As we want the class of functions we consider to be large enough to contain those with compact support, we choose initial function $f(x)$ from the classes of Gevrey functions.

Assumption (A). $A_j(t,x)$, $j = 1, ..., n$, and $V(t,x)$ are real continuous in $(t,x) \in \mathbb{R}^1 \times \mathbb{R}^n$. For every fixed t, $A(t,x)$ and $V(t,x)$ have analytic extensions $A(t,z)$ and $V(t,z)$ into a common complex neighborhood W of $\mathbb{R}^n$ which are continuous in the variables $(t,z) \in \mathbb{R}^1 \times W$.

Assumption (B). The operator $H(t,h)$ with domain $\mathcal{S}(\mathbb{R}^n)$ is essentially selfadjoint in $L^2(\mathbb{R}^n)$ for every $t \in \mathbb{R}^1$ and the equation (1.1) generates a unique unitary propagator $U^h(t,s)$, $-\infty < t, s < \infty$, which preserves $\mathcal{S}(\mathbb{R}^n)$: $U^h(t,s)\mathcal{S}(\mathbb{R}^n) \subset \mathcal{S}(\mathbb{R}^n)$.

We recall the definition of Gevrey functions:

Definition. A function f is called Gevrey function of order $s \geq 1$ on a domain D if for every compact subset K of D, there exist constants A and B such that

$$|\partial_x^\alpha f(x)| \leq AB^{|\alpha|}(\alpha!)^s, \quad x \in K$$

for all α. We write $\mathcal{G}^s(D)$ for the set of all Gevrey functions of order s on D. $\mathcal{G}_0^s(D)$ is the set of $\mathcal{G}^s(D)$ functions with compact supports.

Theorem 1.1. *Let (A) and (B) be satisfied and let $S(x)$ be real analytic. Suppose that $f \in \mathcal{G}_0^s(\mathbb{R}^n)$, $s > 1$. Then, for sufficiently small $0 < t \leq T$,*

$$\|U^h(t,0)(e^{iS(x)/h}f(x))\|_{L^2(\Omega_t(S,supp\, f)^c)} \leq C\exp(-1/C(th)^{1/(2s-1)}) \tag{1.9}$$

where the constant $0 < C$ is independent of $0 < h < 1$ and $0 < t \leq T$.

Theorem 1.2. *Let (A) and (B) be satisfied and $S(x)$ and $f(x)$ be as above. Then, for sufficiently small $0 < t \leq T$ and $D \subset \mathbb{R}^n$ with $dist(D, \Omega_t(S, supp\, f)) > 0$,*

$$\|U^h(t,0)(e^{iS(x)/h}f(x))\|_{L^2(D)} \leq C\exp(-1/Ch^{1/s}), \tag{1.10}$$

where the constant $0 < C$ is independent of $0 < h < 1$ and $0 < t \leq T$.

Note that the right hand side of (1.9) is much larger than that of (1.10) for small $h > 0$. This is natural because (1.9) is taking care of the behavior of the wave function upto the boundary of $\Omega_t(S, supp\, f)^c$ whereas (1.10) strictly inside of $\Omega_t(S, supp\, f)^c$ only.

The estimate as in Theorem 1.2 can in fact be microlocalized in the phase space. For formulating the result in this direction, we introduce the notion of the Gevrey frequency set which generalizes the corresponding notion of C^∞ and analytic frequency set introduced in different contexts by Robert [R] and Martinez [Mar], respectively.

Definition. For a distribution $\phi(x, h) \in \mathcal{D}'(\mathbb{R}^n)$ depending on the parameter $0 < h < 1$, the Gevrey s frequency set $FS_s(\phi) \subset \mathbb{R}^n \times \mathbb{R}^n$, $s > 1$ is defined as follows: $(x_0, \xi_0) \notin FS_s(\phi)$ iff there exists an open neighbourhood U of (x_0, ξ_0) such that

$$\|\chi(x, hD)\phi\|_{L^2(D)} \leq C \exp(-1/Ch^{1/s}), \quad 0 < h < 1 \tag{1.11}$$

for every $\chi(x, \xi) \in \mathcal{G}_0^s(U)$. Here $\chi(x, hD)$ is the standard pseudo-differential operator with the symbol $\chi(x, h\xi)$:

$$\chi(x, hD) f(x) = (2\pi h)^{-n} \int e^{i(x-y)\cdot\xi/h} \chi(x, \xi) f(y)\, dy\, d\xi. \tag{1.12}$$

Remark. In virtue of Lemma 2.7 below, $(x_0, \xi_0) \notin FS_s(\phi)$ iff (1.11) is satisfied for some $\chi(x, \xi) \in \mathcal{G}_0^s(\mathbb{R}^n)$ such that $\chi(x, \xi) = 1$ near (x_0, ξ_0).

We have the following propagation theorem for the $\mathcal{G}^s$ frequency set.

Theorem 1.3. *Let Assumptions (A) and (B) be satisfied. Then*

$$FS_s(U^h(t, 0) f) = \Gamma(t) FS_s(f). \tag{1.13}$$

Finally in this section, we comment, for the free Schrödinger equation, that the wave function $\exp(ith\Delta)(e^{iS(x)/h} f(x))$ never decays like $O(\exp(-1/Ch))$ in any domain.

Theorem 1.4. *Let $f \in L^2(\mathbb{R}^n)$ have compact support and D be any non-empty open subset of $\mathbb{R}^n$. Suppose that*

$$\|\exp(ith\Delta)(e^{iS(x)/h} f(x))\|_{L^2(D)} \leq C \exp(-1/Ch), \quad 0 < h < h_0 \tag{1.14}$$

for some $t \neq 0$ and some constant $C > 0$. Then $f(x) = 0$ almost everywhere.

The rest of this paper is devoted to the proof of the theorems. Our approach is very much elementary in the sense we treat everything by WKB-approximation and elementary estimates. In sectin 2, we construct a local parametrix using WKB-approximation and collect several estimate necessary in subsequent sections and prove Theorem 1.1 as byproduct of this construction. Theorem 1.2 will be proved in section 3. Theorem 1.3 will be proved in section 4 and Theorem 1.4 in the final section 5.

Acknowledgement. The author is grateful to Professor C. Gérard for very helpful discussions. He expresses his thanks to Professor E. Balslev for the hospitality and organizing a stimulating workshop.

§2. $\mathcal{G}^s$-Local Parametrix and Proof of Theorem 1.1.

In this section we collect several preliminary results necessary for proving the theorems. In particular, given $\Psi(x,\xi) \in \mathcal{G}_0^s(\mathbb{R}^{2n})$, $1 < s$, we construct an integral operator $E_\Psi(t,h)$ such that $E_\Psi(0,h) = \Psi(x,hD)$ and such that

$$\|U^h(t,0)\Psi(x,hD) - E_\Psi(t,h)\| \le C\exp(-C/h^{1/(2s-1)}),$$

where $\|\cdot\|$ is the operator norm in $L^2(\mathbb{R}^n)$. We shall call $E_\Psi(t,h)$ the $\mathcal{G}^s$ *local parametrix* for the propagator $U^h(t,0)$ localized on $\Psi(x,\xi)$.

We begin with the following well known lemma. We remark that the Hamilton flow $\Gamma(t,t_0)$ can be defined for complex (y,η) as well under Assumption (A) and $\Gamma(t) = \Gamma(t,0)$. We say that an open set U is a compact open subset of V, if the closure of U is a compact subset of V.

Lemma 2.1. *Let Assumption (A) be satisfied. Let $\tilde{K}_j, j = 1,2$, be a compact open subset of $\mathbb{C}^n$ such that $\tilde{K}_1 \subset W$ and let $S(x,\xi)$ be analytic on a domain containing the closure of $\tilde{K}_1 \times \tilde{K}_2$ and real for real (x,ξ). Then there exists $T > 0$ such that:*
(1) $\Gamma(t)(y,\nabla_y S(y,\xi))$ is analytic in $(y,\xi) \in \tilde{K}_1 \times \tilde{K}_2$, real for reals, and for constants C_0' and C_1'

$$(2.1) \qquad |\partial_y^\alpha \partial_\xi^\beta \Gamma(t)(y,\nabla_y S(y,\xi))| \le C_0' C_1'^{|\alpha+\beta|}\alpha!\beta!, \quad (y,\xi) \in \tilde{K}_1 \times \tilde{K}_2,\ 0 \le t \le T.$$

(2) The map $\tilde{K}_1 \ni y \to x(t,y,\nabla_y S(y,\xi)) \in \Omega_t(S,\tilde{K}_1)$ is diffeomorphic for all $0 \le t \le T$ and $\xi \in \tilde{K}_2$.

Hereafter till the proof of Theorem 1.1, we take and fix a compact open subset $K_j \subset \mathbb{R}^n$, its complex neighborhood $\tilde{K}_j, j = 1,2$ such that $\tilde{K}_1 \subset W$; a compact open subset G_j of K_j, and its complex neighbourhood of $\tilde{G}_j$ which is a compact open subset of $\tilde{K}_j$.

We shall construct the WKB-approximation of $U^h(t,0)u_{0\xi}$ for a family of wave functions $u_{0\xi} = e^{iS(x,\xi)/h}\Psi(x,\xi)$ such that $\Psi(x,\xi)$ and $S(x,\xi)$ satisfy the following conditions. We write $\mathcal{A}(D)$ for the set of all analytic functions on D.

(C1). $\Psi(x,\xi) \in \mathcal{G}_0^s(K_1 \times K_2) \cap \mathcal{A}(\tilde{G}_1' \times \tilde{G}_2')$. where $\tilde{G}_j' \subset \tilde{K}_j$ is an open neighbourhood of the closure of $\tilde{G}_j, j = 1,2$. Moreover it satsfies the estimates

$$(2.2a) \qquad |\partial_y^\alpha \partial_\xi^\beta \Psi(x,\xi)| \le C_0 C_1^{|\alpha+\beta|}(\alpha!)^s(\beta!)^s, \quad (y,\xi) \in K_1 \times K_2.$$

$$(2.2b) \qquad |\partial_y^\alpha \partial_\xi^\beta \Psi(x,\xi)| \le C_0 C_1^{|\alpha+\beta|}\alpha!\beta!, \quad (y,\xi) \in \tilde{G}_1' \times \tilde{G}_2'.$$

We do not, however, exclude the possibility that G_1 and G_2 are empty.

(C2). $S(x,\xi)$ satisfies the condition of Lemma 2.1, that is, $S(x,\xi)$ is analytic on a domain containing the closure of $\tilde{K}_1 \times \tilde{K}_2$ and real for reals.

We take and fix $T > 0$ as in Lemma 2.1. Considering $S(x,\xi)$ and $\Psi(x,\xi)$ as families of functions of x with parameter ξ, we define $S(t,x,\xi)$ and $u_j(t,x,\xi)$, $j = 0,1,\ldots$, by (1.6) and (1.8a-b) corresponding to $S(x,\xi)$ and $\Psi(x,\xi)$, respectively. Note that this definition is possible for $0 \le t \le T$ not only for real

$$(x,\xi) \in W_1(t) \equiv \{(x,\xi) = (x(t,y,\nabla S_y(y,\xi)),\xi) : (y,\xi) \in K_1 \times K_2\}. \tag{2.3a}$$

but also for complex

$$(x,\xi) \in W_2(t) = \{(x(t,y,\nabla S_y(y,\xi)),\xi) : y \in \tilde{G}'_1, \xi \in \tilde{G}'_2\}, \tag{2.3b}$$

thanks to the conditions (C1) and (C2).

We have to estimate the derivatives of $u_j(t,x,\xi)$. For this we set for $j = 0,1,\ldots$,

$$\tilde{u}_j(t,y,\xi) = \{\det(\partial x(t,y,\nabla S_y(y,\xi))/\partial y)\}^{1/2} u_j(t,x(t,y,\nabla S_y(y,\xi))),\xi). \tag{2.4}$$

Then $\tilde{u}_0(t,y,\xi) = \Psi(y,\xi)$ and change of variables shows that $\tilde{u}_j(t,y,\xi)$ satisfies

$$\partial_t \tilde{u}_j(t,y,\xi) = P(t,y,\xi,\partial_y)\tilde{u}_{j-1}(t,y,\xi), \quad \tilde{u}_j(0) = 0, \quad j = 1,2,\ldots \tag{2.5}$$

where, in virtue of Lemma 2.1, the second order differential operator

$$P(t,y,\xi,\partial_y) = \sum_{|\gamma|\le 2} a_\gamma(t,y,\xi)\partial_y^\gamma \tag{2.6}$$

has coefficients $a_\gamma(t,y,\xi)$ which are analytic in (y,ξ), C^1 in (t,y,ξ) and

$$|\partial_y^\alpha \partial_\xi^\beta a_\gamma(t,y,\xi)| \le M_0 M_1^{|\alpha+\beta|}\alpha!\beta!, \quad (y,\xi) \in \tilde{K}_1 \times \tilde{K}_2. \tag{2.6a}$$

We shall often use the following

Lemma 2.2. *Let K be a domain in $\mathbb{R}^m$, $P_j(x,\partial_x) = \sum_{|\gamma|\le k} a_{j\gamma}(x)\partial_x^\gamma$, $j = 1,2,\cdots$, be a sequence k-th order differential operator with $\mathcal{G}^s$-coefficients $a_{j\gamma}(x)$ such that*

$$|\partial_x^\alpha a_{j\gamma}(x)| \le M_0 M_1^{|\alpha|}(\alpha!)^s, \quad x \in K \tag{2.7}$$

and $f(x)$ be a $\mathcal{G}^s(K)$ function:

$$|\partial_x^\alpha f(x)| \le AB^{|\alpha|}(\alpha!)^s, \quad x \in K. \tag{2.8}$$

Then with constants L_0, L_1 and L_2 independent of $j = 0,1,\ldots$ and α, we have

$$|\partial_x^\alpha P_j(x,\partial_x)\cdots P_1(x,\partial_x)f(x)| \le L_0 L_1^j L_2^{|\alpha|}(kj)^{s|\alpha|}(\alpha!)^s((kj)!)^s, \quad x \in K. \tag{2.9a}$$

When $a_{j\gamma}(x), f(x) \in \mathcal{A}(\tilde{K})$, $j = 1,2,\cdots,|\gamma| \le k$ and (2.7) and (2.8) are satisfied in $\tilde{K}$ with $s = 1$, then in any compact open subset $\tilde{G}$ of $\tilde{K}$ we have better estimate

$$|\partial_x^\alpha P_j(x,\partial_x)\cdots P_1(x,\partial_x)f(x)| \le L_0 L_1^j L_2^{|\alpha|}\alpha!(kj)!, \quad x \in \tilde{G}. \tag{2.9b}$$

Proof. By the induction argument with respect to j. we prove

$$(2.10)\quad |\partial_x^\alpha P_j(x,\partial_x)\cdots P_1(x,\partial_x)f(x)| \le L_0(CM_0)^j L_3^{|\alpha|+kj}2^{(m+k)j+|\alpha|}\sup_{|\delta|\le kj}\{(\alpha+\delta)!\}^s.$$

(2.9a) follows from (2.10) by setting $L_1 = 2^{m+k}CM_0L_3^k$ and $L_2 = 2^{s+1}L_3$, since

$$\sup_{|\delta|\le kj}(\alpha+\delta)! \le 2^{|\alpha|}\alpha!(kj)!(kj)^{|\alpha|}.$$

Set $L_0 = A$ and $L_3 = \max(B, M_1)$. Then, (2.10) is certainly satisfied for $j = 0$, by the assumption (2.7) and (2.8). Suppose (2.10) is satisfied up to j and write $f_j(x) = P_j(x,\partial_x)\cdots P_1(x,\partial_x)f(x)$. By Leibniz' rule, we have

$$\partial_x^\alpha(a_{(j+1)\gamma}(x)\partial_x^\gamma f_j(x)) = \sum_{\beta\le\alpha}\frac{\alpha!}{\beta!(\alpha-\beta)!}\partial_x^{\alpha-\beta}a_{(j+1)\gamma}\cdot\partial_x^{\beta+\gamma}f_j(x).$$

It follows by (2.7) and the induction hypothesis that for $|\gamma|\le k$

$$\begin{aligned}|\partial_x^\alpha(a_{(j+1)\gamma}\partial_x^\gamma f_j)| &\le \sum_{\beta\le\alpha}\frac{\alpha!}{\beta!(\alpha-\beta)!}M_0M_1^{|\alpha-\beta|}\{(\alpha-\beta)!\}^s\times\\ &\quad 2^{(m+k)j+|\beta+\gamma|}L_0(CM_0)^jL_3^{|\beta+\gamma|+kj}\sup_{|\delta|\le kj}\{(\beta+\gamma+\delta)!\}^s\\ &\le L_0C^jM_0^{j+1}L_3^{|\alpha|+kj+k}\sup_{|\delta|\le kj+k}\{(\alpha+\delta)!\}^s2^{(m+k)j+k}\sum_{\beta\le\alpha}2^{|\beta|}.\end{aligned}$$

Since $\sum_{\beta\le\alpha}2^{|\beta|} = 2^{|\alpha|+m}$, the right hand side is bounded by

$$(2.11)\qquad L_0C^jM_0^{j+1}L_3^{|\alpha|+k(j+1)}\sup_{|\delta|\le k(j+1)}\{(\alpha+\delta)!\}^s2^{(m+k)(j+1)+|\alpha|}$$

and, if we have had set $C = \sum_{|\gamma|\le k}1 \le (k+1)^m$, we obtain (2.10) for $j+1$ by summing up (2.11) for $|\gamma|\le k$. Thus (2.10) holds for any $j = 0, 1, \ldots$ and α.

Given the argument above, for proving (2.9b) it suffices to note the analyticity of the relevant functions: Setting $s = 1$ in the argument above and working in the complex domain $\tilde{K}$, we obtain (2.9b) for $\alpha = 0$ in the domain $\tilde{K}$. Then, in virtue of Cauchy's estimate, (2.9b) for general α follows in the smaller domain $\tilde{G}$ with possibly different constants. □

Integrating (2.5) by t successively, we have for $j = 1, 2, \cdots$,

$$(2.12)\qquad \tilde{u}_j(t,y,\xi) = \int_0^t dt_{j-1}\cdots\int_0^{t_1}dt_0P(t_{j-1},y,\xi,\partial_y)\cdots P(t_0,y,\xi,\partial_y)\Psi(y,\xi).$$

Applying Lemma 2.2, we obtain the following

Proposition 2.3. *$\tilde{u}_j(t,y,\xi)$ is C^1 in (t,y,ξ) and $\mathcal{G}_0^s(K_1 \times K_2) \cap \mathcal{A}(\tilde{G}_1' \times \tilde{G}_2')$ in (y,ξ) for $j = 0,1,\ldots$. There exist constants L_0, L_1 and L_2 such that for $(y,\xi) \in K_1 \times K_2$,*

$$|\partial_y^\alpha \partial_\xi^\beta \tilde{u}_j(t,y,\xi)| \le \frac{t^j}{j!} L_0 L_1^j L_2^{|\alpha+\beta|} (2j)^{s|\alpha|} ((2j)!)^s (\alpha!)^s (\beta!)^s, \tag{2.13a}$$

and for $(y,\xi) \in \tilde{G}_1 \times \tilde{G}_2$

$$|\partial_y^\alpha \partial_\xi^\beta \tilde{u}_j(t,y,\xi)| \le \frac{t^j}{j!} L_0 L_1^j (2j)! L_2^{|\alpha+\beta|} \alpha! \beta!. \tag{2.13b}$$

Proof. The smoothness and the support property of $\tilde{u}_j(t,y,\xi)$ stated in the proposition are obvious from (2.12). Estimates (2.13a-b) follow by applying (2.9a-b) to (2.12). □

Lemma 2.4. *Let $0 \le t \le T$ and $W_1(t), W_2(t)$ be defined by (2.3a) and (2.3b). Then the following statements are satisfied:*
(1) For $j = 0,1,\ldots$, $u_j(t,x,\xi)$ is C^1 in (t,x,ξ) and is $\mathcal{G}_0^s(W_1(t)) \cap \mathcal{A}(W_2(t))$ in (x,ξ). We have $supp\, u_j(t,\cdot,\xi) = \Omega_t(S, supp\, \Psi(\cdot,\xi))$.
(2) For constants $C_3 \sim C_5$, we have for $(x,\xi) \in \mathbb{R}^n \times \mathbb{R}^n$,

$$|\partial_x^\alpha \partial_\xi^\beta u_j(t,x,\xi)| \le C_3 C_4^{|\alpha+\beta|} (\alpha!)^s (\beta!)^s (2j)^{s|\alpha+\beta|} \cdot t^j C_5^j (j!)^{-1} \{(2j)!\}^s, \tag{2.14a}$$

and for $(x,\xi) \in W_2(t)$ we have better estimate:

$$|\partial_x^\alpha \partial_\xi^\beta u_j(t,x,\xi)| \le C_3 C_4^{|\alpha+\beta|} \alpha! \beta! \cdot t^j C_5^j (j!)^{-1} (2j)!. \tag{2.14b}$$

Here of course the constants are independent of j, α, β and (t,x,ξ).

Proof. Write $y = y(t,x,\xi)$ for the inverse of $y \to x(t,y,\nabla_y S(y,\xi))$ and

$$v_j(t,y,\xi) = \{\det(\partial x(t,y,\nabla_y S(y,\xi))/\partial y)\}^{-1/2} \tilde{u}_j(t,y,\xi).$$

In virtue of (2.4), we have $u_j(t,x,\xi) = v_j(t, y(t,x,\xi), \xi)$. Then the analyticity of $y(t,x,\xi)$ and $\{\det(\partial x(t,y,\nabla_y S(y,\xi))/\partial y)\}^{-1/2}$ and (2.13a-b) imply (2.14a-b). The statement (1) is obvious. □

We set for $N = 0,1,\ldots$ and $0 \le t \le T$,

$$e_N(t,x,\xi,h) = \sum_{j=0}^{N} h^j u_j(t,x,\xi) \tag{2.15a}$$

and

$$e^s(t,x,\xi,h) = e_N(t,x,\xi,h), \quad N = C/(Th)^{1/(2s-1)}, \tag{2.15b}$$

where C is a constant to be specified below.

Lemma 2.5. *Let $e^s(t,x,\xi,h)$ be defined by (2.15a-b) for $0 \le t \le T$. Then, for sufficiently small $\epsilon_0 > 0$, the following statements are satisfied for all $0 < h < \epsilon_0$:*
(1) $e^s(t,x,\xi,h)$ is C^1 in (t,x,ξ) and is $\mathcal{G}_0^s(W_1(t)) \cap \mathcal{A}(W_2(t))$ in (x,ξ). We have $supp\, e^s(t,\cdot,\xi) = \Omega_t(S, supp\,\Psi(\cdot,\xi))$.
(2) There exists a constant $C > 0$ such that for every $\xi \in K_2$ and $0 \le t \le T$,

$$(2.16)\quad \|U^h(t,0)(e^{iS(\cdot,\xi)/h}\Psi(\cdot,\xi)) - e^{iS(t,\cdot,\xi)/h}e^s(t,\cdot,\xi,h)\|_{L^2(\mathbb{R}^n)} \le Ce^{-1/C(Th)^{1/(2s-1)}}$$

(3) There exist constants C_0, C_1 such that for any α and β,

$$(2.17a)\qquad |\partial_x^\alpha \partial_\xi^\beta e^s(t,x,\xi,h)| \le C_0 C_1^{|\alpha+\beta|}\alpha!\beta!, \quad (x,\xi) \in W_2(t).$$

(4) There exist constants $C_{\alpha\beta} > 0$ such that

$$(2.17b)\qquad |\partial_x^\alpha \partial_\xi^\beta e^s(t,x,\xi,h)| \le C_{\alpha\beta}, \quad (x,\xi) \in \mathbb{R}^n \times \mathbb{R}^n.$$

Proof. The statement (1) is obvious from Lemma 2.4. We prove (4) first. In virtue of the Stirling formula $N! = (2\pi)^{1/2}e^{-N-1}N^{N+(1/2)}e^{\theta/12(N+1)}$, $0 \le \theta \le 1$, we have $\{(2j)!\}^s(j!)^{-1} \le C_{s1}C_{s2}^j j^{(2s-1)j}$, $j = 0,1,\cdots$ with constants C_{s1} and C_{s2} depending only on $1 \le s$. Set $C = (1/2C_{s2}C_5)^{1/(2s-1)}$, that is, $N = N_s \equiv 1/(2C_{s2}C_5Th)^{1/(2s-1)}$ in (2.15b) using the constant C_5 in (2.14a-b). Then, we have $C_5^j(th)^j(j!)^{-1}\{(2j)!\}^s \le C_{s1}2^{-j}$ for $0 \le t \le T$ and $0 \le j \le N_s$ and (2.14a) implies the desired estimate:

$$|\partial_x^\alpha \partial_\xi^\beta e^s(t,x,\xi,h)| \le C_3 C_4^{|\alpha+\beta|}(\alpha!)^s(\beta!)^s \sum_{j=0}^{N} C_{s1}(2j)^{s|\alpha+\beta|}2^{-j} \equiv C_{\alpha\beta}.$$

The proof of (3) is almost the same as the one for (4) and the only difference is that we use (2.14b) instead of (2.14a). The result is the desired estimate

$$|\partial_x^\alpha \partial_\xi^\beta e^s(t,x,\xi,h)| \le C_{s1}C_3C_4^{|\alpha+\beta|}\alpha!\beta! \sum_{j=0}^{N} 2^{-j}.$$

Finally we prove (2). Write $E^s(t,x,\xi,h) = e^{iS(t,x,\xi)/h}e^s(t,x,\xi,h)$. A simple computation yields, with $N = N_s$,

$$(ih\partial_t - H^h(t))E^s(t,x,\xi,h) = i2^{-1}e^{iS(t,x)/h}h^{N+2}\Delta u_N(t,x,\xi) \equiv ih^{N+2}g_N(t,x,\xi).$$

Solving the latter equation via Duhamel's formula, we have

$$U^h(t,0)(e^{iS(\cdot,\xi)/h}\Psi(\cdot,\xi))(t,\cdot) - E(t,\cdot,\xi,h) = -h^{N+1}\int_0^t U^h(t,s)g_N(s,\cdot,\xi)ds,$$

and, by the unitarity of $U^h(t,s)$,

$$\|U^h(t,0)(e^{iS(\cdot,\xi)/h}\Psi(\cdot,\xi))(t,\cdot) - E(t,\cdot,\xi,h)\|_{L^2} \le h^{N+1}\int_0^t \|\Delta u_N(s,\cdot,\xi)\|_{L^2}ds.$$

In virtue of (2.14a) the RHS is bounded for $N = N_s$ by a constant times

$$h^{N+1}\int_0^t \tau^N C_5^N (N!)^{-1}\{(2N)!\}^s d\tau = C_5^N (th)^{N+1}((2N)!)^s((N+1)!)^{-1}$$
$$\leq C_{1s}2^{-N} \leq C_{1s}\exp(-(\log 2)/(2ThC_5C_{s2})^{1/(2s-1)}).$$

This proves (4). □

Proof of Theorem 1.1. We apply Lemma 2.5 to the case $\Psi(x,\xi) = f(x)$ and $S(x,\xi) = S(x)$. If we take compact open K_1 such that $supp\, f \subset K_1$ and K_2 arbitrarily, then conditions (C1) and (C2) are clearly satisfied with empty G_j. It follows that there exists $T > 0$ such that the statements of Lemma 2.5 are satisfied for $0 \leq t \leq T$. Since $e^s(t,\cdot,\xi,h)$ which is in fact independent of ξ in this case has support $\Omega_t(S, supp\, f)$, (1.9) for $t = T$ follows from (2.16) immediately . It is obvious that if Lemma 2.5 holds for some T, it does for any smaller $T > 0$. Thus (1.9) is satisfied for any $0 \leq t \leq T$. □

Hereafter in this section we let $S(x,\xi) = x \cdot \xi$ and let $\Psi(x,\xi)$ satisfy the condition (C1). It is obvious that $S(x,\xi)$ satisfies (C2) and we have Lemma 2.5 for this $S(x,\xi)$ and $\Psi(x,\xi)$. Using $S(t,x,\xi)$ and $e^s(t,x,\xi,h)$ defined by (1.6) and (2.15a-b) for this case, we set

$$E_\Psi^s(t,h)u(x) = (2\pi h)^{-n}\int e^{i(S(t,x,\xi)-y\cdot\xi)/h}e^s(t,x,\xi,h)u(y)dyd\xi. \tag{2.18}$$

Since $S(0,x,\xi) = x\cdot\xi$ and $e^s(0,x,\xi,h) = \Psi(x,\xi)$, we have

$$E_\Psi^s(0,h) = \Psi(x,hD) \tag{2.19}$$

and $supp\, e^s(t,\cdot,\xi) = \Omega_t(\xi, supp\,\Psi(\cdot,\xi))$, where with a little abuse of notation $\Omega_t(\xi,K) = \{x(t,y,\xi) : y \in K\}$.

Theorem 2.6. *For sufficiently small $T > 0$ and $\epsilon_0 > 0$, the following statements are satisfied:*
(1) As operators in $L^2(\mathbb{R}^n)$, $\{E_\Psi^s(t,h) : 0 \leq t \leq T,\ 0 < h < \epsilon_0\}$ is uniformly bounded and is strongly continuous in t. Moreover $E_\Psi^s(t,h)$ maps $\mathcal{S}(\mathbb{R}^n)$ into $\mathcal{S}(\mathbb{R}^n)$ continuously.
(2) When $u \in \mathcal{S}(\mathbb{R}^n)$, $t \to E_\Psi^s(t,h)u$ is C^1 in $L^2(\mathbb{R}^n)$.
(3) There exists a constant $C > 0$ such that for $0 < h < \epsilon_0$ and $0 \leq t \leq T$

$$\|U^h(t,0)\Psi(x,hD)u - E_\Psi^s(t,h)\| \leq C\exp(-C/(Th)^{1/(2s-1)}). \tag{2.20}$$

Proof. In virtue of Lemma 2.5, we have $e^s(t,x,\xi,h) \in \mathcal{G}_0^s(W_1(t))$ in (x,ξ) and C_1 in (t,x,ξ). On the other hand, we have $(\partial^2 S/\partial x_j\partial\xi_k)(t,x,\xi) - \delta_{jk} = O(t)$ in $C^\infty(W_1(t))$ topology as $t \to 0$. Hence statements (1) and (2) are results of the well known continuity property of oscillatory integral operators [AF].

It remains to prove (2.20). In virtue of (2.16), we see that

$$F(t,x,\xi,h) = U^h(t,0)(e^{iy\cdot\xi/h}\Psi(y,\xi))(x) - e^{iS(t,x,\xi)/h}e^s(t,x,\xi,h) \tag{2.21}$$

satisfies, for any $0 \le t \le T$ and $0 < h < \epsilon_0$,

$$\|F(t,\cdot,\xi,h)\|_{L^2(\mathbb{R}^n)} \le C\exp(-C/(Th)^{1/(2s-1)}), \quad \xi \in K_2. \tag{2.22}$$

For $u \in \mathcal{S}(\mathbb{R}^n)$, we regard the integral

$$\Psi(x,hD)u(x) = (2\pi)^{-n/2}h^{-n}\int e^{ix\cdot\xi/h}\Psi(x,\xi)\hat{u}(\xi/h)d\xi$$

as the one of $L^2(\mathbb{R}^n_x)$ valued function of ξ with compact support and multiplying both sides by $U^h(t,0)$. Inserting (2.21) into the resulting equation, we obtain

$$U^h(t,0)\Psi(x,hD)u(x) = E^s_\Psi(t,h)u(x) + (2\pi)^{-n/2}h^{-n}\int F(t,x,\xi,h)\hat{u}(\xi/h)d\xi.$$

It follows from (2.22) and the fact that $F(t,x,\xi,h) = 0$ for $\xi \notin K_2$ that

$$\begin{aligned}\|U^h(t,0)\Psi(x,hD)u - E^s_\Psi(t,h)u\|_{L^2} &\le (2\pi)^{-n/2}h^{-n}\int_{K_2}\|F(t,\cdot,\xi,h)\|_{L^2}|\hat{u}(\xi/h)|d\xi \\ \le C\exp(-C/(Th)^{1/(2s-1)})h^{-n}&\int_{K_2}|\hat{u}(\xi/h)|d\xi \le C\exp(-C/(Th)^{1/(2s-1)})\|u\|_{L^2},\end{aligned}$$

which yields (2.20). Here we used Schwartz inequality and absorbed the factor $h^{-n/2}$ into $\exp(-C/h^{1/(2s-1)})$ by changing the constants. □

We shall use the following

Lemma 2.7. *Let $s > 1$ and let $\Psi_1(x,\xi), \Psi_2(x,\xi) \in \mathcal{G}^s_0(\mathbb{R}^n \times \mathbb{R}^n)$ be such that $dist(supp\,(1-\Psi_2), supp\,\Psi_1) = d > 0$. Then for $0 < h < 1$:*

$$\|\Psi_2(x,hD)\Psi_1(x,hD) - \Psi_1(x,hD)\| \le C\exp(-1/Ch^{1/s}), \tag{2.23a}$$

$$\|\Psi_1(x,hD)\Psi_2(x,hD) - \Psi_1(x,hD)\| \le C\exp(-1/Ch^{1/s}), \tag{2.23b}$$

$$\|\Psi_2(x,hD)^*\Psi_1(x,hD) - \Psi_1(x,hD)\| \le C\exp(-1/Ch^{1/s}), \tag{2.23c}$$

$$\|\Psi_1(x,hD)\Psi_2(x,hD)^* - \Psi_1(x,hD)\| \le C\exp(-1/Ch^{1/s}). \tag{2.23d}$$

Proof. We prove (2.23a) only. The proofs for the others are similar. For $u \in \mathcal{S}(\mathbb{R}^n)$, we write $\Psi_2(x,hD)\Psi_1(x,hD)u(x) - \Psi_1(x,hD)u(x)$ in the form of oscillatory integral:

$$(2\pi h)^{-2n}\int e^{i\psi(x,y,z,\xi,\eta)/h}(\Psi_2(x,\xi)-1)\Psi_1(y,\eta)u(z)dzd\eta dyd\xi. \tag{2.24}$$

where $\psi(x,y,z,\xi,\eta) = (x-y)\cdot\xi + (y-z)\cdot\eta$. Using the fact that

$$G = (x-y)^2 + (\xi-\eta)^2 + (y-z)^2 \ge C(x^2+\xi^2+z^2+d^2) \tag{2.25}$$

on the support of $(\Psi_2(x,\xi)-1)\Psi_1(y,\eta)$, we set the first order differential operator L by

$$L=-iG^{-1}((x-y)\cdot\partial/\partial\xi+(\eta-\xi)\cdot\partial/\partial y+(y-z)\cdot\partial/\partial\eta)$$

so that $L(e^{i\psi(x,y,z,\xi,\eta)/h})=he^{i\psi(x,y,z,\xi,\eta)/h}$. Performing integrations by parts N times, we rewrite the integral (2.24) in the form

$$(2\pi h)^{-2n}h^N\int e^{i\psi(x,y,z,\xi,\eta)/h}({}^tL)^N\{(\Psi_2(x,\xi)-1)\Psi_1(y,\eta)\}u(z)dzd\eta dyd\xi.$$

In virtue of Lemma 2.2, (2.25) and the assumption $\Psi_1,\Psi_2\in\mathcal{G}_0^s(\mathbb{R}^{2n})$, we have

$$|({}^tL)^N\{(\Psi_2(x,\xi)-1)\Psi_1(y,\eta)\}|\le C_0(x^2+\xi^2+z^2+d^2)^{-3n}C_1^N(N!)^s.$$

It follows that

$$\|\Psi_2(x,hD)\Psi_1(x,hD)u-\Psi_1(x,hD)u\|\le CC_1^N(N!)^sh^N\|u\|$$

and we obtain (2.23a) by setting $N=(1/C_1h)^{1/s}$.

□

We have so far set the initial time be zero. It is, however, obvious by time translation that the argument above can be generalized for any initial time t_0. In particular, the $\mathcal{G}^s$ local parametric $E_\Psi^s(t,t_0,h)$ of $U^h(t,t_0)$ localized on $\Psi(x,\xi)\in\mathcal{G}_0^s(\mathbb{R}^{2n})$ can be constructed for $|t-t_0|\le T$ with small $T>0$ in the form

$$E_\Psi^s(t,t_0,h)u(x)=(2\pi h)^{-n}\int e^{iS(t,t_0,x,\xi)/h-iy\cdot\xi/h}e^s(t,t_0,x,\xi,h)u(y)dyd\xi \tag{2.26}$$

such that

$$\|U^h(t,t_0)\Psi(x,hD)-E_\Psi^s(t,t_0,h)\|\le C\exp(-1/Ch^{1/(2s-1)}). \tag{2.27}$$

Here the notation should be obvious and $e^s(t,t_0,x,\xi,h)$ satisfies the properties of Lemma 2.6 and Theorem 2.7 corresponding to this case. In particular,

$$supp\, e^s(t,t_0,\cdot,\xi,h)=\{x=(t,t_0,y,\xi):y\in supp\,\Psi(\cdot,\xi)\}. \tag{2.28}$$

§3. Proof of Theorem 1.2.

Since $dist(D,\Omega_t(S,supp\,f))>0$, it is possible to find an open neighbourhood $\tilde D\subset\mathbb{R}^n\times\mathbb{R}^n$ of $\Lambda(S,supp\,f)$ such that $dist(\pi_x\Gamma(t)\tilde D,D)>0$ where π_x is the projection to $\mathbb{R}_x^n$. Take $1<s'<s$ such that $2s'-1<s$ and $\Psi(x,\xi)\in\mathcal{G}_0^{s'}(\tilde D)$ such that $\Psi(x,\xi)=1$ on a neighbourhood of $\Lambda(S,supp\,f)$.

Lemma 3.1. *Let $\Psi(x,\xi) \in \mathcal{G}_0^{s'}(\tilde{D})$ be as above. Then for $0 < h < 1$ we have*

$$\|\Psi(x,hD)u_0 - u_0\| \leq C\exp(-1/Ch^{1/s}), \quad u_0(x) = e^{iS(x)/h}f(x). \tag{3.1}$$

Proof. We apply integration by parts to

$$\Psi(x,hD)u_0(x) - u_0(x) = (2\pi h)^{-n}\int e^{i(x-y)\cdot\xi/h+iS(y)/h}(\Psi(x,\xi)-1)f(y)dyd\xi$$

Since $\Psi(x,\xi) = 1$ on a neighbourhood of $\Lambda(S, supp\, f)$, we have

$$Q \equiv (x-y)^2 + (\nabla S(y)-\xi)^2 \geq C(x^2+\xi^2+y^2+1) \tag{3.2}$$

when $f(y)(1-\Psi(x,\xi)) \neq 0$. Define $M = -iQ^{-1}((x-y)\cdot\partial_\xi + (\nabla S(y)-\xi)\cdot\partial_y)$ so that $Me^{i(x-y)\cdot\xi/h+iS(y)/h} = he^{i(x-y)\cdot\xi/h+iS(y)/h}$. Performing integration by parts N times, we rewrite the integral in the form

$$(2\pi h)^{-n}h^N\int e^{i(x-y)\cdot\xi/h+iS(y)/h}({}^tM)^N\{(\Psi(x,\xi)-1)f(y)\}dyd\xi.$$

Since $(1-\Psi(x,\xi))f(y)$ is $\mathcal{G}^s$ and $\nabla S(y)-\xi$ is analytic, we see by Lemma 2.2 and (3.2) that

$$|h^N({}^tM_y)^N((1-\Psi(x,\xi))f(y))| \leq Ch^NM_0^N(N!)^s(x^2+\xi^2+y^2)^{-3n}.$$

(3.1) follows immediately from this. □

Completion of the proof of Theorem 1.2.

We continue to write $u_0(x) = e^{iS(x)/h}f(x)$. In virtue of (3.1), we have

$$\|U^h(t,0)u_0 - U^h(t,0)\Psi(x,hD)u_0\| \leq C\exp(-1/Ch^{1/s}). \tag{3.3}$$

Now construct the $\mathcal{G}^{s'}$ local parametrix $E_\Psi^{s'}(t,h)$, $0 \leq t \leq T$, localized on $\Psi(x,\xi)$:

$$\|U^h(t,0)\Psi(x,hD) - E_\Psi^{s'}(t,h)\| \leq C\exp(-C/h^{1/(2s'-1)}) \leq C\exp(-C/h^{1/s}).$$

Recall that the amplitude $e^{s'}(t,\cdot,\xi,h)$ in $E_\Psi^{s'}(t,h)$ has support $\Omega_t(\xi, supp\,\Psi(\cdot,\xi)) \subset \Gamma(t)\tilde{D}$ which is disjoint from D. Hence

$$\begin{aligned}(3.4)\qquad &\|U^h(t,0)\Psi(x,hD)u_0\|_{L^2(D)}\\ &= \|U^h(t,0)\Psi(x,hD)u_0 - E_\Psi^{s'}(t,h)u_0\|_{L^2(D)} \leq C\exp(-C/h^{1/s}).\end{aligned}$$

Combining (3.4) with (3.3), we obtain the theorem. □

§4. Proof of Theorem 1.3.

Recall that $\Gamma(t,s)$ is the Hamilton flow associated with (1.3) and $\Gamma(t) = \Gamma(t,0)$. In virtue of time symmetry, it suffices to show

$$\Gamma(t)FS_s(f) \supset FS_s(U^h(t,0)f). \tag{4.1}$$

We suppose $(x_t, \xi_t) = \Gamma(t)(x_0, \xi_0) \notin \Gamma(t)FS_s(f)$ and show that

$$\|\Psi_2(x,hD)U^h(t,0)f\| = O(\exp(-1/Ch^{1/s})) \tag{4.2}$$

for a $\Psi_2 \in \mathcal{G}_0^s(\mathbb{R}^{2n})$ such that $\Psi_2(x,\xi) = 1$ near (x_t, ξ_t).

Since $(x_0, \xi_0) \notin FS_s(f)$, there exist a compact open subset $U \ni (x_0, \xi_0)$ and $\Psi_1 \in \mathcal{G}_0^s(U)$ such that $\Psi_1(x,\xi) = 1$ near (x_0, ξ_0) and such that

$$\|\Psi_1(x,hD)f\| = O(\exp(-1/Ch^{1/s})). \tag{4.3}$$

We suppose $\Psi_1(x,\xi) = 1$ on $V = V_1 \times V_2$. Take subsets $x_t \in V_1' \subset V_1'' \subset G_1 \subset U_1'$ and $\xi_t \in V_2' \subset V_2'' \subset G_2 \subset U_2'$ in such a way that:

(a) Each set is a compact open subset of the succeeding one;
(b) $\bigcup\{x(0,t,V_1'',\xi) : \xi \in U_2'\}$ is a compact open subset of $\bigcap\{x(0,t,G_1,\xi) : \xi \in U_2'\}$;
(c) $\Gamma(0,t)(U_1' \times U_2') \subset V_1 \times V_2$.

It is certainly possible to take these subsets if we choose U_2' sufficiently small around ξ_t after taking $V_1' \subset V_1'' \subset G_1 \subset U_1'$ small enough. We show that (4.2) is satisfied for any $\Psi_2 \in \mathcal{G}_0^s(V_1' \times V_2')$ with $\Psi_2(x,\xi) = 1$ near (x_t, ξ_t).

In virtue of (b), there exists $\phi \in \mathcal{G}_0^s(V_1)$ such that for any $\xi \in U_2'$, we have

$$\phi(x) = 1 \text{ for } x \in x(0,t,V_1'',\xi), \quad \phi(x) = 0 \text{ for } x \notin x(0,t,G_1,\xi). \tag{4.4}$$

Take $1 < s' < s$ such that $2s' - 1 < s$ and take two auxilliary functions $\Psi_3(x,\xi) \in \mathcal{G}_0^s(V_1'' \times V_2'')$ and $\Psi_4(x,\xi) \in \mathcal{G}_0^{s'}(U_1'' \times U_2'')$ such that $\Psi_3(x,\xi) = 1$ on $V_1' \times V_2'$ and $\Psi_4(x,\xi) = 1$ on $G_1 \times G_2$. $\Psi_j(x,hD)$, $j = 1,\cdots,4$ are uniformly bounded operators in $L^2(\mathbb{R}^n)$ for $0 < h < 1$ and, in virtue of Lemma 2.7,

$$\|\Psi_3(x,hD)\Psi_2(x,hD)^* - \Psi_2(x,hD)^*\| = O(\exp(-1/Ch^{1/s})), \tag{4.5}$$

$$\|\Psi_4(x,hD)\chi(hD)\Psi_3(x,hD) - \Psi_3(x,hD)\| = O(\exp(-1/Ch^{1/s})), \tag{4.6}$$

where $\chi(\xi) \in \mathcal{G}_0^{s'}(G_1)$ is such that $\chi(\xi) = 1$ for all $\xi \in V_1''$.

Lemma 4.1. *We have, for $0 < h < 1$,*

$$\|(1-\phi(x))U^h(0,t)\Psi_3(x,hD)\| \le C\exp(-1/Ch^{1/s'}). \tag{4.7}$$

Proof. In virtue of (4.4), $dist(supp\ (1-\phi(x)), \Omega_t(\xi, supp\ \Psi_3(\cdot,\xi))) > d > 0$ for all $\xi \in U_2'$. Hence Theorem 1.2 implies (4.7) as in the proof of Theorem 2.6. □

Lemma 4.2. *Let $\chi(\xi) \in \mathcal{G}_0^{s'}(G_1)$ be such that $\chi(\xi) = 1$ for $\xi \in V_1''$. Then*

$$\|(1 - \Psi_1(x, hD)^*)\phi(x)U^h(0,t)\Psi_4(x, hD)\chi(hD)\| \leq C\exp(-1/Ch^{1/s}) \tag{4.8}$$

Proof. Construct the $\mathcal{G}^{s'}$ local parametrix for $U^h(0,t)$ localized on $\Psi_4(x,\xi)$:

$$E_{\Psi_4}^{s'}(0,t,h)u(x) = (2\pi h)^{-n} \int e^{i(S(0,t,x,\xi)-y\cdot\xi)/h} e^{s'}(0,t,x,\xi,h)u(y)dyd\xi$$

Since $2s' - 1 < s$, we have

$$\begin{aligned} &\|(1 - \Psi_1(x, hD)^*)\phi(x)U^h(0,t)\Psi_4(x, hD)\chi(hD)\| \\ &\leq \|(1 - \Psi_1(x, hD)^*)\phi(x)E_{\Psi_4}^{s'}(0,t,h)\chi(hD)\| + C\exp(-1/Ch^{1/s}) \end{aligned} \tag{4.9}$$

We write $(1 - \Psi_1(x, hD)^*)\phi(x)E_{\Psi_4}^{s'}(0,t,h)\chi(hD)u(x)$ in the form of multi-integral:

$$(2\pi h)^{-2n} \int e^{i((x-z)\eta + S(0,t,z,\xi) - y\xi)/h}(1 - \Psi_1(z,\eta))\phi(z)e^{s'}(0,t,z,\xi,h)\chi(\xi)u(y)dyd\xi dzd\eta$$

and estimate it by integration by parts. Note that the phase function $S(0,t,z,\xi)$ is analytic in (z,ξ) and by the choice of $\phi(x)$ and $\chi(\xi)$ we have the following facts:

(1) $e^{s'}(0,t,z,\xi,h)$ is analytic in $(z,\xi) \in supp\ \phi \times supp\ \chi$ and, by Lemma 2.5,

$$|\partial_z^\alpha \partial_\xi^\beta e^{s'}(0,t,z,\xi,h)| \leq C_1 C_2^{|\alpha+\beta|}\alpha!\beta!, \quad (z,\xi) \in supp\ \phi \times supp\ \chi. \tag{4.10}$$

(2) $R \equiv (x-z)^2 + (\eta - \nabla_z S(0,t,z,\xi))^2 + +(\nabla_\xi S(0,t,z,\xi) - y)^2) \geq C(1 + x^2 + \eta^2 + y^2)$ on the support of integrand because:

i) Since $(z, \nabla_z S(0,t,z,\xi)) \in \Gamma(G_1 \times G_2)$ on the support of $e^{s'}(t,z,\xi,h)$ and $1 - \Psi_1(z,\eta) = 0$ on $V_1 \times V_2$, property (c) implies $(x-z)^2 + (\eta - \nabla_z S(0,t,z,\xi))^2 \geq C(1 + x^2 + \eta^2)$;

ii) $\nabla_\xi S(0,t,z,\xi) \in G_1$ on the support of $e^{s'}(t,z,\xi,h)$.

Thus, performing integration by parts $N = 1/Ch^{1/s}$ times by using the first order operator

$$-iR^{-1}((x-z)\cdot\partial/\partial_\eta - (\eta - \nabla_z S(0,t,z,\xi))\cdot\partial/\partial_z + (\nabla_\xi S(0,t,z,\xi) - y)\cdot\partial/\partial_\xi)$$

and using the fact that $(1 - \Psi_1(z,\eta))$, ϕ and χ are $\mathcal{G}^s$-functions as well as (1) and (2) above, we obtain the desired estimate (4.8) as in the proof of Theorem 1.2. □

We now able to prove (4.2). In virtue of (4.5), we have

$$\begin{aligned} &\|(1 - \Psi_1(x, hD)^*)U^h(0,t)\Psi_2(x, hD)^*\| \\ &\leq C\|(1 - \Psi_1(x, hD)^*)U^h(0,t)\Psi_3(x, hD)\| + C\exp(-1/Ch^{1/s}), \end{aligned}$$

which is bounded, by Lemma 4.1, by

$$\leq C\|(1-\Psi_1(x,hD)^*)\phi(x)U^h(0,t)\Psi_3(x,hD)\| + C\exp(-1/Ch^{1/s}).$$

Then applying (4.6) and Lemma 4.2, we obtain

$$\begin{aligned}&\leq C\|(1-\Psi_1(x,hD)^*)\phi(x)U^h(0,t)\Psi_4(x,hD)\chi(hD)\| + C\exp(-1/Ch^{1/s})\\ &\leq 2C\exp(-1/Ch^{1/s}).\end{aligned}$$

Thus $\|(1-\Psi_1(x,hD)^*)U^h(0,t)\Psi_2(x,hD)^*\| \leq C\exp(-1/Ch^{1/s})$ and taking the adjoint, we have

$$\|\Psi_2(x,hD)U^h(t,0)(1-\Psi_1(x,hD))f\| \leq C\exp(-1/Ch^{1/s}) \tag{4.11}$$

Combining (4.11) with (4.3), we obtain (4.2). This completes the proof of Theorem 1.3.

§5. Proof of Theorem 1.4.

We assume (1.14). By scaling argument we may assume $t = \pm 1$ without losing the generality. We treat the case $t = 1$ only. Set

$$K(\lambda)f(x) = \int_{\mathbb{R}^n} e^{i\lambda(x-y)^2/2+i\lambda S(y)} f(y)dy.$$

It is easy to see that $K(\lambda)f(x)$ is entire analytic in (λ, x) and (1.14) implies that $\|K(\lambda)f\|_{L^2(D)} \leq Ce^{-C\lambda}$ as $\lambda \to \infty$. Let $g \in C_0^\infty(D)$ and $\phi(\lambda) = (K(\lambda)f, g)$. $\phi(\lambda)$ is entire and $|\phi(\lambda)| \leq Ce^{-C\lambda}$ as $\lambda \to \infty$. Hence its Fourier-Laplace transform

$$\hat{\phi}(z) = \int_0^\infty e^{-i\lambda z}\phi(\lambda)d\lambda$$

is analytic in the half plane $\{t : \operatorname{Im} z < C\}$. Let $|S(y)| < R$ on *supp* f which is compact. We see by Fubini's theorem that $\hat{\phi}(z)$ has the expression for $\operatorname{Re} z < -R$

$$\hat{\phi}(z) = i\int_{\mathbb{R}^{2n}} \frac{f(y)\overline{g(x)}}{(x-y)^2/2 + S(y) - z}dydx. \tag{5.2}$$

Here the RHS of (5.2) is obviously analytic in the domain $\mathbb{C}^1 \setminus \{x \in \mathbb{R}^1 : x \geq -R\}$ and (5.2) holds for all z in $\{z \in \mathbb{C}^1 : \operatorname{Im} z < C\} \setminus \{x \in \mathbb{R}^1 : x \geq -R\}$. Thus, for all $t \in \mathbb{R}^1$, we have

$$\begin{aligned}0 &= \lim_{\epsilon\downarrow 0}\frac{\hat{\phi}(t+i\epsilon) - \hat{\phi}(t-i\epsilon)}{2\pi} = \lim_{\epsilon\downarrow 0}\frac{\epsilon}{\pi}\int_{\mathbb{R}^{2n}} \frac{f(y)\overline{g(x)}dydx}{((x-y)^2/2 + S(y) - t)^2 + \epsilon^2}\\ &= \lim_{\epsilon\downarrow 0}\frac{1}{2\pi}\int_{-\infty}^{\infty}\Big(\int_{\mathbb{R}^{2n}} e^{i\lambda((x-y)^2/2+S(y))}f(y)\overline{g(x)}dxdy\Big)e^{-i\lambda t-\epsilon|\lambda|}d\lambda\end{aligned}$$

This implies, as a distribution, hence for any $\lambda \in \mathbb{R}^1$,

$$\int_{\mathbb{R}^{2n}} e^{i\lambda((x-y)^2/2+S(y))} f(y)\overline{g(x)}dxdy = 0. \tag{5.3}$$

Since $g \in C_0^\infty(D)$ is arbitrary, (5.3) implies

$$\int_{\mathbb{R}^n} e^{i\lambda((x-y)^2/2+S(y))} f(y)dy = 0, \quad \text{a.e.} x \in D. \tag{5.4}$$

for any λ. Note, however, that the left hand side of (5.4) is entire analytic in x. Hence (5.4) holds all $x \in \mathbb{R}^n$ and all $\lambda \in \mathbb{R}^1$. Thus, changing x by x/λ, we see that

$$\int_{\mathbb{R}^n} e^{-ixy+i\lambda(y^2/2+S(y))} f(y)dy = 0, \quad x \in \mathbb{R}^n$$

and the Fourier inversion formula shows $f(y) = 0$ almost everywhere. This completes the proof of Theorem 1.4. □

References

[AF] Asada, K. and Fujiwara, D., *On some oscillatory integral transformatiions in $L^2(\mathbb{R}^n)$*, Japan. J. Math. **4** (1978), 299–361.

[H] Hörmander, L., *The analysis of linear partial differential operators I*, Springer, Berlin-Heidelberg-New York-Tokyo, 1983.

[Mar] Martinez, A., *Estimations de l'effect tunnel pour le double puits II, États hautement excites*, Bull. Soc. math. France **116** (1988), 199–229.

[Mas] Maslov, V. P., *Theory of perturbations and asymptotic methods*, in Russian, Moskov. Gos. Univ., Moskow, 1965.

[R] Robert, D., *Autour asymptotique semi-classique*, Birkheuser, Basel, 1988.

[S] Sjöstrand, J., *Singularités analytiques microlocales*, Asterisque **95** (1982).

[Y] Yajima, K., *The quasi-classical limit of quantum scattering theory*, Commun. math. Phys. **69** (1979), 101–129.

GPSR Compliance

The European Union's (EU) General Product Safety Regulation (GPSR) is a set of rules that requires consumer products to be safe and our obligations to ensure this.

If you have any concerns about our products, you can contact us on ProductSafety@springernature.com

In case Publisher is established outside the EU, the EU authorized representative is:

Springer Nature Customer Service Center GmbH
Europaplatz 3
69115 Heidelberg, Germany

Batch number: 10620881

Printed by Printforce, the Netherlands